LATTICE METHODS FOR QUANTUM CHROMODYNAMICS

LATTICE METHODS FOR QUANTUM CHROMODYNAMICS

Thomas DeGrand
University of Colorado, USA

Carleton DeTar
University of Utah, USA

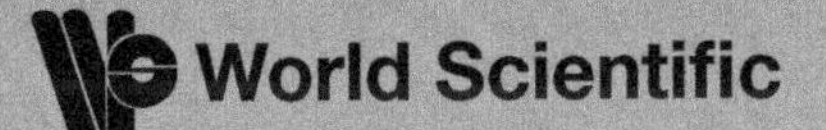

NEW JERSEY · LONDON · SINGAPORE · BEIJING · SHANGHAI · HONG KONG · TAIPEI · CHENNAI

Published by

World Scientific Publishing Co. Pte. Ltd.
5 Toh Tuck Link, Singapore 596224
USA office: 27 Warren Street, Suite 401-402, Hackensack, NJ 07601
UK office: 57 Shelton Street, Covent Garden, London WC2H 9HE

British Library Cataloguing-in-Publication Data
A catalogue record for this book is available from the British Library.

LATTICE METHODS FOR QUANTUM CHROMODYNAMICS

ISBN-13 978-981-256-727-7
ISBN-10 981-256-727-5

Printed in Singapore

to Laurel and Mary Ann

Preface

No book is written in a vacuum. We would like to thank our colleagues for numerous discussions about many of the topics in this book. For many insights we are especially indebted to Claude Bernard, Poul Damgaard, Christof Gattringer, Maarten Golterman, Steve Gottlieb, Anna Hasenfratz, Peter Hasenfratz, Urs Heller, Jim Hetrick, Karl Jansen, Francesco Knechtli, Támas Kovács, Christian Lang, Martin Lüscher, Ferenc Niedermayer, Kostas Orginos, Stefan Schaefer, Yigal Shamir, Steve Sharpe, Rainer Sommer, Bob Sugar, Doug Toussaint, Peter Weisz, Hartmut Wittig, and Ulli Wolff. Any flaws are our own, however. Our research has been supported for many years by the U. S. National Science Foundation and Department of Energy.

Thomas DeGrand, Carleton DeTar

Preface

[illegible]

Contents

Chapter 1

Introduction

Strongly interacting elementary particles such as the proton and neutron are not fundamental particles, but are bound states of spin -1/2 fermions, the quarks, and spin-1 gauge bosons, the gluons. The strong interactions are described by a quantum field theory called Quantum Chromodynamics (QCD). It is believed that quarks and gluons are permanently confined, *i.e.*, it is not possible to separate an isolated quark a macroscopic distance from an antiquark or another quark pair. At short distances the effective coupling constant of QCD becomes small, and it is possible to compare results from experiments with theoretical predictions based on perturbation theory. However, at long distances the strong interactions are truly strong, and perturbation theory breaks down. Some nonperturbative technique is necessary to perform calculations of quantities that are sensitive to the long distance behavior of QCD. One of those techniques is the subject of this book: it is called "lattice QCD."

The approach involves regularizing QCD by introducing a space time lattice. Quarks are restricted to occupy the sites of the lattice; gluons, the links joining the sites. Introducing the lattice spacing converts the Feynman path integral for QCD into an ordinary integral of very large dimensionality. The path integral expectation value of any observable can be computed by evaluating the integral on a computer, typically using importance sampling techniques. Of course, the lattice spacing is itself an unphysical quantity; it was introduced in order to perform the calculation. Physical predictions require an extrapolation from nonzero lattice spacing to zero lattice spacing.

The idea of replacing a field variable, defined at every space-time point, by a variable defined at some location on a lattice is probably older than the idea of a field variable itself. In some physical problems (such as in the description of electrons in a crystal) the lattice is a physical quantity.

However, the particular replacement of a continuum gauge field theory by a lattice theory, from which this field grew, dates only to Wilson (1974). [Lattice gauge theory was in the air in the early 70's: other people working on lattice theories with local gauge invariance included Wegner (1971), J. Smit, and Polyakov (1975)].

Wilson showed that a lattice formulation of QCD exhibited confinement in the strong coupling limit. His work led to an explosive growth of research on the properties of strongly-coupled QCD. At first, all of this research was analytic. The use of Monte Carlo methods for studying statistical systems was already very old [as early as Metropolis *et al.* (1953)], and in 1979 Creutz, Jacobs, and Rebbi (Creutz *et al.*, 1979; Creutz, 1979) adapted these techniques to the study of lattice gauge theories. The first computer simulations involving quarks began two years later (Weingarten and Petcher, 1981; Marinari *et al.*, 1981; Hamber and Parisi, 1981), and by the mid-1980's the field of lattice QCD had been transformed into something like its present form, a mix of analytic calculations and increasingly larger scale numerical simulation. As techniques have advanced, the lattice has become an ever more accurate and important source of calculations of hadronic spectroscopy and matrix elements used in studies of QCD, the Standard Model of elementary particle physics, and beyond.

This book is an introduction to, and motivation for, modern techniques of lattice gauge theory. As we wrote this book, we had several different audiences in mind:

- Graduate students or other scientists beginning to do research in lattice gauge theory. They presumably want to know how quantum field theories are formulated on the lattice, and how to perform calculations in lattice QCD.
- Researchers studying the strong interactions, who might be interested in using lattice calculations of reduced matrix elements in their (non-lattice) calculations. They presumably want to know what lattice people can do, how they do it, and why.
- Experimentalists using results of lattice calculations to interpret measurements. They may want to understand the source of uncertainty in lattice calculations.
- Researchers in computational physics, who might be interested in numerical simulation, and who want to know the physical context in which the calculations are performed.

When we talk about lattice methods or results with a group of particle theorists who do more conventional (non-lattice) calculations, we are struck by how different lattice techniques seem to be from what our colleagues know or do. Certainly, our day to day activities of writing programs, submitting them to remote supercomputers, and analyzing our results, seem very different from what a weak-interaction model builder (for example) does.

And yet, at the bottom, lattice QCD is merely a corner of quantum field theory, and the techniques lattice theorists use are simply decorated versions of techniques used across the board by physicists studying problems with many degrees of freedom in particle, condensed matter, and nuclear physics.

Many of the recent advances in lattice techniques have come when observant researchers recognized a connection between a lattice problem and some other branch of theoretical physics associated with strong interactions. The modern theory of critical phenomena is the source of many important ideas for lattice QCD, followed by the related topic of effective field theories.

While a number of books about lattice gauge theory already exist [two excellent early examples are Creutz (1983) and Montvay and Muenster (1994)], the last five to ten years have seen enormous progress in the development of algorithms and techniques for constructing and simulating lattice discretizations of QCD. Most of these advances are described only in research articles in journals or in occasional summer school lectures. Some of them are, in fact, of a rather fundamental nature. Our goal in writing this book is to bridge the gap between what we think a conventionally-trained theoretical physicist knows and what is written in modern journal articles about topics in lattice QCD for those "who already know it" .

This means that we must describe the rather specialized techniques associated with formulating field theories (mostly QCD) on a lattice, performing numerical simulations, and interpreting their results. It also means that we should attempt to make connections between lattice QCD and other research areas of strong interaction theory and phenomenology.

We have not made any attempt to include results from lattice simulations. To do so would be to invite obsolescence. While the mean values of results from lattice calculations do not vary too rapidly with time, there is a steady decrease in the size of quoted uncertainties as techniques get refined and as data sets get bigger. And then there are the occasional big jumps when a new technique is invented or exploited. We hope we will give our readers enough background, that they can go directly from this little book to the current research literature and understand it. Readers interested in the latest results, and are willing to take those results from very brief articles, should consult the Proceedings of the annual conferences in this research field, "Lattice (year)." These volumes exist as real books through the 2004 conference, and have since been published electronically.

Chapter 2

Continuum QCD and its phenomenology

Since the discovery of strong forces in the atomic nucleus, physicists have sought to understand the strong interactions. Experimental and theoretical efforts during the 1950's and 1960's led to a gradually improved qualitative understanding. Significant developments included the quark model and current algebra. These developments culminated in the 1970's with the formulation of quantum chromodynamics (QCD), the $SU(3)$ gauge theory of interacting quarks and gluons, as the fundamental basis for the strong interactions. Together with its companion gauge theories describing the weak and electromagnetic interactions, QCD became part of the Standard Model.

Perturbative QCD, relying on asymptotic freedom, provided an excellent understanding of hard processes that probe short distances. However, nonperturbative phenomena remained out of reach. The lattice formulation of QCD promised to fill that need. In fact lattice quantum chromodynamics is nearly as old as the continuum formulation itself. But it took more than two decades of improvements in computational and algorithmic power to turn it into an effective tool for answering a wide class of nonperturbative phenomenological questions. Although we are concerned in this text almost entirely with lattice methods, we aim, of course, to explain continuum physics. So we devote this chapter to some of the continuum questions that can be addressed by lattice methods. Continuum models often guide the formulation of problems on the lattice.

We want this chapter to be a bit more than a review, since we will often refer to relevant physics issues we describe here. However, it is less than a full discussion, because the subjects we treat are worthy of (and have received) long discussions in other books.

2.1 The Lagrangian and QCD at short distance

Without further ado, we write the Lagrange density for continuum Euclidean QCD:

$$\mathcal{L} = \sum_{i=1}^{N_f} [\bar{\psi}_i(x)(\not{D} + m_i)\psi_i(x)] - \frac{1}{4}F_{\mu\nu}F^{\mu\nu}. \tag{2.1}$$

Here $\psi_i(x)$ is the fermion field corresponding to a quark of flavor i (with mass m_i); $\not{D} = \gamma^\mu[\partial_\mu - igA_\mu(x)]$, and $A_\mu(x)$ is the gluon field, which can be decomposed into components as

$$A_\mu(x) = \sum_a \lambda^a A_\mu^a(x). \tag{2.2}$$

The field strength tensor is

$$F_{\mu\nu}(x) = \sum_a \lambda^a F_{\mu\nu}^a(x). \tag{2.3}$$

It is related to the vector potential through

$$F_{\mu\nu}^a = \partial_\mu A_\nu^a - \partial_\nu A_\mu^a + g f_{abc} A_\mu^b A_\nu^c. \tag{2.4}$$

The λ^a's are the generators of the color gauge group, *i.e.*, $SU(3)$ in nature, and f_{abc} are the structure constants.

Although one could imagine QCD with any numbers of flavors, in nature there are six. According to the Standard Model the left-hand components $(1-\gamma_5)\psi/2$ form doublets under the weak isospin of the electroweak interactions,

$$\begin{pmatrix} u \\ d' \end{pmatrix} \begin{pmatrix} c \\ s' \end{pmatrix} \begin{pmatrix} t \\ b' \end{pmatrix} \tag{2.5}$$

and the right-hand components $(1+\gamma_5)\psi/2$ are singlets of weak isospin. The $I_3 = 1/2$ (upper entries) all have charge $Q = 2/3$ (in units of the proton's charge); the lower components, $Q = -1/3$. The eigenstates of QCD, which parameterize Eq. (2.1), are not diagonal in the weak interactions; hence, in Eq. (2.5) we have primed the lower entries. The relation between primed and unprimed quarks is given by $q'_i = M_{ij}q_j$ where M_{ij} is called the Cabibbo-Kobayashi-Maskawa (CKM) matrix. The quarks, their charges (in units of the proton's charge), their masses (roughly the $\overline{MS}$ mass at a renormalization point $\mu = 2$ GeV), are displayed in Table 2.1.

Table 2.1 The quarks.

name	charge	mass
u	2/3	5 MeV
d	−1/3	9 MeV
c	2/3	1.5 GeV
s	−1/3	100 MeV
t	2/3	175 GeV
b	−1/3	4.5 GeV

The dominant qualitative feature of QCD seen in perturbation theory is asymptotic freedom. The effective coupling constant of QCD shrinks as the momentum scale Q^2 grows. This means that processes involving only high momentum or short distance are accessible to analysis with perturbative methods. To one loop order the beta function for N_c colors and N_f flavors is

$$\beta(g) = \mu \frac{\partial g}{\partial \mu} = -\frac{g^3}{16\pi^2} \left[\frac{11}{3} N_c - \frac{2}{3} N_f \right] \tag{2.6}$$

so the coupling constant runs as

$$\frac{g^2(Q^2)}{4\pi} = \frac{4\pi}{(11N_c/3 - 2N_f/3)\log(Q^2/\Lambda_{\mathrm{QCD}}^2)}. \tag{2.7}$$

Conversely, asymptotic freedom implies that the effective coupling constant for QCD grows at lower momentum scales or longer distances. The quantity "Λ_{QCD}" expresses, somewhat poetically, the scale at which the strong interactions become strong. (A more precise definition of a Λ parameter is renormalization-prescription dependent.) In qualitative calculations, one might take it to be a few hundred MeV, so that at distances greater than a few tenths of a femtometer, which is the relevant distance scale for spectroscopy or hadronic matrix elements, QCD becomes nonperturbative.

2.2 The nonrelativistic quark model

2.2.1 *Explanations and puzzles*

Before there was QCD, there was the quark model. The multiplet patterns of numerous hadronic resonances in the 1950's and early 1960's was strong evidence for some common substructure. This was provided by the quark model of Gell-Mann and Zweig. Hadrons are composed of fermionic constituents, the quarks. Mesons (integer spin states) are bound states of a quark and an antiquark; baryons (half-integer spin states) are bound states of three quarks. A glance at the intrinsic quantum numbers of Dirac particles shows that S-wave $\bar{q}q$ bound states are pseudoscalars ($J^P = 0^-$) and vectors ($J^P = 1^-$), and a glance at the Reviews of Particle Properties (Eidelman *et al.*, 2004) reveals that the lightest meson states form a pseudoscalar multiplet and a vector multiplet. Three-fermion S-wave bound states have $J^P = \frac{1}{2}^+$ and $\frac{3}{2}^+$ bound states, also in accord with observation.

In the quark model, the mass of a hadron is a sum of the "constituent" quark masses, plus small corrections due, perhaps, to spin effects. At the time of the creation of the quark model, three flavors of quarks were needed: nearly degenerate up and down quarks, (so that ordinary isospin would be a good quantum number) and a somewhat heavier strange quark. The approximate equality of masses and of reduced matrix elements under exchange of up, down, and strange quarks implies an approximate "flavor" $SU(3)$ symmetry of the strong interactions, which combined with the $SU(2)$ of spin to give the so-called "$SU(6)$ quark model."

Elementary angular momentum addition allows one to write down quark wave functions for the mesons and baryons with remarkably accurate consequences for predictions of the ratios of static quantities of the hadrons. The standard textbook quark model result ($-2/3$) for the ratio of neutron to proton magnetic moments is remarkably close to the experimental value (-0.657).

Quarks in neutrons and protons should occupy the same lowest orbital, but, to make the observed pattern of states the spin and flavor, wave functions need to be symmetric under interchange, contrary to the requirements of Fermi statistics. The resolution to this puzzle required the existence of a hidden quantum number, color. Quarks must have three colors to get an antisymmetric three-quark wave function. It seemed plausible (though not required) that color symmetry should also be $SU(3)$. The $\sum_a \bar{q}_a q_a$ meson wave function and $\epsilon_{abc} q_a q_b q_c$ baryon wave function couple the constituents into a color-neutral representation (a color singlet state). Then confinement is equivalent to the statement that all hadrons are color singlets.

Of course, much was unexplained in the quark model. What is the confinement mechanism? And how can one justify a nonrelativistic treatment? Here is how the problem arises: We know that the proton's charge radius is about one femtometer. The uncertainty principle requires that an object confined to a box of this size has a momentum of about h/r, or a few hundred MeV. But the quark model assumes most of the mass of a hadron is just the rest mass of the quarks (so the proton consists of three 300 MeV constituents). With a quark mass comparable to its momentum, a derivation of hadronic properties from the solution of a Schrödinger equation is suspect.

This problem becomes more acute when we look ahead a few pages to a description of the low energy dynamics of the strong interactions in terms of an effective Lagrangian for would-be Goldstone bosons (the pseudoscalar mesons). If the quarks were massless, so would be the pions and kaons. That the pions and kaons are not massless implies that the up and down quarks have masses of a few MeV; the strange quark has a mass of $O(100)$ MeV. How then, does it make sense to talk about a baryon (mass greater than 1000 MeV) as a bound state of three quarks?

Contemporary lattice QCD problems are the descendants of quark model puzzles: What are the masses of the hadrons in terms of the masses of the quarks? What are their static properties? What are their decay amplitudes? What is the origin of the nuclear force? How do strong interactions influence weak decays? What are the properties of strongly interacting matter at high temperature and density? The list goes on.

2.2.2 *Quark model interpolating operators*

Let us continue our description of quark model phenomenology with an eye toward its use in a lattice context. A typical lattice calculation of hadronic masses measures the correlation function between composite operators with the quantum numbers appropriate to the desired states. The correlator is a sum of propagators for all hadrons with these quantum numbers. We generally need some way to construct interpolating operators that couple strongly to one of the states (typically the lightest one, or the next-to-lightest one). Quark model wave functions can be a useful guide to the construction of such interpolating fields.

Let's examine in some detail how nonrelativistic quark model wave functions can help in constructing interpolating operators for baryons. We assume that the quarks each carry spin-1/2, have identical spatial wave functions, and carry three colors labeled a, b, and c. We use fermionic creation operators to assure a properly antisymmetrized state. The lightest states will then have $J = \frac{1}{2} \otimes \frac{1}{2} \otimes \frac{1}{2} = \frac{3}{2} \oplus \frac{1}{2}$. The $J = 3/2$ states are easiest to build. Start with the $J = 3/2, m_J = 3/2$:

$$\Delta^{++}(3/2) |0\rangle = u^\dagger_{a\uparrow} u^\dagger_{b\uparrow} u^\dagger_{c\uparrow} |0\rangle , \tag{2.8}$$

where the notation for the creation operators for quark orbitals identifies flavor, spin projection, and color. Spin and flavor lowering operators then lead to the other states. For example, the isospin- and spin-lowering operators produce the $\Delta^+(1/2)$, given by

$$3\Delta^+(1/2) = \epsilon_{abc}(\frac{1}{2}u^\dagger_{a\uparrow}u^\dagger_{b\uparrow}d^\dagger_{c\downarrow} + u^\dagger_{a\downarrow}u^\dagger_{b\uparrow}d^\dagger_{c\uparrow}) \tag{2.9}$$

Similarly, the flavor $SU(3)$ U- and V-spin raising and lowering operators lead to strange members of the flavor decuplet.

The proton and its relatives have mixed-symmetry spin wave functions and mixed-symmetry flavor wave functions. It is easiest to construct them by considering the proton, a *uud* bound state. The u quarks are identical, so their spin wave function must be symmetric ($J = 1$). Then the spin-up proton interpolating operator is

$$p(1/2) = \epsilon_{abc}(\alpha u^\dagger_{a\uparrow}u^\dagger_{b\uparrow}d^\dagger_{c\downarrow} + \beta u^\dagger_{a\downarrow}u^\dagger_{b\uparrow}d^\dagger_{c\uparrow}) \tag{2.10}$$

Requiring this state to be normalized and orthogonal to the $\Delta^+(1/2)$ gives $\alpha = -\beta = \sqrt{\frac{1}{6}}$. Raising or lowering flavor and spin similarly produces interpolating operators for the other strange and nonstrange members of the baryon octet. See Kokkedee (1969) for a complete list of wave functions.

Meson states are simpler. We have a quark-antiquark pair, so statistics is not an issue. With two spin-1/2 particles in the same spatial orbital and three flavors of quarks, there are $4 \times 9 = 36$ states, a $J = 0$ flavor singlet and octet (9 states) and a $J = 1$ flavor singlet and octet (27 states).

The quark model allows one to construct and classify excited states as well. Mesons are easy: we simply include (and model) orbital angular momentum in addition to spin. In nature, essentially all P-wave states (${}^SL_J = {}^3P_{2,1,0}$, ${}^1P_1$ or $J^{PC} = 2^{++}$, 1^{++}, 0^{++} and 1^{+-}) are seen, along with higher orbital excitations and what are plausibly radial excitations. The first multiplet of radial excitations of the baryon's ground state $\underline{56}$ is largely present, and the lowest P-wave multiplet [a negative parity multiplet, the $L = 1$ $\underline{70}$ of $SU(6)$], is also well known.

2.2.3 *Glueballs, hybrids, and exotics*

The nonrelativistic quark model was designed specifically to explain the well-observed three-quark baryons and quark-antiquark mesons. However, QCD introduced yet another player, namely, gluons, opening the possibility for "hybrid" states, consisting of combinations of quarks and gluons, and "glueball" states, consisting entirely of gluons. If the conserved quantum numbers agree, these states would be expected to mix strongly with the purely quark states. But some "exotic" combinations of quantum numbers cannot be obtained with only a quark and antiquark, *e.g.* the spin, parity, and charge conjugation combinations $J^{PC} = 0^{+-}$, 0^{--}, and 1^{-+}. The experimental candidates for the 1^{-+} state are in the 1400 to 1600 MeV range. In the language of the constituent quark model, the gluon would then have an effective mass of several hundred MeV, but as a gauge particle it must be strictly massless.

2.2.4 *Flavor and glueball mixing example: the anomaly*

A particularly noteworthy example of flavor and glueball mixing occurs in the three pseudoscalar quark-antiquark meson states. Unbroken flavor $SU(3)$ requires three states, two of them in the octet, namely the $I = 1$, $I_3 = 0$ pion formed as $(\bar{u}^\dagger u^\dagger - \bar{d}^\dagger d^\dagger)/\sqrt{2}$ and the unmixed $I = 0$ $\eta_8 = (\bar{u}^\dagger u^\dagger + \bar{d}^\dagger d^\dagger - 2\bar{s}^\dagger s^\dagger)/\sqrt{6}$, and the third, the unmixed flavor singlet $\eta_0 = (\bar{u}^\dagger u^\dagger + \bar{d}^\dagger d^\dagger + \bar{s}^\dagger s^\dagger)/\sqrt{3}$. A pseudoscalar glueball (presumably much higher in mass) mixes with the flavor singlet eta. This effect is alternatively understood as a consequence of the $U(1)$ axial anomaly reviewed in Sec. 2.6 below. The splitting of the quark masses breaks the flavor symmetry, thereby involving both the isospin singlet etas in the mixing, resulting, finally, in the observed η and η'. Isospin conservation protects the pion from mixing.

2.2.5 *Quark model hadron masses*

Having built states, we proceed to dynamics. We could begin by assuming that the mass of a hadron is additive in the number of quarks, so from a strictly nonrelativistic point of view

$$M = \sum_{i=u,d,s} n_i m_i. \tag{2.11}$$

The quantity m_i is called a "constituent quark mass" and corresponds roughly to the energy of the quark in the state. It should be roughly the same parameter for baryons as for mesons. Then we would predict for the decuplet that

$$\begin{aligned} M(\Omega) - M(\Xi^*) &= M(\Xi^*) - M(\Sigma^*) = M(\Sigma^*) - M(\Delta) \\ 140 \text{ MeV} &= 145 \text{ MeV} \quad = 153 \text{ MeV} \end{aligned} \tag{2.12}$$

which seems to be very successful, but the predictions

$$\begin{aligned} M(\Xi) - M(\Sigma) &= M(\Sigma) - M(p) \\ 125 \text{ MeV} &= 250 \text{ MeV} \end{aligned} \tag{2.13}$$

and

$$\begin{aligned} M(\Lambda) &= M(\Sigma) \\ 1115 \text{ MeV} &= 1189 \text{ MeV} \end{aligned} \tag{2.14}$$

do not work so well.

Clearly, there should also be spin-dependent forces. The octet and decuplet states with the same quark content are not degenerate. To suggest a way to model them, we recall the hyperfine interaction of atomic physics

$$V(r) = \sum_{ij} \lambda_i^a \lambda_j^a \vec{\sigma}_i \cdot \vec{\sigma}_j f(m_i, m_j, r). \tag{2.15}$$

We have replaced the product of charges from the electromagnetic hyperfine interaction with a product of color matrices λ, which would be appropriate for the exchange of a gluon. For nonrelativistic quarks we can copy the formula from a quantum mechanics book,

$$f(m_i, m_j, r) \sim \frac{1}{m_i m_j r^3}. \tag{2.16}$$

The inverse masses arise from the magnetic moments of the quarks. In this model the $\Sigma^0 - \Lambda$ splitting is, then, due to the difference between the magnetic moments of the strange and nonstrange quarks. For light quarks we construct a model by substituting the constituent quark mass in the formula. Hadrons are color singlet objects, and an interaction of this form immediately suggests a connection between baryon and meson mass splittings. For a meson, $(\lambda_i + \lambda_j)^2 = 0$ so $\lambda_i \lambda_j = -\lambda^2$ where λ^2 is the square of any generator. In baryons, $(\lambda_1 + \lambda_2 + \lambda_3)^2 = 0$ so $2\lambda_i \lambda_j = -\lambda^2$. The intrinsic fine structure splitting is thus twice as big in mesons as in baryons. Computing matrix elements of the hyperfine interaction between mesons and baryons, we see that pseudoscalar masses are driven down, vector meson masses are driven up, and octets and decuplets are split apart as observed. Ratios of the predicted shifts for the pion, the rho, the nucleon, and the Delta are $-6 : 2 : -3 : 3$ for a net meson to baryon mass splitting ratio of $8 : 6$. This is an underestimate of what is seen in nature, where the baryon splitting is 1232 MeV - 938 MeV = 294 MeV, whereas the meson splitting is 770 MeV-140 MeV = 630 MeV.

If one assumes that the hyperfine interaction varies smoothly with the quark mass [so that $f(m_i, m_j, r) = f_0 + n_s f_s$] one can obtain phenomenologically interesting mass formulas such as the equal-spacing rule for the decuplet, the Gell-Mann Okubo formula for the baryon octet, $2m(N)+2m(\Xi) = 3m(\Lambda) + m(\Sigma)$, and the $SU(6)$ relation, $m(\Sigma^*) - m(\Sigma) = m(\Xi^*) - m(\Xi)$.

The first model with a confining potential and a hyperfine interaction of the form of Eqs. (2.15-2.16) was written down by De Rujula *et al.* (1975) in 1975. The most successful quark model of baryon spectroscopy of which we are aware is due to Isgur and Karl (Isgur and Karl, 1978, 1979). It uses a harmonic oscillator potential to confine the quarks and the color magnetic dipole interaction plus a suppressed spin-orbital interaction to split the states. As a phenomenological description of low lying hadronic states, it works enormously well. However, it is not QCD, nor can it be derived from QCD.

We could continue describing quark model phenomenology, but that is not the purpose of this book. We believe we have presented enough of an introduction to orient the reader. We hope she is dissatisfied with our discussion, since one of the goals of lattice QCD is to address questions of hadron spectroscopy in a more controlled way.

2.3 Heavy quark systems

2.3.1 *Quarkonium*

When the quarks become heavy, much of their dynamics simplifies. Bound states of a heavy quark-antiquark pair can plausibly be modeled using a nonrelativistic Schrödinger equation. The charm quark is the lightest "heavy" quark. Its observed spectrum reveals a family of two S-wave radial excitations, including both the ${}^3S_1$ and ${}^1S_0$ level for each, and a set of ${}^3P$ levels in between. (It is easiest to make these states in e^+e^- annihilation, which produces the ${}^3S_1$ states directly. They can decay by electric dipole radiation to the ${}^3P_{2,1,0}$ levels, or by magnetic dipole radiation to the ${}^1S_0$ states.) Above the higher ${}^3S_1$ state, the decay widths of the observed resonances grow, signaling that higher states are above threshold for light quark-antiquark pairs to be produced. The state can then "fall apart" into D and $\bar{D}$ mesons, e.g. $c\bar{c} \rightarrow (c\bar{u})(\bar{c}u)$. The bottom quark system is richer, with three ${}^3S_1$ states and two radial recurrences of ${}^3P$ states below $B\bar{B}$ threshold.

Unspecified in the Schrödinger phenomenology is the shape of the potential $V(r)$. At short distances asymptotic freedom suggests that it take a Coulomb-like form, $V(r) \sim g^2/r$, and one might expect, also from asymptotic freedom, that the coefficient g^2 would also show logarithmic length dependence. If quarks are confined, the potential must rise monotonically at large distance. It is easy to imagine that the potential might rise linearly, $V(r) \sim \sigma r$, with "string tension" σ. Such behavior has been justified variously in models and in phenomenology. In bag models the color electric field is focused into a flux tube with a stable radius (Chodos *et al.*, 1974; DeGrand *et al.*, 1975). The observed linearly rising Regge trajectories suggest a linearly rising potential (Regge, 1959). A linearly rising potential gives a good fit to the observed energy levels for heavy quark systems (Eichten *et al.*, 1975). Of course, these statements are merely ones of plausibility; to determine truly the shape of the heavy quark potential from first principles requires a nonperturbative approach.

A second phenomenological question would be to explain the quarkonium fine and hyperfine structure. In potential models these are quasi-relativistic effects, requiring additional assumptions about the relativistic behavior of the confining term.

2.3.2 *Heavy quark symmetries*

Mesons and baryons containing a single heavy quark also simplify as m_Q goes to infinity. The momentum transfer between the heavy quark and the rest of the system is typically of the scale of the light quark masses, *i.e.*, $\mathcal{O}(\Lambda_{\rm QCD})$, and thus remains bounded as the heavy quark mass goes to infinity. Hence, in that limit the heavy quark is on shell and at rest relative to the hadron. In this limit the QCD dynamics becomes independent of the heavy quark mass. All heavy quark mass dependence can be extracted analytically leading to a power series expansion in $1/m_Q$. New symmetry relations appear. These can be exploited to provide relations between matrix elements, especially when the kinematics are arranged so that the velocity of the initial and final heavy quarks are unchanged. (For a review, see Manohar and Wise (2000).)

Unstated in the heavy quark effective theory are the specific values of reduced matrix elements. Also unstated is the minimum value of the heavy quark mass for which the quark can be considered to be "heavy." A nonperturbative treatment is needed to resolve these questions, either within the heavy quark limit or without it.

2.4 Chiral symmetry and chiral symmetry breaking

2.4.1 *Chiral symmetry*

The low energy limit of QCD is much more than a phenomenology; it can be described by an effective field theory with its own reasonably high predictive power. Let us consider the Lagrangian for QCD, given in Eq. (2.1), truncated to two flavors, up and down. We introduce left- and right-handed spinors,

$$u_L = \frac{1}{2}(1-\gamma_5)u \qquad u_R = \frac{1}{2}(1+\gamma_5)u$$

and similarly for the down quark. Then the truncated Lagrangian is

$$\mathcal{L}_{ud} = (\bar{u}_L\ \bar{d}_L)\begin{pmatrix} \not{D} & 0 \\ 0 & \not{D} \end{pmatrix}\begin{pmatrix} u_L \\ d_L \end{pmatrix} + (\bar{u}_L\ \bar{d}_L)\begin{pmatrix} m_u & 0 \\ 0 & m_d \end{pmatrix}\begin{pmatrix} u_R \\ d_R \end{pmatrix} + (L \leftrightarrow R). \tag{2.17}$$

If $m_u = m_d = 0$, the Lagrangian Eq. (2.17) is invariant under separate $U(2)$ rotations of the left- and right-handed fields,

$$\begin{pmatrix} u_i \\ d_i \end{pmatrix} \to V_i \begin{pmatrix} u_i \\ d_i \end{pmatrix} \tag{2.18}$$

with $V_i \in U(2)$ and $i = L, R$. In other words gluon interactions do not change the helicity of the quarks. However, the terms in the Lagrangian proportional to quark masses do not remain invariant under the transformations of Eq. (2.18).

From Noether's theorem, each continuous parameter of the symmetry group has a corresponding conserved current. The group $U(2)$ has four real parameters, and so there should be eight conserved currents. We focus on six of the conserved currents, namely, the vector and axial vector currents.

$$J^a_\mu = \bar{q}\gamma_\mu \frac{\tau^a}{2} q \qquad J^{5a}_\mu = \bar{q}\gamma_\mu\gamma_5 \frac{\tau^a}{2} q. \tag{2.19}$$

Here $q = \begin{pmatrix} u \\ d \end{pmatrix}$, $a = 1, 2, 3$, and τ^a is a generator of the group $SU(2)$.

If the quark masses were set to zero, the six vector and axial vector charges $Q^a_{V,A}$ corresponding to the axial and vector currents would commute with the Hamiltonian, $[H_{\text{QCD}}, Q^a_V] = [H_{\text{QCD}}, Q^a_A] = 0$, and would be conserved. Let us consider the consequences. Suppose we had eigenstates of H_{QCD}, $H_{\text{QCD}}|\psi\rangle = E|\psi\rangle$. The opposite parity states $Q^a_V|\psi\rangle$ and $Q^a_A|\psi\rangle$ would be degenerate. Such parity doubling does not occur in nature. It must be that nature does not respect the symmetry of the Hamiltonian in this (Wigner) way. The alternative possibility, that the symmetry is "spontaneously broken," is widely accepted. We present the classic example of symmetry breaking below, namely, the linear sigma model. It makes a number of important predictions for QCD: Although the vacuum is annihilated by vector charges, $Q_V^a|0\rangle = 0$, it is not annihilated by axial charges, $Q_A^a|0\rangle \neq 0$. As a consequence, the spectrum of H_{QCD} contains three massless pseudoscalar particles, Goldstone bosons. The axial charges act on any state in the Hilbert space to create additional Goldstone boson states, with no resemblance to the single particle states created by the vector currents.

The order parameter associated with this spontaneous symmetry breakdown is the quark condensate

$$\langle 0|\bar{q}q|0\rangle = \langle 0|\bar{q}_L q_R + \bar{q}_R q_L|0\rangle \neq 0. \tag{2.20}$$

Physically, it signals the instability of the vacuum against production of a condensate of quark-antiquark pairs. The pairs must have net zero total momentum and total angular momentum, so they must involve a pairing of left-handed quarks with right-handed anti-quarks, and vice versa. The vacuum expectation value of Eq. (2.20) signals the spontaneous breakdown of the $U(2) \times U(2)$ symmetry of Eq. (2.18) down to the subgroup of vector symmetries with $V_L = V_R$.

Since the up and down quarks actually have a small mass, chiral symmetry is not only spontaneously broken, but also explicitly broken. Thus there are no truly massless Goldstone bosons. However, the π^+, π^-, and π^0, are Goldstone bosons candidates. Indeed, they are the the lightest of all hadrons.

2.4.2 *Linear sigma model*

The classic example of spontaneous symmetry breaking is the $O(4)$ linear sigma model (Gell-Mann and Levy, 1960). It is a toy model for chiral symmetry breaking in QCD. This is a field theory for four real scalar fields, bundled into a multiplet $\vec{\phi} = (\phi_0, \phi_1, \phi_2, \phi_3)$, with a Lagrangian

$$\mathcal{L} = \frac{1}{2}\partial^\mu \vec{\phi}\partial_\mu \vec{\phi} - \frac{g}{4}(\vec{\phi}^2 - v^2)^2 + h\phi_0. \tag{2.21}$$

When the explicit symmetry breaking term h vanishes, the Lagrangian is invariant under transformations $\phi_i \to R_{ij}\phi_j$ where R is an element of the symmetry group $O(4)$. Then there are six Noether currents, which we may choose to be

$$J^a_\mu = \epsilon^{abc}\phi^b\partial_\mu\phi^c \qquad J^{5a}_\mu = -\phi^0 \overleftrightarrow{\partial}_\mu \phi^a, \tag{2.22}$$

where $a, b, c = 1, 2, 3$. With our choice for a potential the $O(4)$ symmetry is spontaneously broken. For small h the minimum of the potential occurs at $\phi_G = (v + h/2gv^2, \vec{0})$. Thus the vacuum expectation value $\langle\phi\rangle$ has a nonzero value that survives the limit $h \to 0$. So the bare vacuum state breaks the $O(4)$ symmetry.

Masslessness is the first property of Goldstone bosons. To identify them, we consider small fluctuations about the minimum for small h, expand the fields as $\vec{\phi} = (v + \sigma + h/2gv^2, \vec{\pi})$, and rewrite the Lagrangian as

$$\begin{aligned}\mathcal{L} = &\frac{1}{2}[\partial_\mu\sigma\partial_\mu\sigma - (2gv^2 + 3h/v)\sigma^2] + \frac{1}{2}\partial_\mu\vec{\pi}\partial_\mu\vec{\pi} \\ &-h\pi^2/2v - (gv + h/2v^2)\sigma(\sigma^2 + \pi^2) - \frac{g}{4}(\sigma^2 + \pi^2)^2 - hv.\end{aligned} \tag{2.23}$$

The bare masses of the σ and π fields are

$$M^2_\pi = h/v \quad M^2_\sigma = 2gv^2 + 3h/v. \tag{2.24}$$

At zero h the three components of the π field are the massless Goldstone bosons, the pions.

We have already alluded to the fact that in nature, (probably) no quark is massless. Nonzero up and down quark mass terms explicitly break the chiral symmetry, just as the symmetry-breaking term does in the linear sigma model. (In QCD the corresponding statement is $M^2_\pi \propto m_u + m_d$.)

The second property of Goldstone bosons, that the axial current creates them from the vacuum, follows from the canonical commutation relations and the properties of field creation and annihilation operators:

$$\langle 0|J^{5a}_\mu(0)|\pi^b(p)\rangle = -ip_\mu\delta^{ab}\langle 0|\phi_0|0\rangle = ip_\mu\delta^{ab}\left(v + \frac{h}{2gv^2}\right). \tag{2.25}$$

Next, let us consider scattering of Goldstone bosons in the limit of zero h. We compute the T matrix for the process $\pi^i(p_1) + \pi^j(p_2) \to \pi^k(p_3) + \pi^l(p_4)$. In lowest order in g, this comes from the direct $(\pi^2)^2$ term, and in order g^2 it occurs through exchange of a σ meson. We find

$$T_{k,l;i,j} = \delta_{ij}\delta_{kl}A(s,t,u) + \delta_{ik}\delta_{jl}A(t,s,u) + \delta_{il}\delta_{kj}A(u,s,t) \tag{2.26}$$

where s, t, and u are the Mandelstam invariants $[s = (p_i + p_j)^2$, $t = (p_i - p_k)^2$, $u = (p_i - p_l)^2]$ and

$$A(s,t,u) = \frac{4g^2v^2}{m^2 - s} - 2g. \tag{2.27}$$

Note that in the low energy limit $s \to 0$ the scattering amplitude vanishes, since $m^2 = 2gv^2$. Expanding about $s = 0$ gives

$$A(s,t,u) \to \frac{s}{v^2} + O(s^2). \tag{2.28}$$

The coefficient of s is given in terms of the axial current matrix element, Eq. (2.25).

2.4.3 *Nonlinear sigma model*

The elementary fields of the linear sigma model are massless Goldstone bosons and a massive scalar particle. A more economical effective field theory, the "nonlinear sigma model", treats only pions as the elementary fields. Since we are interested in the theory only as a low-energy imitation of QCD, we are content to dispense with the high mass elementary sigma meson. To carry out the imitation we must tune the parameters of the nonlinear sigma model so it gives the same set of low energy predictions as QCD. So in this section, as a prelude to carrying out the corresponding exercise with QCD, we will tune the parameters of the nonlinear model so it matches the predictions of the linear model of the previous section at energies well below the threshold for producing the massive scalar. Since the exercise is only pedagogical, we do this only to the extent of reproducing the tree graphs of the linear sigma model.

We see from Eqs. (2.27-2.28) that the scattering matrix contains arbitrarily high powers of momenta. To construct the effective Lagrangian, we must then keep terms with arbitrarily high powers of pion momenta or derivatives of the pion field

$$\mathcal{L}_{\text{eff}} = \mathcal{L}_0 + \mathcal{L}_2 + \mathcal{L}_4 + \mathcal{L}_6 + \ldots \tag{2.29}$$

The term $\mathcal{L}_n$ contains n derivatives of the pion fields.

Except for explicit symmetry-breaking contributions, we require that the terms in the Lagrangian be invariant under the $SU(2)_L \times SU(2)_R$ chiral symmetry. A simple way to realize this is to introduce the elementary fields through the unitary two by two matrix U, written either in the form

$$U = \frac{\sigma'}{f} + i\frac{\tau^k \pi^{k\prime}}{f} \tag{2.30}$$

or

$$U = \exp\left(\frac{i}{f}\tau^k \pi^k\right) \tag{2.31}$$

with $\sigma'^2 + \vec{\pi}'^2 = f^2$, and the τ^k's are again Pauli matrices. Please note we are using the same notation for the field π^k as in the linear sigma model.

For the leading term $\mathcal{L}_2$, we take

$$\mathcal{L}_2 = \frac{f^2}{4}\mathrm{Tr}(\partial_\mu U \partial_\mu U^\dagger). \tag{2.32}$$

This expression is indeed invariant under $U \to V_R U V_L^\dagger$, where V_L and V_R are elements of $SU(2)$. So it has the desired symmetry.

This simple contribution to the full Lagrangian can be expanded in a Taylor series in the π's as

$$\mathcal{L}_2 = \frac{1}{2}\partial_\mu \vec{\pi} \partial_\mu \vec{\pi} + \frac{1}{8f^2}\partial_\mu(\vec{\pi}^2)\partial_\mu(\vec{\pi}^2) + O(\pi^6). \tag{2.33}$$

The expansion parameter is the ratio of the pion momentum to f.

To match the nonlinear and linear models at tree level we require the pion decay amplitudes to agree. The conserved vector and axial vector Noether currents in this model are

$$J_\mu^a = \frac{if^2}{4}\mathrm{Tr}\left(\tau_a[U, \partial_\mu U^\dagger]\right); \qquad J_\mu^{5a} = \frac{if^2}{4}\mathrm{Tr}\left(\tau_a\{U, \partial_\mu U^\dagger\}\right). \tag{2.34}$$

The pion decay amplitude is, then,

$$\langle 0|J_\mu^{5a}(0)|\pi^b(p)\rangle = ip_\mu \delta^{ab} f, \tag{2.35}$$

so we require $f = v$ at tree level.

Next we introduce an explicit symmetry breaking term

$$\mathcal{L}_0 = mBf^2 \mathrm{Re}\mathrm{Tr}(U) \approx -mB\vec{\pi}^2 \tag{2.36}$$

to complete the lowest order Lagrangian. Notice that from the standpoint of the polynomial expansion Eq. (2.33), we recognize a bare pion mass $M_\pi^2 = 2mB$, so we require $mB = h/(2v)$ to match the linear and nonlinear models at tree level. The symmetry breaking term assures that the ground state occurs at $U = 1$, or $\sigma' = f$.

We have thus determined the two low energy constants f and B at tree level. As a consistency check we can compute pi-pi scattering using $\mathcal{L}_2$. Notice that this effective theory only involves the pions. To lowest order in the momenta there is only one diagram (corresponding to the second term in Eq. (2.33)). Indeed, the scattering amplitude reproduces Eq. (2.28).

We would continue the construction of the effective Lagrangian by adding higher order terms to $\mathcal{L}_{\text{eff}}$, whose coefficients would be tuned to match the scattering amplitudes of the linear sigma model at still higher order in the momenta.

2.4.4 *Nonlinear effective chiral Lagrangian for QCD*

So far we have been matching results for the linear and nonlinear models only at tree level. Our intention for QCD is to compute quantities such as the pion mass, pion decay constant and the low-energy pion scattering amplitude nonperturbatively. In that case we compute the corresponding quantities in the effective theory up to as many orders in the expansion parameter p^2/f^2 as needed and match results between the two theories to fix the parameters of the effective theory.

We write the effective nonlinear sigma model Lagrangian as before [Eq. (2.29)], but modify the parameterization of the symmetry breaking term Eq. (2.36) slightly, leading to the Lagrangian

$$\mathcal{L}_2 = \frac{f^2}{4}\text{Tr}[\partial_\mu U \partial_\mu U^\dagger] + B\frac{f^2}{2}\text{Tr}[M(U + U^\dagger)] + \text{higher order}, \qquad (2.37)$$

where the symmetry breaking term now includes a "mass" matrix [specializing to $SU(2)$]

$$M = \begin{pmatrix} m_u & 0 \\ 0 & m_d \end{pmatrix}, \qquad (2.38)$$

leading to a tree-level squared pion mass

$$M_\pi^2 = B(m_u + m_d). \qquad (2.39)$$

As before the constant f in the first term can be fixed by matching the pion decay constant in QCD to the pion decay constant in the nonlinear theory.

The symmetry-breaking term is related to the quark mass term $\bar{q}q$ in QCD. A quick way to see this is to observe that the vacuum energy density of QCD, $\langle 0| -\mathcal{L} |0\rangle$ at small m_u and m_d is $m_u \langle \bar{u}u\rangle + m_d \langle \bar{d}d\rangle$). At small m_u and m_d, isospin symmetry gives $\langle \bar{u}u\rangle = \langle \bar{d}d\rangle$. In the effective field theory at small M, the vacuum configuration is $U = 1$ and the vacuum energy is $\langle 0| \mathcal{H} |0\rangle = -f^2 B(m_u + m_d)$. Equating these quantities fixes

$$\langle \bar{u}u \rangle = -f^2 B. \tag{2.40}$$

Combining this result with Eq. (2.39) gives

$$f^2 M_\pi^2 = -(m_u + m_d)\langle \bar{u}u \rangle. \tag{2.41}$$

This is called the "Gell-Mann, Oakes, Renner (GMOR) relation" (Gell-Mann *et al.*, 1968).

We close this section with a couple of remarks.

- This discussion of an effective field theory for QCD is on a very different footing from the one we gave for the linear sigma model. Even if a perturbative expansion of QCD made sense, we would have to match many Feynman graphs in QCD against a few processes in the effective theory. But it is clear that the dynamics of QCD responsible for chiral symmetry breaking must be nonperturbative. The formation of a bound state (like the pion) is intrinsically nonperturbative as is the formation of the chiral condensate.
- From the point of view of the effective Lagrangian, the parameters f and B are simply bare parameters; there is no short distance structure for them to encode. However, from the point of view of QCD, Eqs. (2.25) and (2.40) involve matrix elements of quark operators, and should be computable. One of the goals of the lattice program is to perform that calculation, and a test of QCD would be whether those determinations would agree with the values of f and B, as determined by a fit of experimental data to the effective Lagrangian.

2.5 A technical aside: Ward identities

Conservation laws are reflected in relations between correlation functions, called Ward identities. We will use Ward identities repeatedly, so we need to provide a brief introduction. Suppose we consider an expectation value of any operator

$$\langle O(x_1, x_2, \ldots, x_m)\rangle = \frac{1}{Z}\int dU d\psi d\bar{\psi}\, O(x_1, x_2, \ldots, x_m) \exp\left(-\int d^4x\, \mathcal{L}\right). \tag{2.42}$$

If we perform a local change of the variables of functional integration, for example, $\psi(x) \to \psi(x) + \alpha(x)\delta\psi(x)$, $\bar{\psi}(x) \to \bar{\psi}(x) + \alpha(x)\delta\bar{\psi}(x)$, the operator, the Lagrangian, and the functional integration measure changes, but the functional integral does not. If at the same time the functional Jacobian of the transformation is unity, the measure is also unchanged and we must have

$$\left\langle \delta O - O \int d^4x\, \delta\mathcal{L}\right\rangle = 0. \tag{2.43}$$

This is the Ward identity. Seemingly innocuous, it expresses a relation between two vacuum expectation values. Let us look at some examples.

First, consider a free Dirac particle and an ordinary $U(1)$ rotation $\delta\psi = i\alpha(x)\psi(x)$, $\delta\bar{\psi}(x) = -i\alpha(x)\bar{\psi}(x)$. The functional Jacobian is $\det[1 - i\alpha(x)\delta(x-x')]\det[1 + i\alpha(x)\delta(x-x')] = 1 + \mathcal{O}(\alpha^2)$, so the functional integration measure is unchanged to first order in α. From Noether's theorem the change in the Lagrangian is $\delta\mathcal{L} = i\partial^\mu\alpha(x)J_\mu(x)$, with $J_\mu = \bar{\psi}\gamma_\mu\psi$. Suppose in addition, that the operator O is just the correlator $\psi(y_1)\bar{\psi}(y_2)$. Then the Ward identity is

$$0 = \langle\psi(y_1)\bar{\psi}(y_2)\rangle[\alpha(y_1) - \alpha(y_2)] + \left\langle \int d^4x\, \alpha(x)\partial^\mu J_\mu(x)\psi(y_1)\bar{\psi}(y_2)\right\rangle. \tag{2.44}$$

Taking the functional derivative with respect to $\alpha(x)$ gives

$$0 = \langle\psi(x)\bar{\psi}(y_2)\rangle[\delta(x-y_1) - \delta(x-y_2)] + \langle\partial^\mu J_\mu(x)\psi(y_1)\bar{\psi}(y_2)\rangle. \tag{2.45}$$

Translational invariance requires that all correlators be functions of relative distances $y_1 - x$ and $x - y_2$. We can Fourier transform this expression with a weighting function $\exp[ip(y_1 - x)]\exp[ip'(x - y_2)]$. The first and second terms of Eq. (2.45) are just the fermion propagators $\Delta(p')$ and $\Delta(p)$ and so we derive the Ward identity relating their difference to the unamputated vertex (the term on the rhs):

$$(p' - p)_\mu\hat{\Gamma}_\mu = \Delta(p') - \Delta(p). \tag{2.46}$$

By pre-multiplying by $\Delta(p)^{-1}$ and post-multiplying by $\Delta(p')^{-1}$, we can re-express this result as a constraint on the amputated vertex (the usual vertex with no external legs),

$$(p' - p)_\mu \Gamma_\mu = \Delta(p)^{-1} - \Delta(p')^{-1}. \tag{2.47}$$

In free field theory, the identity is trivially satisfied, $\Delta(p')^{-1} = \not p' + m$, and $\Gamma_\mu = \gamma_\mu$, but it is true for interacting theories as well. Its consequence is that the vertex and propagator renormalizations are related.

A second example is a bit less straightforward. We work with a flavor multiplet of quarks with different masses. The symmetry transformation is a flavor nonsinglet chiral rotation

$$\psi(x) \to \left[1 + i\epsilon(x)^a \frac{\lambda^a}{2}\gamma_5\right]\psi(x), \qquad \bar\psi(x) \to \bar\psi(x)\left[1 + i\epsilon(x)^a \frac{\lambda^a}{2}\gamma_5\right]. \tag{2.48}$$

The λ's form a Lie algebra appropriate to the flavor group with commutator $[\lambda^a, \lambda^b] = 2if^{abc}\lambda^c$. The conserved vector current is

$$J^a_\mu(z) = \bar\psi(x)\frac{\lambda^a}{2}\gamma_\mu\psi(x) \tag{2.49}$$

and at zero quark mass the conserved axial vector current is

$$J^{5a}_\mu(z) = \bar\psi(x)\frac{\lambda^a}{2}\gamma_\mu\gamma_5\psi(x). \tag{2.50}$$

If the quarks are massive, the axial current is not conserved, and in addition to the $J^{a5}_\mu(x)\partial^\mu\epsilon^a(x)$ term there is the term $i\epsilon^a(x)\bar\psi(x)\{\frac{\lambda^a}{2}, M\}\gamma_5\psi(x)$, where M is the flavor mass matrix. The Ward identity becomes

$$\left\langle \frac{\delta O}{\delta\epsilon^a(x)} + O\int d^4x\left[\partial^\mu J^{a5}_\mu(x) - \bar\psi(x)\left\{\frac{\lambda^a}{2}\gamma_5, M\right\}\gamma_5\psi(x)\right]\right\rangle = 0. \tag{2.51}$$

If $O = 1$ the Ward identity for the axial current is

$$\langle\alpha|\partial^\mu J^{a5}_\mu(x)|\beta\rangle = \langle\alpha|\bar\psi\left\{\frac{\lambda^a}{2}, M\right\}\gamma_5\psi|\beta\rangle. \tag{2.52}$$

This is called the "PCAC" (partial conservation of axial current) relation.

If $O = J^{5b}_\rho(y)J^c_\sigma(z)$ we can get δO from the transformations of the currents under a chiral rotation:

$$\begin{aligned}\delta J^c_\mu(y) &= \epsilon^a(y)f^{acd}J^{5d}_\mu(y)\\ \delta J^{5c}_\mu(y) &= \epsilon^a(y)f^{acd}J^d_\mu(y).\end{aligned} \tag{2.53}$$

The Ward identity is then

$$\langle[\partial^\mu J_\mu^{5a}(x) - \bar\psi(x)\left\{\frac{\lambda^a}{2}, M\right\}\psi(x)]J_\rho^{5b}(y)J_\sigma^c(z)\rangle = \\ if^{abd}\delta^4(x-y)\langle J_\rho^d(y)J_\sigma^c(z)\rangle \\ +if^{acd}\delta^4(x-z)\langle J_\rho^{5b}(y)J_\sigma^{5d}(z)\rangle \tag{2.54}$$

[plus terms proportional to $\delta^4(x-y)\delta^4(y-z)$]. The integral over x of this expression is also interesting: the divergence of the axial current integrates to zero, leaving

$$-\int d^4x\langle\bar\psi(x)\left\{\frac{\lambda^a}{2}, M\right\}\psi(x)J_\rho^{5b}(y)J_\sigma^c(z)\rangle = \\ if^{abd}\langle J_\rho^d(y)J_\rho^c(z)\rangle + if^{acd}\langle J_\rho^{5b}(y)J_\sigma^{5d}(z)\rangle. \tag{2.55}$$

Suppose that the quark mass were somehow hidden in other parameters in the Lagrangian. One could determine the quark mass from a calculation of the correlators, and the massless limit would be the value of m where the vector and axial correlators were equal in magnitude and opposite in sign.

2.6 The axial anomaly and instantons

2.6.1 *Nonconservation of the flavor-singlet axial current*

The last Ward identity example is a bit less straightforward and deserves its own section. Let us consider a flavor-singlet chiral rotation, defined by $\psi(x) \to [1 + i\epsilon(x)\gamma_5]\psi(x)$, $\bar\psi(x) \to \bar\psi(x)[1 + i\epsilon(x)\gamma_5]$. At zero quark mass the Noether current is $J_\mu^5 = \bar\psi\gamma_\mu\gamma_5\psi$. At nonzero quark mass the variation of the Lagrangian includes both $J_\mu^5(x)\partial^\mu\epsilon(x)$ and $\epsilon(x)\bar\psi(x)\{\gamma_5, M\}\psi(x)$.

There is a crucial change in the Ward identity in this case, because the functional Jacobian is anomalous (Peskin and Schroeder, 1995). To first order in $\epsilon(x)$ it is

$$J = \exp\left[-i\int d^4x\, \epsilon(x)Q_T(x)\right] \tag{2.56}$$

(per quark flavor) where

$$Q_T(x) = \frac{g^2}{16\pi^2}\epsilon^{\alpha\beta\mu\nu}F^a_{\alpha\beta}(x)F^a_{\mu\nu}(x). \tag{2.57}$$

The Ward identity (with functional derivative included) thus becomes

$$\left\langle \frac{\delta O}{\delta\epsilon(x)} - iO\int d^4x[-\partial^\mu J^5_\mu(x) + \bar{\psi}(x)\{\gamma_5, M\}\psi(x) - N_f Q_T(x)]\right\rangle = 0, \tag{2.58}$$

implying for $\mathcal{O} = 1$ and vanishing quark masses that

$$\partial_\mu J^{\mu 5} = -N_f Q_T(x). \tag{2.59}$$

We see that even at zero quark mass the flavor singlet current is not conserved due to the presence of quantum fluctuations, and therefore the flavor singlet pseudoscalar meson is not a Goldstone boson. The η and (especially) the η' mesons have a large singlet component, thus qualitatively accounting for their mass.

(We note that there is no anomaly for an isospin axial current: its strength would be proportional to $\mathrm{Tr}t^a\{\lambda^b, \lambda^c\}$, where t^a is an isospin generator and the λ's are color matrices. The trace is the product of a color piece and a flavor piece; the latter vanishes since $\mathrm{Tr}\; t^a = 0$. As a corollary, a theory containing many fermions with axial charges summing to zero also has a vanishing anomaly. We will encounter this result on the lattice in Sec. 6.1.)

2.6.2 *Witten-Veneziano formula*

The first attempt to understand the mass of the η' from the anomaly was made by Witten and Veneziano in 1979 (Witten, 1979b; Veneziano, 1979). Using anomalous Ward identities plus additional assumptions, they found

$$\frac{f_\pi^2}{2N_f}(m_\eta^2 + m_{\eta'}^2 - 2m_K^2) = \int d^4x \langle Q_T(x) Q_T(0)\rangle \equiv \chi_{\mathrm{top}}. \tag{2.60}$$

This is not an identity, since it requires some mild model assumptions. It relates quantities in full QCD on the left side to the topological susceptibility calculated in the pure gluon theory, which is, of course, not measurable in experiments. Nevertheless, if we assume a topological susceptibility appropriate to a random distribution of independent instantons with a density of one per fm^4, *i.e.*, $\chi_{\mathrm{top}} \approx (180\ \mathrm{MeV})^4$, we can account approximately for a large η' mass. Lattice simulations, on the other hand, can determine the susceptibility with or without quarks.

2.6.3 *Suppression of the topological susceptibility*

The integral of Eq. (2.59) gives the change in axial charge over the time extent of integration:

$$\Delta Q_5 = \int dt \frac{dQ}{dt} = \int d^4 x \partial_\mu J^{\mu 5} = -N_f \int d^4 x \, Q_T(x). \tag{2.61}$$

The integral would seem superficially to be zero (since it is the integral over a divergence), but the surface term at infinity does not vanish rapidly enough. Instead the integral is an integer, the so-called winding number or topological charge of the gauge configuration ν. It is possible to construct specific solutions of the Yang-Mills field equations with $\nu = \pm 1$. These solutions are called instantons and anti-instantons. They can be used to build multi-instanton solutions, with arbitrary ν. The precise form of these solutions need not concern us here.

One can solve the Dirac equation in the background of an instanton gauge configuration. 't Hooft (1976) discovered that the Dirac operator has a chiral zero mode $i\not{D}\psi = 0$, whose chirality is correlated with the topological charge ν of the underlying gauge configuration. This correlation is encoded in the Atiyah-Singer "index theorem," $\nu = n_- - n_+$, where $n_\pm$ counts the number of zero modes of positive or negative chirality.

The index theorem implies that light quarks suppress fluctuations in the topological charge, so they reduce the topological susceptibility. As we will find in Ch. 3, the probability of finding a particular set of gauge field variables in the generating functional for gauge fields is proportional to the determinant of the Dirac operator. In a gauge configuration of winding number ν the massless Dirac operator has ν zero modes. Nonzero modes of the massless Dirac operator obey $\not{D}\psi_r = i\lambda_r \psi_r$, and since γ_5 anticommutes with $\not{D}$, $\not{D}\gamma_5\psi_r = -i\lambda_r\gamma_5\psi_r$, and all $\lambda_r \neq 0$ modes are paired. With a quark mass, the fermionic determinant in the background gauge configuration becomes

$$\det D = m^\nu \prod_{\lambda_r \neq 0} (\lambda_r^2 + m^2). \tag{2.62}$$

As the fermion mass vanishes, the contribution of gauge configurations with $\nu \neq 0$ to the functional integral is suppressed, and χ_{top} must also vanish.

2.6.4 *QCD vacuum*

There is an extensive phenomenology of the QCD vacuum based on instantons. For a review of this subject, see Schafer and Shuryak (1998). This phenomenology has had a fair amount of success, mainly in predictions for correlation functions that are sensitive to chiral symmetry breaking. It is based on a description of the long distance structure of the QCD vacuum derived from sum rules, as developed by Shifman, Vainshtein, and Zakharov (Shifman *et al.*, 1979b,a) in the late 1970's.

The formal results about the anomaly and models of the QCD vacuum we have listed here are based on general principles, but all examples of these results involve semi-classical (instanton) solutions of the field equations. Are they then relevant in the real world of strongly-coupled QCD? We will see that lattice fermion actions with exact chiral symmetry allow us to revisit these questions in a more controlled context. Of course, instanton phenomenology, like quark model phenomenology, is an uncontrolled description for QCD. Can it be justified directly from QCD by *ab initio* lattice calculations?

2.7 The large N_c limit

QCD with massless quarks does not have an obvious expansion parameter. The coupling constant is itself not really a free parameter, because, through the renormalization group, its size is connected to the distance scale. When we do perturbation theory, we are restricted to scales for which the effective coupling is small, that is, to short distance scales. There does not seem to be any way to organize a calculation without requiring a dimensional scale. However, as first pointed out by 't Hooft (1974) the inverse of the number of colors $1/N_c$ can be used as an expansion parameter. It is surprising that the large N_c limit bears a striking resemblance to $N_c = 3$ QCD. To be precise, the coupling is decreased in this limit so $g^2 N_c = g_0^2$ is constant as $N_c \to \infty$. This is "large-N_c QCD" ('t Hooft, 1974).

In the large-N_c limit the number of gluons scales as N_c^2, whereas the number of quarks scales only as N_c. This has interesting consequences for Feynman diagrams. It is easiest to introduce large-N_c QCD for mesons, which we will briefly do.

We consider a Feynman diagrammatic expansion of QCD. We cannot hope to perform a complete calculation of all the Feynman diagrams, but we can hope to identify the particular class of diagrams that dominate in the large N_c limit. Propagating quarks carry their colors from one vertex to the next. A gluon propagating between two vertices carries a color factor $(t^a)_{ij}(t^a)_{kl}$. The group theory identity of $SU(N)$

$$(t^a)_{ij}(t^a)_{kl} = \frac{1}{2}\left(\delta_{il}\delta_{kj} - \frac{1}{N}\delta_{ij}\delta_{kl}\right) \tag{2.63}$$

means that in the large-N_c limit we can neglect the last term, keep the first term, and replace the color factor in a gluon propagator by a double line. (A good mnemonic is to write the color indices in the gluon as $A^i_{\mu j}$, the quark and antiquark as q^i and $\bar{q}_j$. In traces, we sum pairs of raised and lowered indices: $\mathrm{Tr}\bar{q}q = \bar{q}_i q^i$, $\mathrm{Tr}\bar{q}Aq = \bar{q}_i A^i_j q^j$ and so on. A trilinear gluon vertex is $\mathrm{Tr}A_\mu A_\nu \partial_\mu A_\nu \sim A^i_{\mu j} A^j_{\nu k} \partial_\mu A^k_{\nu i}$. These vertices (and the four-gluon vertex) have quark-line decompositions shown in Fig. 2.1.

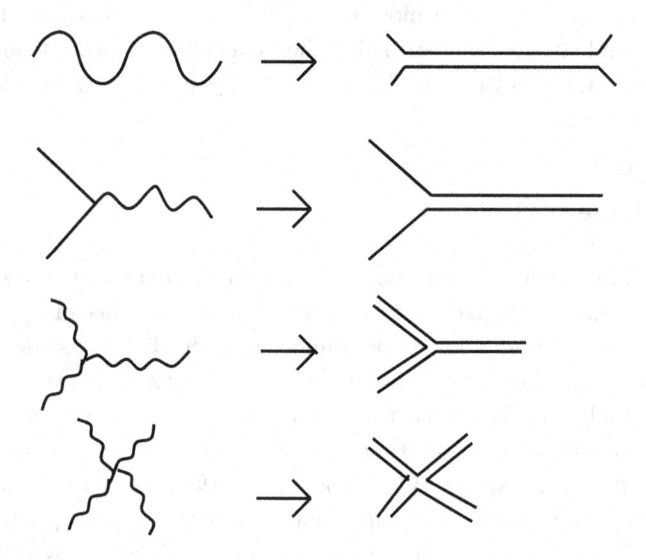

Fig. 2.1 Color factor decomposition into quark lines for vertices in large-N_c QCD.

Now we consider a Feynman diagram containing an arbitrary number of quarks and gluons. We replace all gluon propagators by their double lines. Color traces are performed by counting a factor of N_c for every closed loop. We associate a factor of $1/\sqrt{N_c}$ to every vertex (because $g = g_0/\sqrt{N_c}$). Finally, if we consider a graph involving quarks coupled to an external color singlet meson, there is an extra $1/\sqrt{N_c}$ for every such vertex, from the color wave function of the meson.

For purely mesonic processes (involving the creation of $\bar{q}q$ pairs at every space time point), we discover the following remarkable results:

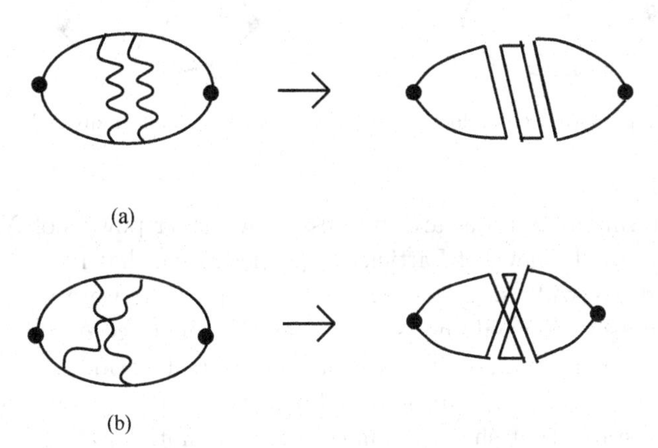

Fig. 2.2 A planar diagram (a) and a nonplanar diagram (b) and their double line color decomposition. The color factor of (a) is $(N_c^3)(1/N_c^2)$ while for (b) it is $(N_c)(1/N_c^2)$

- Planar diagrams dominate nonplanar diagrams by $O(N_c^2)$. An example of this dominance is shown in Fig. 2.2, where we have shown the graphs first with gluons, and then with the color double lines.
- Planar diagrams with quarks on the outside dominate ones with outer gluons, by $1/N_c^2$. Compare Fig. 2.3.
- Graphs with internal quark loops are down by $1/N_c$ per quark line compared with planar diagrams with only gluons.
- Color factors in the leading planar diagrams are such that no combination of propagating particles form color singlets. This follows from the decomposition of vertices into quark lines. For example, in Fig. 2.2a, the quark and gluon color factors are coupled in the pattern $\bar{q}_l A^l_k A^k_j q^j$. This means that, while the four fields couple to a color singlet, no smaller (planar) combination of them is a singlet: not $\bar{q}_l A^l_k$ nor $\bar{q}_l A^l_k A^k_j$. The singlet part of $A^l_k A^k_j$ is a $1/N_c$ effect.
- Amplitudes involving three mesons are suppressed by $1/\sqrt{N_c}$ compared with two-body ones, and ones involving p mesons are suppressed by $1/N_c^{p-1/2}$.

The consequences of these results are that mesons are $\bar{q}q$ states as in the quark model and do not have a $\bar{q}q\bar{q}q$ or higher quark number content. They are noninteracting in leading order in N_c, with decay widths of order $1/N_c$.

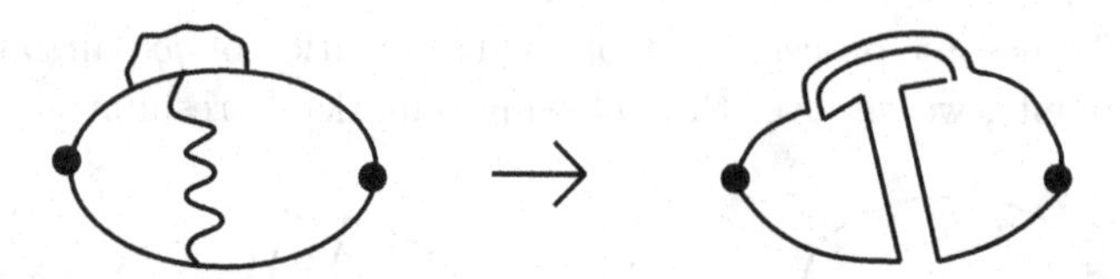

Fig. 2.3 A graph with a gluon loop outside the quark loop, and its quark line decomposition.

Meson scattering amplitudes are suppressed by higher powers of N_c. The summaries in the Reviews of Particle Properties reveal that many hadrons have small decay widths: in fact, if their decay widths were not smaller than their masses, we would not see them at all. The large N_c limit seems to capture some of the qualitative features of the real world.

Zweig's rule is the phenomenological observation that decay widths of hadrons into states that share no quarks with the initial state are smaller than decay widths in processes where quarks can flow from the initial to final state hadrons. In the large-N_c limit, these processes are down by $1/N_c$ compared with direct ones, as the reader can check by comparing Figs. 2.4a and b.

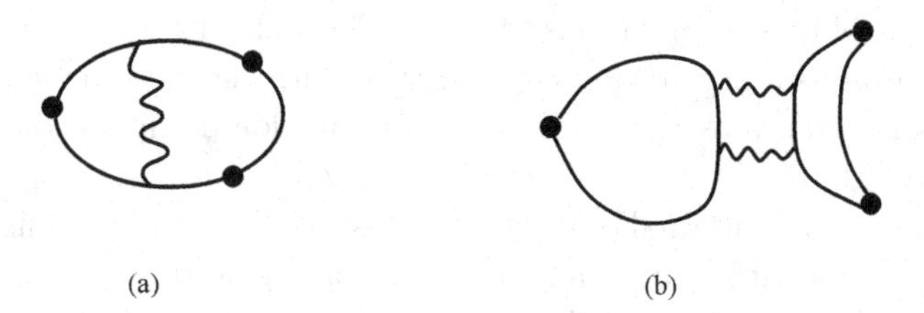

Fig. 2.4 A decay process that shares quarks between the initial and final states, and a "disconnected" diagram, which would be forbidden by Zweig's rule.

The divergence of the flavor singlet axial current is given by the anomaly, Eqs. (2.57-2.59), which is proportional to g^2. In the large-N_c limit, this factor is replaced by g_0^2/N_c, so that the anomaly disappears in the large-N_c limit. This implies that the eta-prime meson becomes an ordinary Goldstone boson in that limit.

In the large-N_c limit, baryons are very different objects from mesons. They contain N_c quarks and presumably are heavy objects. A variety of approaches to baryons as heavy, semiclassical objects appear in the literature (Witten, 1979a; Skyrme, 1961; Witten, 1983). It is possible to compute ratios of matrix elements relevant to baryon masses and couplings as a systematic expansion in $1/N_c$. (For early work, see (Dashen and Manohar, 1993a,b; Dashen *et al.*, 1994).)

Of course, like all the other approaches to QCD we have so far discussed, much is missing: there is no way to compute the sizes of any of the matrix elements we have described. Some nonperturbative approach (obviously, the lattice, again) is needed. And is $N_c = 3$ large or small? The qualitative discussion we have made here argues that it is large, but of course nothing prevents a leading order matrix element from being numerically small, while its $1/N_c$ correction might happen to be multiplied by a big factor.

Much of the discussion in this chapter has been phenomenological. One could ask, why do we need phenomenological models? We have the QCD Lagrangian in hand. The rest of the book shows how to use it to do calculations of hadronic properties from first principles without any need for models. We think such an attitude is too narrow. We have mentioned these models of QCD because each gives us a particular insight into how QCD works. Models suggest questions to address with numerical simulation, and they guide the interpretation of numerical results. Furthermore, limitations to numerical simulation often require recourse to models to extend our understanding of QCD.

Chapter 3

Path integration

3.1 Lattice Schwinger model

Lattice field theories are usually defined directly from the lattice action via the Feynman path integral. A fundamental requirement is that the formal continuum limit of the lattice action reproduces the correct continuum action. The continuum Feynman path integral is, in turn, justified from a Hamiltonian formulation of the continuum field theory through methods that usually challenge mathematical rigor. Because they involve countable degrees of freedom, lattice field theories require less sleight of hand. Thus it is instructive to derive the Feynman path integral from the lattice Hamiltonian. Moreover, it is important for students of lattice field theory to be able to speak both languages. We often return to the Hamiltonian language when we build interpolating operators for correlation functions. So we begin our introduction to lattice gauge theory by considering a simple model, namely, the lattice version of the massive Schwinger model — quantum electrodynamics in one spatial dimension. Despite its simplicity, it anticipates almost all of the technical features of more complicated gauge theories with fermions in higher dimensions.

We will allow ourselves some leeway in making the connection between the lattice Hamiltonian and its corresponding action. Since we intend always to take the continuum limit, it is acceptable to sacrifice precision for simplicity, provided that the resulting theories agree in the continuum limit.

3.1.1 *Free fermions in one dimension*

We consider first the free fermion field on a one-dimensional lattice. Later we will introduce the electromagnetic field. We will derive the spectrum of the free fermion and then show in the next subsection that it has twice the number of fermions as expected.

In one spatial dimension we use a two-component Dirac spinor and define Dirac gamma matrices as follows:

$$\psi_x(t) = \begin{pmatrix} \psi_{1x}(t) \\ \psi_{2x}(t) \end{pmatrix} \quad \gamma_0 = \begin{pmatrix} 1 & 0 \\ 0 & -1 \end{pmatrix} \quad \gamma_1 = \begin{pmatrix} 0 & 1 \\ -1 & 0 \end{pmatrix} \tag{3.1}$$

where $x = 0, 1, \ldots N-1$ labels the lattice points and $\psi_x(t)$ is the two-component fermion field operator.

The continuum Dirac energy density is $\bar{\psi}(x)(-\gamma_1 d/dx + m)\psi(x)$. We will choose a simple nearest-neighbor discretization of the derivative for our lattice Hamiltonian. For the free fermion of mass m it is

$$H_F = \frac{-i}{2a} \sum_{x=0}^{N-1} \bar{\psi}_x \gamma_1 (\psi_{x+1} - \psi_{x-1}) + \sum_{x=0}^{N-1} m \bar{\psi}_x \psi_x. \tag{3.2}$$

The lattice spacing is a and we have used a finite difference formula for the lattice derivative. We used the central difference in order to get a hermitian Hamiltonian. As usual $\bar{\psi}_x = \psi_x^\dagger \gamma_0$. We take periodic boundary conditions $\psi_0 = \psi_N$. In one dimension spin has no meaning. The two fermion spinor components produce a particle and an antiparticle state.

The field operators satisfy the usual anticommutation relations

$$\{\psi_{x\alpha}, \psi_{x'\alpha'}^\dagger\} = \delta_{\alpha,\alpha'} \delta_{x,x'}. \tag{3.3}$$

From them one can derive the Heisenberg equations of motion:

$$i\dot{\psi}_x = [\psi_x, H_F] = \frac{-i\gamma_0\gamma_1}{2a} (\psi_{x+1} - \psi_{x-1}) + m\gamma_0\psi_x. \tag{3.4}$$

The Hamiltonian is diagonalized in momentum space

$$\psi_k = \frac{1}{\sqrt{N}} \sum_{x=0}^{N-1} \exp(2\pi i k x/N) \psi_x \tag{3.5}$$

for $k = 0, 1, \ldots, N-1$. It becomes

$$H_F = \frac{-i}{2a} \sum_{k=0}^{N-1} \psi_k^\dagger \begin{pmatrix} m & \frac{1}{a}\sin(\frac{2\pi k}{N}) \\ \frac{1}{a}\sin\left(\frac{2\pi k}{N}\right) & -m \end{pmatrix} \psi_k. \tag{3.6}$$

The eigenenergies are

$$E = \pm\sqrt{m^2 + \frac{1}{a^2}\sin^2\left(\frac{2\pi k}{N}\right)}. \tag{3.7}$$

The zero mass Hamiltonian has a chiral symmetry

$$\psi_x \rightarrow e^{i\gamma_5\theta}\psi_x \tag{3.8}$$

where $\gamma_5 = \gamma_0\gamma_1$. This symmetry mixes upper and lower components, which corresponds to mixing particles and antiparticles.

3.1.2 *Species doubling*

The continuum limit of the model takes $a \rightarrow 0$ with Na fixed. For a finite energy we require k to be close to 0, $N/2$, or N. Notice that the dispersion relation, like the discrete Fourier transform, is periodic, so the point $k = N$ is equivalent to $k = 0$. Thus in the continuum limit there are two finite energy solutions, both of the form $E = \pm\sqrt{m^2 + p^2}$, where the momentum p is defined to be

$$p = \begin{cases} 2\pi k/aN; & k > 0 \\ 2\pi(k - N)/aN; & k < N \end{cases} \tag{3.9}$$

or

$$p = 2\pi(k - N/2)/aN;\ k \approx N/2. \tag{3.10}$$

Because of periodicity the first two cases cover positive and negative p continuously, so count as one particle of mass m, and the third case counts as a second particle of mass m. Both particles have their associated antiparticles. This doubling of fermion species is an awkward consequence of lattice formulations of relativistic fermions. In d spatial dimensions one gets 2^d species where only one is intended. Switching from a central difference to a one-sided difference of the form $(\psi_{x+q} - \psi_x)/a$ would eliminate the problem, but then the Hamiltonian would not be Hermitian. Later on we will see that the lattice action with a similar discrete time derivative gives rise to one further doubling, resulting in 16 species in four space-time dimensions.

In one dimension Susskind (1977) found a simple remedy to the doubling problem, namely, a field redefinition that diagonalizes the Hamiltonian in coordinate space. Let N be even and define

$$\psi_x = (\gamma_0\gamma_1)^x\psi'_x. \tag{3.11}$$

Then

$$H_F = \frac{-i}{2a}\sum_{x=0}^{N-1}\psi'^{\dagger}_x(\psi'_{x+1} - \psi'_{x-1}) + \sum_{x=0}^{N-1} m\psi'^{\dagger}_x\bar{\psi}'_x(-)^x. \tag{3.12}$$

This is the sum of two identical Hamiltonians, one for the upper and one for the lower component of the spinor. If we select only the upper component $\chi_x = \psi_{x1}$, then we have the one-component Hamiltonian,

$$H_F = \frac{-i}{2a} \sum_{x=0}^{N-1} \chi_x^\dagger (\chi_{x+1} - \chi_{x-1}) + \sum_{x=0}^{N-1} m \chi_x^\dagger \chi_x (-)^x . \tag{3.13}$$

The same trick can be used in higher dimensions to reduce doubling. It is the basis for the "staggered fermion" scheme. In momentum space, because $(-)^x = \exp(i\pi x)$, the one-component Hamiltonian becomes

$$H_F = \frac{-i}{2a} \sum_{k=0}^{N/2-1} \begin{pmatrix} \chi_k^\dagger & \chi_{k+N/2}^\dagger \end{pmatrix} \begin{pmatrix} \frac{1}{a}\sin\left(\frac{2\pi k}{N}\right) & m \\ m & \frac{-1}{a}\sin\left(\frac{2\pi k}{N}\right) \end{pmatrix} \begin{pmatrix} \chi_k \\ \chi_{k+N/2} \end{pmatrix} . \tag{3.14}$$

We have bundled the Fourier modes at k and $k+N/2$ into a two component spinor. Setting the mass to zero to make the argument, we can compare Eqs. (3.6) and (3.14): the first Hamiltonian has two-component zero energy modes at $k = 0$ and at $k = N/2$, while the latter Hamiltonian has one-component zero energy modes at $k = 0$ and $N/2$, which we have interpreted as a system with a single two-component zero mode at $k = 0$. To do this, we have restricted the summation over k in Eq. (3.14) to cover only half the interval in Eq. (3.6).

Cutting the number of degrees of freedom in half in this way solves the doubling problem, but we lose chiral symmetry in the process for the simple reason that the lower fermion component is no longer available for mixing. This impossibility of formulating a hermitian, local, chirally symmetric theory on the lattice without species doubling is codified in the Nielson-Ninomiya "no-go" theorem (Nielsen and Ninomiya, 1981c,a,b).

3.2 Hamiltonian with gauge fields

We now want to add gauge fields and consider interactions. We want to find a lattice-regularized version of the Schwinger model, a single fermion in one spatial dimension coupled to a $U(1)$ gauge field. The continuum Hamiltonian in temporal gauge is

$$H = \int dx \left\{ \frac{1}{2} E_1^2(x) + \bar\psi(x)\gamma_1 \left[-i\frac{d}{dx} + gA_1(x) \right] \psi(x) + m\bar\psi(x)\psi(x) \right\} . \tag{3.15}$$

The vector potential in the spatial direction is $A_1(x)$, and $E_1(x)$ is the electric field. In temporal gauge the time component of the vector potential is fixed to be zero. The canonically conjugate variables $E_1(x)$ and $A_1(x)$ satisfy the commutation relation

$$[A_1(x), E_1(x')] = i\delta(x - x'). \tag{3.16}$$

The electric field is coupled to the fermionic charge density ρ through the Gauss' law constraint $\vec{\nabla} \cdot \vec{E} = \rho$ or $\partial E_1(x)/\partial x = \rho$. We will define the charge density carefully, below.

Passing to the lattice involves replacing the derivative by a lattice difference (as we have already done) while maintaining local gauge invariance. The lattice fermion is $\psi_x = \sqrt{a}\psi(ax)$. A local gauge transformation is a local change of phase of ψ:

$$\psi_x \to e^{i\lambda_x}\psi_x. \tag{3.17}$$

Thus in the finite difference, a term like $\bar{\psi}_x\psi_{x+1}$ will pick up a phase $\exp[i(\lambda_x - \lambda_{x'})]$. We compensate for this change by replacing the vector potential by a variable that lives on the link between the points x and $x+1$,

$$gaA_1(ax) \to \theta_x \tag{3.18}$$

where $\theta_x \in [0, 2\pi]$ and define the local gauge transformation to be

$$\psi_x \to e^{i\lambda_x}\psi_x \tag{3.19}$$

$$\theta_x \to \theta_x - \lambda_{x+1} + \lambda_x. \tag{3.20}$$

Then the combinations $\bar{\psi}_x \exp(i\theta_x)\psi_{x+1}$ and $\bar{\psi}_{x+1} \exp(-i\theta_x)\psi_x$ are gauge invariant. It is also convenient to rescale the electric field into

$$E_1(ax) = g\ell(x). \tag{3.21}$$

The lattice Hamiltonian is then

$$H = \frac{1}{2}g^2 a \sum_{x=0}^{N-1} \ell_x^2 - \frac{i}{2a}\sum_{x=0}^{N-1} \left[\bar{\psi}_x e^{i\theta_x}\gamma_1\psi_{x+1} - \bar{\psi}_{x+1}e^{-i\theta_x}\gamma_1\psi_x\right] \tag{3.22}$$

$$+ \sum_{x=0}^{N-1} m\bar{\psi}_x\psi_x.$$

The new gauge variables obey the commutation relation

$$[\theta_x, \ell_{x'}] = i\delta_{x,x'}. \tag{3.23}$$

We assume periodicity in the links: $\ell_{-1} = \ell_{N-1}$. Since all quantities have period 2π in the link angle θ_x, ℓ_x must have only integer quantum numbers, in analogy with a quantized angular momentum.

Temporal gauge also requires that physical states satisfy Gauss' law:

$$\ell_x - \ell_{x-1} = \rho_x =: \psi_x^\dagger \psi_x : . \tag{3.24}$$

That is to say, the total flux leaving a site must equal the total charge on that site. The notation :: represents a suitably chosen normal ordering. Our periodic boundary condition on the electric flux then requires that the net charge be zero. We usually adjust the charge so that the vacuum is neutral. In perturbation theory the vacuum is the state obtained by occupying all the negative energy states. With two-component fermions the average vacuum occupation has one fermion per lattice site. So if we define

$$\rho_x =: \psi_x^\dagger \psi_x := \psi_x^\dagger \psi_x - 1, \tag{3.25}$$

the vacuum is neutral. In atomic lattices the vacuum state consists of a half-filled band with static ions of charge -1 on each site.

3.3 Feynman path integral

For numerical simulation it is desirable to reformulate the quantum theory in terms of ordinary numbers. The Hamiltonian formalism works with noncommuting operators, of course, so it is less amenable to direct computation. Feynman showed that the quantum partition function

$$Z(\beta) = \mathrm{Tr}\exp(-\beta H) \tag{3.26}$$

for bosons can be written as an integral involving ordinary numbers. As we will see, for fermions the integration involves anticommuting Grassmann variables, which might also present a computational challenge. Fortunately, for the models we are concerned with, that integration can be done by hand, leaving only ordinary numbers.

Of course, we are not merely interested in a partition function. We want to calculate particle masses, decay constants, and a host of other physically important observables. All such quantities start from the expectation value of a quantum observable $\mathcal{O}$:

$$\langle \mathcal{O} \rangle = \mathrm{Tr}[\mathcal{O}\exp(-\beta H)]/Z(\beta). \tag{3.27}$$

The Feynman technique allows us to compute both numerator and denominator from a multidimensional integral over ordinary numbers.

The steps leading to the Feynman path integration are particularly transparent for a lattice Hamiltonian. We begin by considering the noninteracting gauge field and fermions separately and finally give the result for the interacting theory.

3.3.1 *Pure gauge theory*

The Hamiltonian for the one-dimensional pure gauge theory is

$$H_G = \frac{1}{2} g^2 a \sum_{x=1}^{N-1} \ell_x^2 \tag{3.28}$$

and with no matter charges, physical states must satisfy $\ell_x - \ell_{x-1} = 0$. Periodicity requires $\ell_{-1} = \ell_{N-1}$. That makes this a very simple system. In the eigenbasis of ℓ_x the electric flux is a constant integer ℓ throughout.

The partition function is trivially calculated:

$$Z_G(\beta) = \operatorname{Tr} \exp(-\beta H_G) = \sum_{\ell=-\infty}^{\infty} \exp\left(-\frac{1}{2} g^2 a \beta \ell^2 N\right). \tag{3.29}$$

This was too easy. Let's recalculate the partition function following steps that generalize to the interacting case and higher dimensions.

We recall that the time evolution operator for a time-independent Hamiltonian is $\exp(iHt)$. So the partition function is the trace of the time evolution operator over an imaginary (often called Euclidean) time interval $t = i\beta$. We break this imaginary time interval into small intervals $\Delta\tau$ so that $M\Delta\tau = \beta$ and write

$$Z_G(\beta) = \operatorname{Tr} T^M \tag{3.30}$$

where $T = \exp(-H_G \Delta\tau)$. The operator T is called a transfer matrix. Acting on a state in the Hilbert space, it evolves the state through an imaginary time interval $\Delta\tau$.

The trace is conventionally done on the basis in which the θ_x are diagonal. However, such a state vector does not automatically satisfy Gauss' law. We introduce the law-enforcing projection operator

$$P = \prod_{x=0}^{N-1} \int_0^{2\pi} \frac{d\omega_x}{2\pi} \exp[i(\ell_x - \ell_{x-1})\omega_x]. \tag{3.31}$$

Then the trace over physical states is obtained by integrating freely over the set of all $\{\theta_x\}$ as follows:

$$Z(\beta) = \int \prod_{x=0}^{N-1} d\theta_x \left\langle \{\theta_x\} \right| \exp(-\beta H_G) P \left| \{\theta_x\} \right\rangle . \tag{3.32}$$

The projection P commutes with H_G so

$$Z(\beta) = \mathrm{Tr}(TP)^M . \tag{3.33}$$

We write the matrix product on the θ_x basis. There are M factors and a trace, requiring that we distinguish M sets of variables of integration. We do this with an extra subscript $\theta_{x,\tau}$ for $\tau = 0, \ldots, M-1$. We have, then,

$$\begin{aligned} Z(\beta) = \int \prod_x d\theta_{x,M-1} \prod_x d\theta_{x,M-2} \cdots \prod_x d\theta_{x,0} \left\langle \{\theta_{x0}\} \right| TP \left| \{\theta_{x,M-1}\} \right\rangle \\ \left\langle \{\theta_{x,M-1}\} \right| TP \left| \{\theta_{x,M-2}\} \right\rangle \cdots \left\langle \{\theta_{x,1}\} \right| TP \left| \{\theta_{x,0}\} \right\rangle . \end{aligned} \tag{3.34}$$

To evaluate the factors in the integrand from (3.28) and (3.31), we first evaluate

$$J = \left\langle \{\theta'_x\} \right| \exp\left(-\frac{\epsilon}{2}\ell_x^2\right) \exp\left[\sum_x i\omega_x(\ell_x - \ell_{x-1})\right] \left| \{\theta_x\} \right\rangle \tag{3.35}$$

for $\epsilon = g^2 a \Delta\tau$ and postpone the integration over the $\{\omega_x\}$. The translation operator $\exp(i\omega\ell_x)$ shifts θ_x to $\theta_x - \omega$. Thus

$$J = \left\langle \{\theta'_x\} \right| \exp\left(-\frac{\epsilon}{2}\sum_{x=0}^{N-1} \ell_x^2\right) \left| \{\theta''_x\} \right\rangle = \prod_{x=0}^{N-1} \left\langle \{\theta'_x\} \right| \exp\left(-\frac{\epsilon}{2}\ell_x^2\right) \left| \{\theta''_x\} \right\rangle \tag{3.36}$$

where the shifted angles are

$$\theta''_x = \theta_x - \omega_x + \omega_{x+1} . \tag{3.37}$$

for $x = 0, \ldots, N-1$ and $\omega_N \equiv \omega_0$. Each factor in J is similar to the standard quantum mechanical Green's function for a free nonrelativistic particle, except that the time interval is imaginary and the result must be periodic in the coordinate θ. The evaluation is done for small ϵ as follows:

$$\begin{aligned}\langle\theta'_x|\exp\left(-\frac{\epsilon}{2}\ell_x^2\right)|\theta''_x\rangle &= \sqrt{4\pi\epsilon}\,\langle\theta'_x|\int_{-\infty}^{\infty} d\theta\,\exp[i\ell_x\theta-\theta^2/2\epsilon]\,|\theta''_x\rangle \\ &= \sqrt{4\pi\epsilon}\sum_{n=-\infty}^{\infty}\exp\left[-(\theta'_x-\theta''+2\pi n)^2/2\epsilon\right] \quad (3.38)\\ &\sim \sqrt{4\pi\epsilon}\exp\left\{\frac{1}{\epsilon}[\cos(\theta'_x-\theta''_x)-1]\right\}.\end{aligned}$$

In the second step we used the translation property again and the identity

$$\langle\theta'_x|\theta''_x+\theta\rangle = \sum_{n=-\infty}^{\infty}\delta(\theta'_x-\theta''_x-\theta+2\pi n). \quad (3.39)$$

The third step is valid in the limit of small ϵ (continuum limit).

So, up to an irrelevant constant factor, the matrix element we need is

$$J\sim\exp\left\{\frac{1}{g^2a\Delta\tau}\sum_{x=0}^{N-1}[\cos(\theta'_x-\theta_x+\omega_x-\omega_{x+1})-1]\right\}, \quad (3.40)$$

and, finally, the partition function is

$$Z(\beta)\sim\int\prod_{\tau=0}^{M-1}\prod_{x=0}^{N-1}d\theta_{x,\tau}d\omega_{x,\tau}\,\exp(-S_G) \quad (3.41)$$

where

$$S_G=\frac{1}{g^2a\Delta\tau}\sum_{\tau=0}^{M-1}\sum_{x=0}^{N-1}[1-\cos(\theta_{x,\tau+1}-\theta_{x,\tau}+\omega_{x,\tau}-\omega_{x+1,\tau})]\,. \quad (3.42)$$

The trace requires periodicity in τ: $\theta_{xM}=\theta_{x0}$.

This important result (3.41) is the Feynman path integral with the Wilson action (3.42) for a $U(1)$ gauge field in one space and one Euclidean time dimension. The angle variables $\theta_{x\tau}$ and $\omega_{x,\tau}$ correspond to the space and time components of the vector potential:

$$\theta_{x,\tau}=gaA_1(ax,\tau\Delta\tau) \quad (3.43)$$

$$\omega_{x,\tau}=g\Delta\tau A_0(ax,\tau\Delta\tau). \quad (3.44)$$

The angle variables are associated with links on a space-time lattice as shown in Fig. 3.1. It is often convenient to represent the lattice link variables by the corresponding $U(1)$ group elements, $U_x(x,t)=\exp(i\theta_{x,\tau})$ and

$U_t(x,t) = \exp(i\omega_{x,\tau})$. Then the action is

$$S_{G-U(1)} = \frac{1}{g^2 a \Delta\tau} \sum_{\tau=0}^{M-1} \sum_{x=0}^{N-1} (1 - \mathrm{Re} U_{x\tau,01}) \tag{3.45}$$

where the plaquette variable,

$$U_{x\tau,01} = \exp(i\omega_{x,\tau}) \exp(i\theta_{x,\tau+1}) \exp(-i\omega_{x+1,\tau}) \exp(-i\theta_{x,\tau}), \tag{3.46}$$

is the product of the $U(1)$ link variables encountered in a circuit around a unit lattice square (plaquette) with a corner at x, τ and the other corners in a positive direction from there. The links are oriented according to the arrows in Fig. 3.1. When a link is traversed backwards we take the complex conjugate.

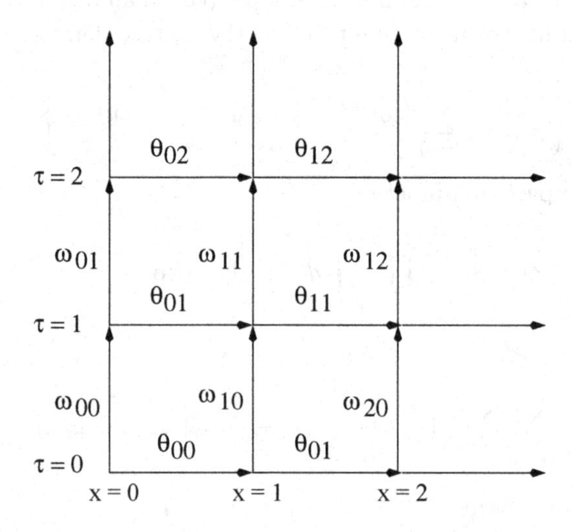

Fig. 3.1 Space-time lattice.

We make some important observations about this result:

- In the formal continuum limit $g\Delta\tau \to 0$ and $ga \to 0$ with A_0 and A_1 fixed, the action becomes

$$S_G \sim \int dx_0 \int_0^\beta d\tau \frac{1}{4} F_{\mu\nu}^2 \tag{3.47}$$

with $F_{\mu\nu} = \partial_\mu A_\nu - \partial_\nu A_\mu$.

- The action S_G is a Euclidean (imaginary time) classical action on a periodic space-time. Imaginary time is inherited from the quantum thermal partition function. Periodicity in time comes from the trace that defines the partition function. The extent of the time domain defines the inverse temperature β. Zero temperature corresponds to an infinite imaginary time extent.
- Periodicity in space is a feature of the model and could be replaced by other boundary conditions.
- The integration is over all classical values of the vector potential. This is called a "path integration", since a particular point in the space of integration variables describes a possible evolution (in discrete steps) of the classical vector potential through the imaginary time domain.
- A general $U(1)$ gauge transformation is specified by a $U(1)$ field $\exp(i\lambda_{x,\tau})$ on the lattice sites. The gauge links transform according to

$$e^{i\theta_{x,\tau}} \to e^{i\lambda_{x,\tau}} e^{i\theta_{x,\tau}} e^{-i\lambda_{x+1,\tau}} \tag{3.48}$$

$$e^{i\omega_{x,\tau}} \to e^{i\lambda_{x,\tau}} e^{i\theta_{x,\tau}} e^{-i\lambda_{x,\tau+1}}, \tag{3.49}$$

where $\lambda_{x,\tau} = \lambda_{x+N,\tau} = \lambda_{x,\tau+M}$. The action S_G is invariant under such a transformation.
- Although we started in temporal gauge, we ended with a path integration over all possible gauges. This freedom came from the temporal gauge projection operator that was needed to enforce Gauss's law. The auxiliary variable $\omega_{x,\tau}$ in that projection took on a new significance as the missing time component of the vector potential.

3.3.2 *Generalization to SU(3) pure gauge theory*

It is straightforward to generalize the result of the previous section to higher dimensions and other gauge groups. We do not take space here to derive everything from a Hamiltonian formalism. Instead we start from the lattice action (3.45).

In three spatial dimensions and one time dimension we define a symmetrical four-dimensional hypercubic lattice with lattice constant a. Each lattice site is labeled by a Cartesian four-vector $x = (\mathbf{x}, \tau)$ and has four links connecting nearest neighbors in the four positive coordinate directions. Let these four directions be denoted by unit vectors $\hat{\mu}$ for $\mu = 0, 1, \ldots, 3$. We put an $SU(3)$ link matrix $U_{x,\mu}$ on the link between sites x and $x + \hat{\mu}$.

The $SU(3)$ link field is related to the color vector potential field $A^c_{x,\mu}\lambda_c$ through

$$U_{x,\hat{\mu}} = \exp(igaA^c_{x,\mu}\lambda_c) \tag{3.50}$$

where λ_c for $c = 1, \ldots, 8$ are the eight generators of $SU(3)$ and g is the gauge coupling constant.

Wilson's generalization to $SU(3)$ (or any other gauge group) replaces the product of $U(1)$ variables by the trace of a product of $SU(3)$ matrices:

$$U_{x,\mu\nu} = \mathrm{Tr}\left(U_{x,\hat{\mu}}U_{x+\hat{\mu},\nu}U^\dagger_{x+\hat{\nu},\mu}U^\dagger_{x,\nu}\right). \tag{3.51}$$

Here x refers to the four-coordinate. The $U(1)$ complex conjugate has been replaced by the matrix hermitian conjugate.

In two dimensions we have only the 01 plane. In higher dimensions we need to distinguish the plane of the plaquette. The coupling constant becomes dimensionless in three dimensions and we have a prefactor proportional to $1/(g^2\Delta\tau/a)$. It is customary to set $\Delta\tau = a$. Then the full $SU(3)$ action in conventional normalization becomes

$$S_{G-SU(3)} = \frac{6}{g^2}\sum_x\sum_{\mu<\nu}(1 - \mathrm{Re}U_{x,\mu\nu}/3). \tag{3.52}$$

To complete the leap to lattice $SU(3)$ Yang-Mills theory we replace the $U(1)$ integration measure by the invariant $SU(3)$ Haar measure:

$$Z(\beta) = \int\prod_{x,\mu} dU_{x,\mu}\exp(-S_{G-SU(3)}). \tag{3.53}$$

This choice completes all the requirements for gauge invariance. A gauge transformation is specified by an $SU(3)$ matrix field W_x on the lattice sites. An $SU(3)$ link variable transforms according to

$$U_{x,\mu} \to W_x U_{x,\mu} W^\dagger_{x+\hat{\mu}}. \tag{3.54}$$

By definition the Haar measure is invariant under such a transformation. We leave as an easy exercise for the student to show that the action is also invariant.

In the formal continuum limit $a \to 0$ the lattice action becomes the continuum Yang-Mills action:

$$S_{G-SU(3)} \to \int_0^\beta d\tau \int d\mathbf{x} \frac{1}{4}(F^c_{\mu\nu})^2, \tag{3.55}$$

where

$$F^c_{\mu\nu} = \partial_\mu A^c_\nu - \partial_\nu A^c_\mu + g f_{abc} A^a_\mu A^b_\nu. \tag{3.56}$$

3.3.3 *Static quark potential*

In this section we show how to determine the potential between two point charges in $U(1)$ gauge theory and indicate its generalization to $SU(3)$.

3.3.3.1 *Polyakov loops and potential energy of point charges*

First, we consider how to introduce a point source in the two-dimensional $U(1)$ theory. Because of our periodic spatial boundary condition, we have to introduce charges in positive-negative pairs. A positive point charge of unit strength at coordinate $x = x_1$ and negative charge at x_2 modifies Gauss' law according to

$$\ell_x - \ell_{x-1} = \delta_{x,x_1} - \delta_{x,x_2}. \tag{3.57}$$

Now consider what this small change does to our derivation of the Feynman path integral. The projection operator (3.31) must be rewritten as

$$P(x_1, x_2) = \int_0^{2\pi} \left\{ \prod_{x=0}^{N-1} \frac{d\omega_x}{2\pi} \exp[i(\ell_x - \ell_{x-1})\omega_x] \right\} \exp(i\omega_{x_1}) \exp(-i\omega_{x_2}). \tag{3.58}$$

We have acquired two extra phase factors because of the point charge contribution.

Proceeding as before to the path integral, but with the modified projection operator, we obtain the partition function in the presence of the point charges:

$$Z(\beta, x_1, x_2) \sim \int \prod_{\tau=0}^{M-1} \prod_{x=0}^{N-1} d\theta_{x,\tau} d\omega_{x,\tau} \exp(-S_G) \mathcal{P}(x_1) \mathcal{P}^*(x_2) \tag{3.59}$$

where the Polyakov loop observable is

$$\mathcal{P}(x) = \prod_{\tau=0}^{M-1} \exp(i\omega_{x,\tau}). \tag{3.60}$$

To generalize to $SU(3)$ pure gauge theory in three spatial dimensions, replace the $U(1)$ point charges by spatially fixed color sources in the fundamental representation of the gauge group (fixed quarks). The Polyakov loop variable becomes

$$\mathcal{P}(x) = \mathrm{Tr} \prod_{\tau=0}^{M-1} U_{\mathbf{x},\tau}. \tag{3.61}$$

This result leads us to our first interesting physical observable. The free energy of separation of two point charges at inverse temperature β is obtained from the ratio

$$\exp[-\beta F(\mathbf{x_1}, \mathbf{x_2})] = Z(\beta, \mathbf{x_1}, \mathbf{x_2})/Z(\beta). \tag{3.62}$$

If we take the zero temperature limit $\beta \to \infty$ we obtain the potential energy between two static quarks.

3.3.3.2 *Wilson loops and point charge potential*

Wilson used a different approach to obtain the same quantity. He considered, instead, the process of creating a static pair of charges at imaginary time τ_1 and then annihilating them at time τ_2.

Such a process is a special case of the general thermal propagator expectation value as defined by (3.27):

$$C(\tau_2, \tau_1) = \langle \mathcal{A}_{\tau_2} \mathcal{B}_{\tau_1} \rangle . \tag{3.63}$$

Here the operators are time dependent. We make the time dependence explicit by going from the Heisenberg to the Schrödinger picture in imaginary time as follows:

$$\mathcal{O}_\tau = e^{\tau H} \mathcal{O} e^{-\tau H}. \tag{3.64}$$

So the thermal propagator becomes

$$C(\tau_2, \tau_1) = \mathrm{Tr} \left\{ e^{(-\beta+\tau_2)H} \mathcal{A} e^{(-\tau_2+\tau_1)H} \mathcal{B} e^{-\tau_1 H} \right\} / Z(\beta). \tag{3.65}$$

In the eigenbasis of the Hamiltonian, the result is reexpressed as

$$C(\tau_2,\tau_1) = \sum_n e^{(-\beta+\tau_2-\tau_1)E_n} \langle n|\,\mathcal{A}\,|m\rangle\, e^{(-\tau_2+\tau_1)E_m} \langle m|\,\mathcal{B}\,|n\rangle \,/\, \sum_n e^{-\beta E_n}. \tag{3.66}$$

If there is a gap between the vacuum and first excited state, this expression simplifies in the limit of low temperature and large time separation $|\tau_2 - \tau_1|$:

$$C(\tau_2,\tau_1) = \langle 0|\,\mathcal{A}\,|1\rangle \langle 1|\,\mathcal{B}\,|0\rangle\, e^{(-\tau_2+\tau_1)(E_1-E_0)} \tag{3.67}$$

where the state $|1\rangle$ is the lowest state reached from the vacuum by both operators.

A measurement of the exponential decay of the propagator correlation function $C(\tau_2,\tau_1)$ gives the excitation energy. This is the standard method for determining hadron masses.

We return, now, to the operator that creates a pair of static charges. We first consider the $U(1)$ theory. As we have seen, the creation of a pair of charges changes Gauss' law. Consistency then requires that when we create the pair of charges, we must also create the accompanying electric flux. The operator that increases the flux on a single link by one unit is the $U(1)$ link variable:

$$e^{i\theta_x}\,|\ell_x\rangle = |\ell_x + 1\rangle\,. \tag{3.68}$$

So the operator that increases the flux by one unit on all links between the two charges is

$$L = \prod_{x=x_1}^{x_2-1} e^{i\theta_x}\,. \tag{3.69}$$

Likewise the operator L^* decreases the flux by one unit. We must use the modified form of Gauss' law while the charges are present.

The Wilson loop is defined as the correlator

$$\langle W\rangle = \left\langle L^*_{\tau_2} L_{\tau_1} P\right\rangle = \left\langle L^*_{\tau_2} P(x_1,x_2) L_{\tau_1}\right\rangle. \tag{3.70}$$

The path integral representation becomes

$$\langle W\rangle = \int \prod_{\tau=0}^{M-1}\prod_{x=0}^{N-1} d\theta_{x\tau} d\omega_{x\tau}\, \exp(-S_G) W(x_1,\tau_1,x_2,\tau_2)/Z \tag{3.71}$$

where

$$W(x_1, \tau_1, x_2, \tau_2) = \prod_{x=x_1}^{x_2-1} e^{i\theta_{x,\tau_1}} \prod_{\tau=\tau_1}^{\tau_2-1} e^{i\omega_{x_2,\tau}} \prod_{x=x_1}^{x_2-1} e^{-i\theta_{x,\tau_2}} \prod_{\tau=\tau_1}^{\tau_2-1} e^{-i\omega_{x_1,\tau}}. \tag{3.72}$$

This observable is a generalization of the plaquette operator. The path traversed is a rectangle with opposite corners at (x_1, τ_1) and (x_2, τ_2).

The generalization of this result to $SU(3)$ is straightforward. The product of $U(1)$ links is replaced by a product of $SU(3)$ link matrices arranged in the order the links are traversed, and we take a trace over color. Both operators are gauge invariant under their respective gauge transformations.

3.4 Free fermions

We turn now to the fermion sector without the gauge field. The Hamiltonian formulation uses creation and annihilation operators. The corresponding path integral formulation requires Grassmann numbers. So we digress with a brief introduction to Grassmann calculus.

3.4.1 *Grassmann calculus*

Let $\eta_1, \ldots, \eta_n$ be a set of independent Grassmann numbers. They satisfy

$$\eta_i \eta_k + \eta_k \eta_i = \{\eta_i, \eta_k\} = 0 \tag{3.73}$$

so that $\eta_i^2 = 0$. A function of these variables can be expressed as a Taylor series, but the series terminates:

$$f(\eta) = a + \sum_i b_i \eta_i + \sum_{i<j} c_{ij} \eta_i \eta_j + \sum_{i<j<k} d_{ijk} \eta_i \eta_j \eta_k + \ldots + z \eta_1 \eta_2 \ldots \eta_n. \tag{3.74}$$

If f is a c-number-valued function, the even coefficients a, $c_{i,j}$, etc. are c-numbers and the odd coefficients b_i, d_{ijk}, etc., are independent Grassman numbers. Left partial differentiation with respect to the Grassmann numbers is defined as follows:

$$\frac{\partial}{\partial \eta_j} \eta_i = \delta_{ij} \quad ; \quad \frac{\partial}{\partial \eta_k} a = 0 \tag{3.75}$$

$$\frac{\partial}{\partial \eta_k} \eta_i \eta_j = \delta_{ik} \eta_j - \delta_{jk} \eta_i. \tag{3.76}$$

Note that the Grassman derivative anticommutes with an independent Grassmann number. Grassmann integration is the same as Grassmann differentiation, but for notational reasons it has been given a separate symbol:

$$\int d\eta = 0 \;\; ; \;\; \int d\eta_i \, \eta_j = \delta_{ij} \tag{3.77}$$

$$\int d\eta_k \, \eta_i \eta_j = \delta_{ik}\eta_j - \delta_{jk}\eta_i. \tag{3.78}$$

The Grassmann delta function is

$$\delta(\eta - \eta') = \eta - \eta' \tag{3.79}$$

so that

$$\int d\eta \, f(\eta)\delta(\eta - \eta') = f(\eta') \tag{3.80}$$

where $f(\eta)$ is a c-number valued function independent of η'. This result is easily demonstrated from a Taylor expansion of $f(\eta)$.

An important property of Grassmann integration is that if $\eta_1, \ldots, \eta_n$ and $\eta_1^*, \ldots, \eta_n^*$ are $2n$ independent Grassmann numbers and if M is an n-dimensional square matrix,

$$\int [d\eta^* d\eta] \exp(-\eta^\dagger M \eta) = \det M \tag{3.81}$$

where

$$[d\eta^* d\eta] = [d\eta_1^* d\eta_1 \ldots d\eta_n^* d\eta_n]. \tag{3.82}$$

Please note that η_k^* does not mean the complex conjugate of η_k! These labels characterize a second set of fermionic variables, which are written this way for notational convenience. The notation $\eta^\dagger$ means the transpose of η^*. As we will see, this identity is used to carry out the Grassmann path integration completely, leaving only an integration over ordinary numbers. An important related identity states that

$$\int [d\eta^* d\eta] \eta_j \eta_k^* \exp(-\eta^\dagger M \eta) = (M^{-1})_{jk} \det M, \tag{3.83}$$

which we need to evaluate operator correlations involving fermions. We also have

$$\int [d\eta^* d\eta] \exp(-\eta^\dagger M \eta) \exp(\chi^\dagger \eta + \eta^\dagger \chi) = \exp(\chi^\dagger M^{-1} \chi) \det M. \tag{3.84}$$

We next turn to the fermion Fock space. We use Grassmann numbers to parameterize the Fock basis. This is done through Fermi-coherent states, defined as follows. Let η be a Grassmann number and ϕ and $\phi^\dagger$ be fermion annihilation and creation operators. Let $|0\rangle$ be the vacuum state defined so that

$$\phi\,|0\rangle = 0. \tag{3.85}$$

Then define the state vector

$$|\eta\rangle = \exp(\phi^\dagger\eta)\,|0\rangle = (1+\phi^\dagger\eta)\,|0\rangle\,. \tag{3.86}$$

This is a vector in an enlarged Hilbert space in which the coefficients of the vectors can be Grassmann numbers. This vector is "coherent" in the same sense as a bosonic coherent state. It is an eigenstate of the annihilation operator with eigenvalue η:

$$\phi\,|\eta\rangle = \eta\,|\eta\rangle\,. \tag{3.87}$$

This result follows trivially from the definition. The dual vector is

$$\langle\eta| = \langle 0|\exp(\eta^*\phi) = \langle 0|\,(1+\eta^*\phi) \tag{3.88}$$

where η^* is independent of η. We can then compute the inner product of two vectors:

$$\langle\eta|\eta'\rangle = 1+\eta^*\eta' = \exp(\eta^*\eta'). \tag{3.89}$$

Finally, we have the completeness relation

$$1 = \int d\eta^* d\eta\,|\eta\rangle\langle\eta|\exp(-\eta^*\eta) \tag{3.90}$$

and the trace relation for an operator $\mathcal{O}$ on the Fock space

$$\mathrm{Tr}\mathcal{O} = -\int d\eta^* d\eta\,\langle\eta|\,\mathcal{O}\,|\eta\rangle\exp(\eta^*\eta). \tag{3.91}$$

Notice the sign change in the exponent compared with the sign in the completeness relation.

3.4.2 *Grassmann path integral for a two-level system*

We now consider the simplest fermion theory, namely, a two-level system with Hamiltonian

$$H_F = m\phi^\dagger\phi, \tag{3.92}$$

where $\phi^\dagger$ and ϕ are fermion creation and annihilation operators. The two-level Fock space consists of the vacuum, which we take to have zero energy, and a single excited state of energy m of maximum occupation 1. Direct substitution into the trace gives the partition function

$$Z(\beta) = \mathrm{Tr}\exp(-\beta H_F) = 1 + \exp(-\beta m). \tag{3.93}$$

To generalize this result to QCD it is convenient to derive a Grassmann functional integral representation. Proceeding in analogy with the pure $U(1)$ gauge theory we write the partition function as

$$Z(\beta) = \mathrm{Tr} T^M \tag{3.94}$$

where $M\Delta\tau = \beta$ and the transfer matrix is

$$T = \exp(-H_F \Delta\tau). \tag{3.95}$$

Using the trace and completeness relations for the Grassmann coherent states (3.90, 3.91) we obtain

$$\begin{aligned} Z(\beta) = \int & d\eta_M^* d\eta_M d\eta_{M-1}^* d\eta_{M-1} \dots d\eta_1^* d\eta_1 \, e^{-\eta_M^* \eta_0} \\ & \langle \eta_M | T | \eta_{M-1} \rangle \, e^{-\eta_{M-1}^* \eta_{M-1}} \langle \eta_{M-1} | \dots | \eta_1 \rangle \, e^{-\eta_1^* \eta_1} \langle \eta_1 | T | \eta_0 \rangle . \end{aligned} \tag{3.96}$$

where $\eta_M \equiv -\eta_0$, as required to produce the trace. This trace requirement is the origin of the fermion antiperiodic boundary condition in Euclidean time.

We evaluate the matrix element of T in two steps. First, it is easy to show that

$$\exp(-m\phi^\dagger\phi\Delta\tau) \left| \eta \right\rangle = \left| e^{-m\Delta\tau} \eta \right\rangle . \tag{3.97}$$

The transfer matrix element is then just the inner product of this result with the dual state

$$\langle \eta' | T | \eta \rangle = \langle \eta' \left| e^{-m\Delta\tau} \eta \right\rangle = \exp\left(\eta'^* \eta e^{-m\Delta\tau} \right) . \tag{3.98}$$

Therefore

$$Z(\beta) = \int [d\eta^* d\eta] \exp(-S_F), \tag{3.99}$$

where $[d\eta^* d\eta] = d\eta_0^* d\eta_0 \dots d\eta_{M-1}^* d\eta_{M-1}$,

$$S_F = \sum_{\tau,\tau'=0}^{M-1} \eta_{\tau'}^* M_{F\tau'\tau} \eta_\tau, \tag{3.100}$$

and the fermion matrix is

$$M_{F\tau'\tau} = \delta_{\tau'\tau} - \delta^F_{\tau'+1,\tau} e^{-m\Delta\tau}. \tag{3.101}$$

The antiperiodic Kronecker symbol δ^F is the ordinary Kronecker symbol on the interval $\tau, \tau' \in [0, M-1]$ and otherwise satisfies

$$\delta^F_{\tau',\tau+kM} = (-)^k \delta^F_{\tau',\tau} \tag{3.102}$$

for integer k.

From (3.81) and a direct evaluation of the determinant we can evaluate the partition function exactly:

$$Z(\beta) = \det M_F = 1 + \exp(-\beta m). \tag{3.103}$$

For small $\Delta\tau$ the action S_F becomes the continuum action for this problem:

$$S_F \sim \int_0^\beta d\tau\, \eta^*(\tau) \left(\frac{\partial}{\partial \tau} + m \right) \eta(\tau) \tag{3.104}$$

with $\eta(\beta) = -\eta(0)$ and $\eta^*(\beta) = -\eta^*(0)$.

3.4.3 *Fermion propagator*

It is instructive to evaluate the Euclidean (thermal) fermion propagator for the simple two-level fermion theory. In Hamiltonian language it is

$$G(\tau_2 - \tau_1) = \langle \phi(\tau_2) \phi^\dagger(\tau_1) \rangle . \tag{3.105}$$

where the Euclidean time dependence of any operator $\mathcal{O}$ is given by (3.64) with $H = H_F$, and $\langle\rangle$ is the thermal expectation value. Thus,

$$G(\tau_2 - \tau_1) = \mathrm{Tr}\left[e^{-\beta H_F} \phi(\tau_2) \phi^\dagger(\tau_1) \right] / \mathrm{Tr}\, e^{-\beta H_F}. \tag{3.106}$$

Therefore

$$G(\tau_2 - \tau_1) = \mathrm{Tr}\left[e^{-(\beta-\tau_2)H_F} \phi e^{-(\tau_2-\tau_1)H_F} \phi^\dagger e^{-\tau_1 H_F} \right] / \mathrm{Tr}\, e^{-\beta H_F}. \tag{3.107}$$

In this trivial model the propagator can be calculated immediately in the two-level Fock space basis, with the result

$$G(\tau_2 - \tau_1) = \theta(\tau_2 - \tau_1)\exp[-m(\tau_2 - \tau_1)\Delta\tau]/[1 + \exp(-\beta m)] \quad (3.108)$$

To deal with cases that cannot be solved so easily in closed form, we need to reexpress the thermal expectation value as a Grassmann functional integral. This is easily done following the same steps as before. The only difference is the appearance of

$$\begin{aligned} \langle\eta'|\,\phi^\dagger T\,|\eta\rangle &= \eta'^* \exp(\eta'^* \eta e^{-m\Delta\tau}) \\ \langle\eta'|\,T\phi\,|\eta\rangle &= \exp(\eta'^* \eta e^{-m\Delta\tau})\eta. \end{aligned} \quad (3.109)$$

Therefore

$$G(\tau_2 - \tau_1) = \int [d\eta^* d\eta]\eta_{\tau_2}\eta^*_{\tau_1} \exp(-S_f)/Z(\beta), \quad (3.110)$$

which, from the identities (3.81) and (3.83) is

$$G(\tau_2 - \tau_1) = M^{-1}_{F\tau_2\tau_1}. \quad (3.111)$$

for $\tau_2 \geq \tau_1$. This result generalizes to more complicated actions. The fermion propagator is just the inverse of the fermion matrix.

To evaluate the inverse of the matrix is trivial if we diagonalize it first, using a discrete Fourier transform,

$$\eta_\omega = \frac{1}{\sqrt{M}} \sum_{\tau=0}^{M-1} e^{2\pi i\omega\tau/M}\eta_\tau \quad (3.112)$$

for $\omega = 1/2, 3/2, \ldots, M - 1/2$. These values of omega give the required antiperiodicity in Euclidean time. On this basis the fermion matrix is just

$$M_{F\omega} = 1 - \exp(-m\Delta\tau + 2\pi i\omega/M) \quad (3.113)$$

so in the Euclidean time basis

$$M^{-1}_{F,\tau'\tau} = \frac{1}{M} \sum_{\omega=\frac{1}{2}}^{M-\frac{1}{2}} \frac{e^{2\pi i\omega(\tau'-\tau)/M}}{1 - \exp(-m\Delta\tau + 2\pi i\omega/M)}. \quad (3.114)$$

The sum can be carried out via contour integration in $z = e^{2\pi i\omega/M}$. There is one pole at $E = 2\pi i\omega/\beta = m$. Thus the dispersion relation is trivial, as we would expect. The result of the summation is (3.108).

3.4.4 *Path integral with particles and antiparticles*

For the more interesting relativistic case, we need to include antiparticles. So we consider the simple Hamiltonian

$$H = m\bar{\psi}\psi = m\psi_+^\dagger\psi_+ - m\psi_-^\dagger\psi_-, \tag{3.115}$$

where ψ and $\bar{\psi}$ are the two-component spinors introduced in Sec. 3.1.1 for the lattice Schwinger model, but for now living on only one spatial site. The lowest energy state (conventionally the vacuum) has an occupied negative energy level. The operator ψ_- is then interpreted as a creation operator for a hole state (antiparticle) and $\psi_-^\dagger$ as the antiparticle annihilation operator. The vacuum is assigned zero energy by normal ordering the Hamiltonian with respect to the newly defined creation and annihilation operators,

$$H =: m\bar{\psi}\psi := m\psi_+^\dagger\psi_+ + m\psi_{c+}^\dagger\psi_{c+}. \tag{3.116}$$

The operators $P_\pm = (1 \pm \gamma_0)/2$ project out the particle and antiparticle states. It is convenient to define particle-antiparticle Grassmann coherent states in terms of these projection operators as follows:

$$|\eta\bar{\eta}\rangle = \exp(\bar{\psi}P_+\eta + \bar{\eta}P_-\psi)\,|0\rangle \tag{3.117}$$

where

$$\eta = \begin{pmatrix}\eta_1\\ \eta_2\end{pmatrix} \quad \bar{\eta} = (\eta_1^*, -\eta_2^*). \tag{3.118}$$

The corresponding dual vector is

$$\langle\eta\bar{\eta}| = \langle 0|\exp(\bar{\eta}P_-\psi + \bar{\psi}P_+\eta). \tag{3.119}$$

The inner product is easily shown to be

$$\langle\eta'\bar{\eta}'\,|\eta\bar{\eta}\rangle = \exp(\bar{\eta}'P_+\eta + \bar{\eta}P_-\eta'). \tag{3.120}$$

The completeness relation is

$$1 = \int d\bar{\eta}d\eta\;|\eta\bar{\eta}\rangle\,\langle\eta\bar{\eta}|\exp(-\bar{\eta}\eta). \tag{3.121}$$

where $d\bar{\eta}d\eta = d\eta_1^* d\eta d\eta_2 d\eta_2^*$. The time evolution of the state is then given by

$$e^{-H\Delta\tau}\,|\eta\bar{\eta}\rangle = \left|e^{-m}\eta e^{-m}\bar{\eta}\right\rangle. \tag{3.122}$$

For future reference, we need the analogous identity involving the charge density (3.25):

$$e^{i\omega\rho}\left|\eta\bar{\eta}\right\rangle = \left|e^{i\omega}\eta e^{-i\omega}\bar{\eta}\right\rangle. \tag{3.123}$$

It is then easy to get the transfer matrix element

$$\langle\eta'\bar{\eta}'|\, e^{-H\Delta\tau}\left|\eta\bar{\eta}\right\rangle = \exp\left[e^{-m\Delta\tau}\left(\bar{\eta}'P_{+}\eta + \bar{\eta}P_{-}\eta'\right)\right]. \tag{3.124}$$

Thus the partition function is

$$Z(\beta) = \int [d\bar{\eta}d\eta]\exp(-S_F) \tag{3.125}$$

where

$$S_F = \sum_{\tau,\tau'=0}^{M-1} \bar{\eta}_{\tau'} M_{F\tau'\tau}\eta_{\tau}, \tag{3.126}$$

and

$$M_{F\tau'\tau} = \delta_{\tau'\tau} - \delta^F_{\tau',\tau+1}\frac{1+\gamma_0}{2}e^{-m\Delta\tau} - \delta^F_{\tau'+1,\tau}\frac{1-\gamma_0}{2}e^{-m\Delta\tau}. \tag{3.127}$$

In the continuum limit the action becomes

$$S_F \sim \int_0^\beta d\tau\,\bar{\eta}(\tau)\left(\gamma_0\frac{\partial}{\partial\tau} + m\right)\eta(\tau). \tag{3.128}$$

Again the lattice fermion matrix can be diagonalized in the time coordinate through a Fourier transform, leading to

$$M_{F\omega} = 1 - P_{+}\exp(-m\Delta\tau + 2\pi i\omega/M) - P_{-}\exp(-m\Delta\tau - 2\pi i\omega/M). \tag{3.129}$$

As before we identify the real energy $E = 2\pi i\omega/\beta$. The determinant vanishes for $E = \pm m$ as expected. The projection operators in (3.127) assure that the particle of positive energy moves forward in time and the particle of negative energy moves backward, which we interpret as an antiparticle moving forward.

A simpler action with the same continuum limit is

$$M'_{F\tau'\tau} = \delta_{\tau'\tau}\sinh(m\Delta\tau) + \frac{1}{2}\gamma_0 e^{-m\Delta\tau}\left[\delta^F_{\tau'+1,\tau} - \delta^F_{\tau',\tau+1}\right]. \tag{3.130}$$

But this action suffers from species doubling. It has two particle and two antiparticle states. This is easily seen by carrying out the Fourier transform and examining the energies that give a vanishing determinant. It is easy to show that the condition is $\sinh(E\Delta\tau)/\Delta\tau = \pm\sinh(m\Delta\tau)$. For the particle state (plus sign) there are two choices relating the energy to the discrete frequency, namely $E = 2\pi i\omega/\beta = m$ and $E = 2\pi i\omega/\beta + i\pi/\Delta\tau = -m$.

3.4.5 *Generalization to higher dimensions*

This simple example captures most of the important technology of fermion functional integration. A quick generalization to higher dimensions starts from the lattice functional integral, rather than the Hamiltonian. So in three spatial dimensions and one time dimension on a hypercubic lattice with spacing a, a relativistic free fermion of mass m could be described by the partition function

$$Z(\beta) = \int [d\psi^* d\psi] \exp(-S_F), \tag{3.131}$$

where

$$S_F = \sum_{x',x} \bar{\psi}_{x'} M_{F,x',x} \psi_x, \tag{3.132}$$

and where we might take the simplest discretization

$$M_F = \frac{1}{2a} \sum_{\mu=0}^{3} \left(\delta^F_{x'+\hat{\mu},x} - \delta^F_{x'-\hat{\mu},x} \right) \hat{\gamma}_\mu + \delta_{x',x} m. \tag{3.133}$$

The four-dimensional hermitian Euclidean gamma matrices are related to the conventional Minkowski gamma matrices through $\hat{\gamma}_0 = \gamma_0$ and $\hat{\gamma}_j = -i\gamma_j$ for $j = 1, 2, 3$. In the formal continuum limit the action becomes the familiar continuum Euclidean Dirac action

$$S_F \sim \int d^4x\, \bar{\psi}(x)(\partial_\mu \hat{\gamma}_\mu + m)\psi(x). \tag{3.134}$$

Here $\bar{\psi}_x$ and ψ_x are independent four-component Grassmann fields. The lattice dispersion relation is obtained by first diagonalizing the action with a Fourier transform.

$$S_F = \sum_p \bar{\psi}(p) \left[i\hat{p}_\mu \hat{\gamma}_\mu + m \right] \psi(p). \tag{3.135}$$

where $\hat{p}_\mu = \sin(p_\mu a)/a$. The zeros of the determinant $\det(\hat{p}_\mu \hat{\gamma}_\mu + m)$ as a function of $p_0 = iE$ at fixed $\mathbf{p}$ give the dispersion relation:

$$\sinh(Ea)/a = \pm\sqrt{\sum_{i=1}^{3} [\sin(p_i a)^2/a^2] + m^2}. \tag{3.136}$$

In the limit $a \to 0$ we recover the familiar relativistic relation $E = \pm\sqrt{\mathbf{p}^2 + m^2}$.

The central difference representation of the derivative is needed here to assure a real dispersion relation. This free relativistic fermion action is called "naive", because it is straight forward, but suffers severely from the fermion doubling problem – there are 16 equivalent fermion states of mass m. We return to this problem and its cures in Ch. 6.

3.5 The interacting theory

3.5.1 *Functional integral representation for the lattice Schwinger model*

The previous sections showed how to construct a functional integral representation of the partition function for a free gauge field and a free fermion field. We are now ready to combine these ingredients to obtain the functional integral representation for the interacting theory. We first return to the lattice Schwinger model (3.22).

There are a few new features. First, the Gauss' Law projection operator (3.31) must now include the fermion charge density:

$$P = \prod_{x=0}^{N-1} \int_0^{2\pi} \frac{d\omega_x}{2\pi} \exp[i(\ell_x - \ell_{x-1} - \rho_x)\omega_x]. \tag{3.137}$$

where ρ_x is given by (3.25). We need the particle-antiparticle coherent states of Sec. 3.4.4 with two components for every spatial site $|\{\eta\}\{\bar{\eta}\}\rangle$. So we have the relations

$$\begin{aligned}
P_+\psi_x \left|\{\eta\}\{\bar{\eta}\}\right\rangle &= P_+\eta_x \left|\{\eta\}\{\bar{\eta}\}\right\rangle \\
\bar{\psi}_x P_- \left|\{\eta\}\{\bar{\eta}\}\right\rangle &= \bar{\eta}_x P_- \left|\{\eta\}\{\bar{\eta}\}\right\rangle \\
\left\langle\{\eta\}\{\bar{\eta}\}\right| \bar{\psi}_x P_+ &= \left\langle\{\eta\}\{\bar{\eta}\}\right| \bar{\eta}_x P_+ \\
\left\langle\{\eta\}\{\bar{\eta}\}\right| P_-\psi_x &= \left\langle\{\eta\}\{\bar{\eta}\}\right| P_-\eta_x.
\end{aligned} \tag{3.138}$$

These identities are useful for evaluating matrix elements, as for example,

$$\langle\{\eta'\}\{\bar\eta'\}|\,\bar\psi_x(P_+ + P_-)\gamma_1\psi_{x+1}\,|\{\eta''\}\{\bar\eta''\}\rangle = \\ \langle\{\eta'\}\{\bar\eta'\}|\,\bar\eta'_x P_+\gamma_1\eta_{x+1} + \bar\eta''_x P_-\gamma_1\eta''_{x+1}\,|\{\eta''\}\{\bar\eta''\}\rangle\,, \tag{3.139}$$

where we have used $P_\pm\gamma_1 = \gamma_1 P_\mp$. States are then labeled by both the Grassmann variables and the link angles θ_x. The evaluation of the functional integral follows the same steps as for the free fermion and pure gauge case. We need to evaluate the transfer matrix element

$$\langle\{\eta'\},\{\theta'\}|\,e^{-H\Delta\tau}P\,|\{\eta\rangle\}\{\theta\}. \tag{3.140}$$

As in the pure gauge case we substitute the integral expression for the projection operator (3.137) and postpone integration over $\{\omega\}$. As before those angles are proportional to the time component of the vector potential. The integrand of the projection operator in (3.137) shifts the link angles θ in exactly the same way as in the pure gauge case. The action of the new factors $\exp(i\omega_x\rho_x)$ on the state follow (3.123):

$$\prod_{x=0}^{N-1}\exp(i\omega_x\rho_x)\,|\{\eta\}\{\bar\eta\}\rangle = |\{\eta''\}\{\bar\eta''\}\rangle \tag{3.141}$$

where $\eta''_x = \exp(i\omega_x)\eta_x$ and $\bar\eta''_x = \exp(-i\omega_x)\bar\eta_x$.

We are left with the matrix element

$$J = \langle\{\eta'\}\{\bar\eta'\},\{\theta'\}|\,e^{-H\Delta\tau}\,|\{\eta''\}\{\bar\eta\}''\{\theta''\}\rangle\,. \tag{3.142}$$

An exact evaluation is complicated by the fact that the Hamiltonian now involves noncommuting terms. For our purposes we seek a result that agrees with the transfer matrix element in the limit of small $\Delta\tau$. So we are allowed to discard terms of order $\Delta\tau^2$ for which noncommutation matters. In this spirit we obtain

$$\begin{aligned} J \sim \exp[-S_G - \frac{i\Delta\tau}{2a}\sum_{x=0}^{N-1}&(\bar\eta'_x P_+ e^{i\theta_x}\gamma_1\eta'_{x+1} + \bar\eta''_x P_- e^{i\theta_x}\gamma_1\eta''_{x+1} \\ &- \bar\eta'_{x+1}P_+e^{-i\theta_x}\gamma_1\eta'_x - \bar\eta''_{x+1}P_-e^{-i\theta_x}\gamma_1\eta''_x) \\ &+ (1 - m\Delta\tau\sum_{x=0}^{N-1}(\bar\eta'_x P_+\eta''_x + \bar\eta''_x P_+\eta'_x))] \end{aligned} \tag{3.143}$$

$$S_G = \frac{1}{g^2 a\Delta\tau}\sum_{x=0}^{N-1}[1-\cos(\theta'_x - \theta''_x)] \tag{3.144}$$

where we have dropped some irrelevant normalization constants. So we have, finally

$$Z(\beta) \sim \int [d\omega d\theta] \exp(-S) \tag{3.145}$$

where

$$S = S_G + \sum_{x',\tau',x,\tau} \bar{\eta}_{x'\tau'} M_{x'\tau',x\tau} \eta_{x\tau}, \tag{3.146}$$

S_G is given by (3.42), and

$$\begin{aligned} M_{x'\tau',x\tau} &= \delta_{x'x}(1 + m\Delta\tau)\left(P_+ e^{i\omega_{x\tau}} \delta_{\tau',\tau+1} + P_- e^{-i\omega_{x\tau}} \delta_{\tau,\tau'+1}\right) \\ &- \frac{i\Delta\tau}{2a}\gamma_1 \left(\delta_{x'+1,x} e^{i\theta_{x\tau}} - \delta_{x+1,x'} e^{-i\theta_{x\tau}}\right)\delta_{\tau',\tau}. \end{aligned} \tag{3.147}$$

The action, like the Hamiltonian, describes two fermion species. With the reduction of (3.13) we would have only one species.

The action is invariant under the gauge transformation (3.20) together with

$$\begin{aligned} \eta_{x\tau} &\to e^{i\lambda_{x\tau}} \eta_{x\tau} \\ \bar{\eta}_{x\tau} &\to e^{-i\lambda_{x\tau}} \bar{\eta}_{x\tau}. \end{aligned} \tag{3.148}$$

The continuum limit of the action is found by making the replacements (3.44) and $\eta_{x\tau} = \sqrt{a}\chi(ax, \tau\Delta\tau)$. It is necessary to take $\Delta\tau/a \to 0$ and $m\Delta\tau \to 0$ first and then $ga \to 0$ with A_μ and χ fixed. Then

$$S \sim \int dx \int_0^\beta d\tau \left[\frac{1}{4}F_{\mu\nu}^2 + \bar{\chi}(\hat{\gamma}_\mu \partial_\mu + igA_\mu \hat{\gamma}_\mu + m)\chi\right]. \tag{3.149}$$

where the two-dimensional gamma matrices coming from (3.1) are $\hat{\gamma}_0 = \gamma_0$ and $\hat{\gamma}_1 = -i\gamma_1$.

The more common "naive Wilson action" has the more symmetric form

$$S = S_G + \sum_{x',\tau',x,\tau} \bar{\eta}_{x'\tau'} \hat{M}_{x'\tau',x\tau} \eta_{x\tau} \tag{3.150}$$

where

$$\begin{aligned} \hat{M}_{x'\tau',x\tau} &= \frac{1}{2}\hat{\gamma}_0 \left[\delta_{\tau',\tau+1} e^{-i\omega_{x\tau}} - \delta_{\tau'+1,\tau} e^{i\omega_{x\tau}}\right]\delta_{x'x} \\ &+ \frac{\Delta\tau}{2a}\hat{\gamma}_1 \left[\delta_{x'+1,x} e^{i\theta_{x\tau}} - \delta_{x+1,x'} e^{i\theta_{x\tau}}\right]\delta_{\tau'\tau} \\ &+ m\Delta\tau \delta_{x',x}\delta_{\tau'\tau}, \end{aligned} \tag{3.151}$$

which has the same formal continuum limit, but it has four species of fermions. The additional species comes from eliminating the projection operators in the time derivative as discussed in Sec. 3.4.4.

In this chapter we have shown how the Feynman path integral is obtained from the lattice Hamiltonian for the Schwinger model, and we indicated how the result generalizes to QCD. The resulting Euclidean lattice action reduces to the familiar continuum Euclidean action in the limit of small lattice spacing. This criterion is the basis for generalizations and improvements to be discussed in later chapters. As we will see, the discretized lattice action that emerges takes on a new significance as a means of regulating ultraviolet divergences in the continuum action.

Chapter 4

Renormalization and the renormalization group

The renormalization group plays a central role in determining the relationship between the lattice spacing and coupling in QCD, in translating results obtained with lattice regularization to results obtained in other regularization schemes, and in fixing the anomalous dimensions of a variety of observables. In this chapter we give an brief, general overview of renormalization theory with applications to the lattice.

4.1 Blocking transformations

Suppose that we are interested in physical phenomena at some characteristic energy scale, and we happen to know the Lagrangian for the physical system at a deep fundamental level. For example, as particle physicists we might want to compute the properties of weak interactions at low energy, and we happen to know the Theory of Everything, or as atomic physicists we might understand the physics of electrons in atoms and be interested in studying Bose-condensed atomic gases. Also, we assume that we know that even the fundamental theory is incomplete; there is some shortest distance or highest momentum scale Λ beyond which we cannot calculate. The theory is then said to possess a cutoff. How can we extract information efficiently about low-energy-scale physics from such a fundamental theory?

The most efficient approach is not to compute directly with the fundamental theory: rather, we should identify the degrees of freedom of the fundamental theory that are relevant to the energy scale of interest and work with them, "thinning out" the more fundamental degrees of freedom to give us a simpler, effective Lagrangian. This thinning procedure is the basis of the "renormalization group."

For QCD a cutoff is required in order to regulate ultraviolet divergences.

The renormalization group describes the asymptotic removal of the cutoff. Critical behavior in condensed matter is characterized by an infinite correlation length. The renormalization group describes the long-distance behavior of the system at the critical point. Despite the fact that the cutoff in QCD is artificial and in solids, real, the mathematics is the same. So for simplicity, we introduce the renormalization group in the context of a lattice theory appropriate to a solid.

In the statistical mechanics literature, the object in the exponential of the Boltzmann factor is usually called the Hamiltonian H, and $Z = \mathrm{Tr}\exp(-H)$. The inverse temperature is absorbed into the parameters used to characterize H. We will retain field theoretic vocabulary and refer to this quantity as the action, S. Also in statistical mechanics, the techniques we describe apply to equilibrium systems, so the dimensionality of the system is the number of spatial dimensions. For a field theory in Euclidean space, the dimensionality is the sum of spatial and temporal dimensions (and of course in Euclidean space there is no distinction between them).

Consider a classical scalar field ϕ_x defined on discrete lattice points x and an action $S(\phi_x)$ in D dimensional Euclidean space. Our thinning procedure might be done in coordinate space: we might define a new block-averaged field by averaging over a volume $V = s^D$ around a point x,

$$\phi'_x = s^{-D} \sum_{x \in V} \phi_x. \tag{4.1}$$

The probability of finding a field configuration $\{\phi_x\}$ in our ensemble is given by Boltzmann weighting, $P(\{\phi\}) = \exp[-S(\phi)]$ for some action S. Now write the partition function as

$$Z = \int [d\phi_x] \exp[-S(\phi)] \tag{4.2}$$

and introduce the blocked variable through

$$Z = \int [d\phi_x][d\phi'_x] \prod_{x'} \delta\left(\phi'_{x'} - s^{-D} \sum_{x \in V_{x'}} \phi_x\right) \exp[-S(\phi)]. \tag{4.3}$$

Integrating over the original variables, we find

$$Z = \int [d\phi'_{x'}] \exp[-S'(\phi')] \tag{4.4}$$

where

$$\exp[-S'(\phi')] = \int [d\phi_x] \prod_{x'} \delta\left(\phi'_{x'} - s^{-D} \sum_{x \in V_{x'}} \phi_x\right) \exp[-S(\phi)]. \quad (4.5)$$

This thinning process defines a new action $S'(\phi')$ such that $P(\{\phi'\}) = \exp[-S'(\phi')]$.

Alternatively, we could introduce a momentum cutoff Λ/s and divide the field variables ϕ into low and high frequency parts, $\phi = \phi_H + \phi_L$, where ϕ_H has $\Lambda > \omega > \Lambda/s$ and ϕ_L has $\omega < \Lambda/s$. Thinning can be done by integrating over the high frequency part,

$$\int [d\phi_H][d\phi_L] \exp[-S(\phi_L, \phi_H)] = \int [d\phi_L] \exp[-S'(\phi_L)] \quad (4.6)$$

with

$$\exp[-S'(\phi_L)] = \int [d\phi_H] \exp[-S(\phi_L, \phi_H)]. \quad (4.7)$$

Either way, we can formally write the result as

$$S' = K_s S \quad (4.8)$$

where K_s is an operator that transforms the action. The subscript s reminds us that we have combined s^D spins, or reduced the effective size (degrees of freedom) of the system by a scale factor s. Clearly, the blocking can be done again with scale factor s', or both blockings can be done in one step with a scale factor ss' with the same result:

$$S'' = K_{s'} K_s S = K_{ss'} S. \quad (4.9)$$

Typically the actions have the form

$$S(\phi) = \sum_i g_i S_i(\phi) \quad (4.10)$$

where the S_i are coordinate space integrals (sums) over local functions of the field variables and their derivatives and the g_i are coupling constants. For example for a spin model, $S(\phi) = -J/kT \sum_{xy} \phi_x^* \phi_y - h/kT \sum_x \mathrm{Re}\phi_x$, where J/kT and h/kT are interpreted as coupling constants. In principle the transformation K_s preserves the form of the action. That is

$$S'(\phi') = \sum_i g'_i S_i(\phi'). \quad (4.11)$$

New operators may arise in the blocking process, but they can be included formally in the starting action with zero couplings. In this way repeated blocking can lead to very complicated actions.

It is often convenient to include a rescaling of the coordinates and fields with each blocking step. If, as is typically the case, the operators S_i are homogeneous in ϕ and the coordinates, the action still preserves its form under the combination of blocking and rescaling. Suppose that the original lattice spacing was b. After the blocking transformation, the new fields are defined on a grid of spacing sb; *i.e.*, there are grid points with no fields defined on them. We rename the lattice spacing label to remove these empty sites. It is also convenient to redefine the blocked field with an overall scale factor λ_s, so

$$\phi'_{x'} = \lambda_s \Phi_{x=x'/s} \tag{4.12}$$

and

$$\sum_i g'_i S_i[\phi'_{x'}] \equiv \sum_i g_{is} S_i[\Phi_x]. \tag{4.13}$$

Relabeling Φ as ϕ gives us a right hand side which looks identical to the original action (the fields are defined on every site), except that the couplings have changed.

We require that $\lambda_s \lambda_{s'} = \lambda_{ss'}$, so $\lambda_s = s^a$ for some as-yet undetermined exponent a. (Picking a is part of the choice of the renormalization group transformation.)

In this way we can characterize an action by listing all its coupling constants g_i: $S = S(g_1, g_2, \ldots)$, and the blocking transformation of the action is now specified by the formal relation

$$g_s = R_s(g). \tag{4.14}$$

Of course, we must have the composition rule

$$R_s * R_{s'} = R_{ss'}. \tag{4.15}$$

The word "renormalization" refers to the change in size of the g_i's and the (possible) rescaling of the ϕ_i's and x's. The composition relation Eq. (4.15) is the equation for a semi-group (not a group – in general R has no inverse). The equation Eq. (4.14) is called a "renormalization group equation" for the couplings.

Under a renormalization group transformation, a starting action that was originally specified by a small number of nonzero couplings can evolve into one characterized by many couplings. We will then have to find a way to organize the calculation. To begin doing that, let us consider an artificial case where there is only one g, and under a renormalization group transformation the new theory is also characterized by a single $g_s = f(g)$ for some function f. Let's assume that the function $f(g) - g$ has the behavior shown in Fig. 4.1. For $g > g_1^*$ (segment A), we get $g_s > g$. Here g grows under blocking, and in the limit of many blocking steps, it becomes infinite. This means that the long distance behavior of this theory is strongly coupled. If $g_1^* > g > g_2^*$ (segment B), then under repeated blocking, g shrinks until it reaches the point g_2^*. If $g_2^* > g$, the coupling grows until it reaches g_2^*. This is called an "infrared attractive fixed point." The physics at long distance scales is described by a theory whose coupling constant is equal to g_2^*. Point C, $g = g_1^*$, is also a fixed point. It is called an "infrared repulsive" fixed point, because the long distance behavior of the theory is dramatically different on either side of $g = g_1^*$.

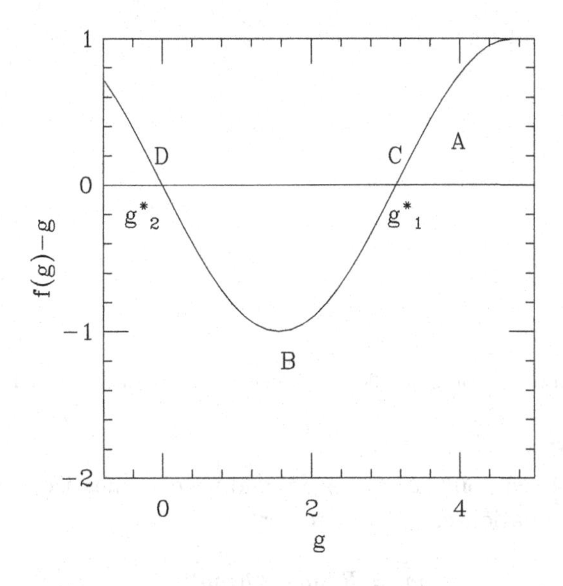

Fig. 4.1 A toy renormalization group flow.

When we consider flows in higher dimensional spaces of coupling constants, fixed points can be simultaneously attractive and repulsive. If we block by a series of small steps, we can map out "flows" or "streamlines" in coupling constant space showing the variation of couplings under blocking.

For example Fig. 4.2 shows a possible flow in a two dimensional coupling constant space. In this example the fixed point has one attractive direction and one repulsive direction. Notice that there is a line on which the flow converges to the fixed point: if the starting theory has coupling constants on that "critical" line it blocks to the same fixed point theory. With more couplings we could have a "critical surface." The set of all theories that lie in the same critical surface is called a "universality class." Typically, universality classes are characterized by the dimensionality of space and the internal symmetries of the bare actions. Thus, all three-dimensional isotropic ferromagnets form a universality class.

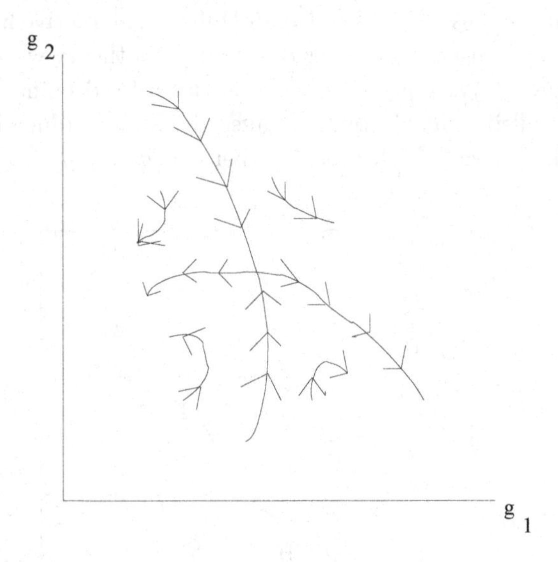

Fig. 4.2 A renormalization group flow with one attractive direction and one repulsive direction.

Close to a fixed point, $g = g^* + \delta g$, and we can linearize the renormalization group equation $g_s = R_s(g)$ to get

$$\delta g_s = R_s^L \delta g + \mathcal{O}((\delta g)^2) \tag{4.16}$$

where the constant matrix R_s^L is

$$(R_s^L)_{\alpha\beta} = \frac{\partial g_{s,\alpha}}{\partial g_\beta}|_{g=g^*}. \tag{4.17}$$

The eigenvectors e_j of R_s^L define directions in coupling constant space for which a renormalization group transformation is a pure scale change. The corresponding eigenvalues ρ_s must satisfy $\rho_s \rho_{s'} = \rho_{ss'}$. So $\rho(s)_j = s^{y_j}$. We expand an arbitrary displacement of the coupling as

$$\delta g = \sum_j t_j e_j. \tag{4.18}$$

Under a blocking transformation, $\delta g_s = \sum_j t'_j e_j$ with $t'_j = t_j s^{y_j}$. Clearly, if $y_j > 0$, t'_j grows with s, and the distance from the fixed point grows with s. If $y_j < 0$, the distance shrinks. The subspace of e_j's with $y_j < 0$ defines the critical surface near g^*.

Let us introduce some terminology. Coupling parameters $t'_i = t_i s^{y_i}$ that grow under a blocking transformation are called "relevant parameters" and the operators they multiply are called "relevant operators." In general, relevancy is only defined with respect to the fixed point. If R_s drives the operator away from the fixed point, it is relevant, and its value in the original short distance theory must be "tuned" (in particle physics contexts, we speak of "fine tuning") to produce critical physics. We must do something special to the coupling of a relevant operator in order to sit on the critical surface. Operators with $y < 0$ naturally flow into the fixed point. It does not matter what their values were in the original theory. These operators are called "irrelevant." If R_s^L does not change the operator, $y = 0$, and the operator is called "marginal." Thus, the low energy or long distance behavior of systems in the same universality class can be described by the same set of relevant and marginal operators.

Let's suppose that we have only one relevant operator, which we will label by the subscript 1. We tune the bare parameters in our theory to move on or off the critical surface; in fact, we can imagine that there is a single parameter that carries us on or off the critical surface. Again, by analogy with a ferromagnet, we call this single bare parameter T and suppose that when $T = T_c$ we are on the critical surface. Under a renormalization group transformation, the distance of coupling g_α from the critical surface, δg_α, shifts:

$$\delta g' = R_s \delta g = t_1(T) s^{y_1} e_1 + \mathcal{O}(s^{y_2}). \tag{4.19}$$

By definition the coefficient $t_1(T)$ vanishes at the fixed point, so we expect $t_1 \sim A(T - T_c)$, and

$$R_s \delta g \sim A(T - T_c) s^{y_1} e_1 + \ldots, \tag{4.20}$$

or

$$R_s \delta g \sim \left(\frac{s}{\xi}\right)^{y_1} e_1 \tag{4.21}$$

where we have defined $\xi = |A(T - T_c)|^{-y_1}$. The parameter ξ is called the "correlation length".

Earlier in this section we introduced two blocking methods: coordinate space and momentum space. The two blocking methods are not identical, but we want them to be approximately consistent. So we motivate the momentum space blocking prescription by taking the Fourier transform of the coordinate space blocking prescription. In D dimensions with lattice spacing b we have the following Fourier transforms of the original and blocked fields. The spatial volume L^D is unchanged, but the lattice spacing goes from b to sb. In momentum space the cutoff, then, goes from $\Lambda = 1/b$ to $\Lambda' = \Lambda/s$, and the couplings become g_i'.

$$\begin{aligned}\phi_k &= L^{-D/2} b^D \sum_x e^{ikx} \phi_x \\ \phi_k' &= L^{-D/2} (sb)^D \sum_{x'} e^{ikx'} \phi_{x'}' \\ &= L^{-D/2} (sb)^D \sum_{x'} e^{ikx'} s^{-D} \sum_{x \in V_{x'}} \phi_x.\end{aligned} \tag{4.22}$$

For low momentum the variation of the phase over the volume $V_{x'}$ is negligible and we have

$$\phi_k' \approx \phi_{L,k}. \tag{4.23}$$

So, before we do any rescaling, the low momentum components of the field are unchanged under blocking. In the momentum blocking scheme Eq. (4.7) this relation was a definition. Next, we rescale according to Eq. (4.12). The spatial volume is rescaled to $(L/s)^D$ and the lattice spacing is restored to b. The momentum cutoff is similarly restored to Λ. We have

$$\begin{aligned}\Phi_k &= (L/s)^{-D/2} b^D \sum_x e^{ikx} \Phi_x \\ &= b^D \sum_{x'} e^{ikx'/s} s^{-a} \phi_{x'}' = s^{-D/2-a} \phi_{k/s}'.\end{aligned} \tag{4.24}$$

So in the momentum blocking scheme, scaling is defined to be

$$\phi_{L,k} = \phi_k' = s^{a+D/2} \Phi_{sk}. \tag{4.25}$$

How are expectation values of observables altered in a blocking transformation? It is easiest to introduce the relevant formula for the case of momentum blocking, Eq. (4.25). Consider a two-point Green's function involving the low momentum modes. From Eq. (4.25) we have

$$\begin{aligned} G(k) &= \langle |\phi_{L,k}|^2 \rangle_P = \langle |\phi'_k|^2 \rangle_{P'} \\ &= s^{2a+D} \langle |\phi_{sk}|^2 \rangle_{P'} \end{aligned} \tag{4.26}$$

where P and P' indicate averaging with the unblocked and blocked probability distributions. So, reinserting the implicit dependence of the correlator on the coupling constants, we get

$$G(k, g_i) = s^{2a+D} G(sk, g'_i), \tag{4.27}$$

showing that the correlator picks up an overall scale prefactor under the blocking.

Let us see how this function behaves near the critical surface:

$$G[k, g(T)] = s^{2a+D} G[sk, g^* + R_s \delta g(T)] = s^{2a+D} G[sk, g^* + \left(\frac{s}{\xi}\right)^{y_1} e_1 + \ldots]. \tag{4.28}$$

This is true for any s as long as s is large, so in particular, we evaluate it for $s = \xi$,

$$G[k, g(T)] = \xi^{2a+D} G[\xi k, g^* + e_1 + \mathcal{O}(\xi^{y_2})] \sim \xi^{2a+D} g(\xi k). \tag{4.29}$$

We can also evaluate the expression at the fixed point $T = T_c$, where ξ diverges. We keep $k \neq 0$ and set $s = 1/k$:

$$G(k, g^*) = k^{-(2a+D)} G[1, g^* + \mathcal{O}(e^{y_2})] = k^{-(2a+D)} \overline{g}(1). \tag{4.30}$$

It is customary to define a correlation length through the coordinate-space correlator, which in free field theory is $G(r) \sim \exp(-r/\xi)/r^{D-2}$. In momentum space, the corresponding free field correlator is $G(k) \sim 1/(k^2 + 1/\xi^2)$. In the latter formula, we recognize that $1/\xi$ is the mass of the exchanged particle. When this mass vanishes, or when the correlations length diverges, $G(k)$ shows pure power law dependence on k, $G(k) \sim k^{-2}$. This is the same kind of behavior as shown in Eq. (4.30). This behavior is called "scaling." We associate scaling behavior with the presence of a massless excitation. Interactions cause the exponent $2a + D$ to deviate its free field value of 2.

It is easy to see that physics on the critical surface is universal: at long distances the effective theory is the action at the fixed point, to which all the couplings have flowed, regardless of their original bare values.

Next, imagine tuning bare parameters close to the critical surface, but not on it. The system will flow toward the fixed point, then away from it. The flow lines in coupling constant space will asymptotically approach a particular trajectory, called the renormalized trajectory, which connects (at $\xi = \infty$) with the fixed point. Along the renormalized trajectory, ξ is finite. However, because it is connected to the fixed point, it shares the scaling properties of the fixed point–in particular, the complete absence of cutoff effects in the spectrum and in Green's functions. (To visualize this remarkable result, imagine doing QCD spectrum calculations with the original bare action with a cutoff equal to the Planck mass and then coarse graining. Now exchange the order of the two procedures. If this can be done without making any approximations the answer should be the same.)

A useful analogy of this picture is think of the critical surface as the top of a high mountain ridge. The fixed point is a saddle point on the ridge. A stone released on the ridge will roll to the saddle and come to rest. If it is not released exactly on the ridge, it will roll near to the saddle, then go down the gully leading away from it. That gully is the renormalized trajectory. For a cartoon, see Fig. 4.3.

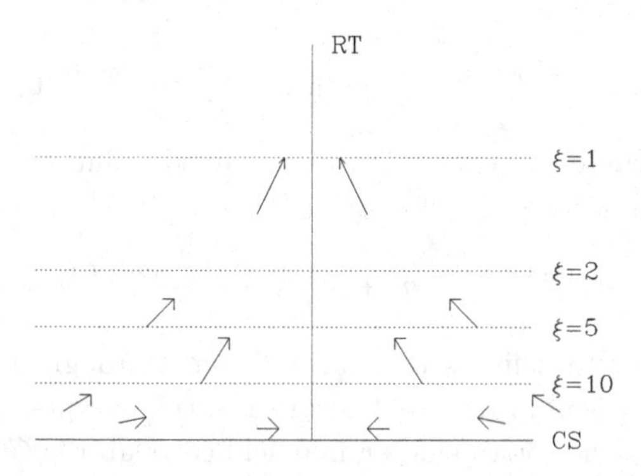

Fig. 4.3 A schematic picture of renormalization group flows along a one-dimensional critical surface, with the associated renormalized trajectory, and superimposed contours of constant correlation length.

The bare action for an asymptotically free theory like QCD can be characterized by one marginally relevant operator, $F_{\mu\nu}^2$, and its associated coupling g^2, plus many irrelevant operators. In lattice language, our bare actions are described by one overall factor of $\beta = 2N/g^2$ and arbitrary weights of various closed loops,

$$\beta S = \frac{2N}{g^2} \sum_j c_j S_j. \tag{4.31}$$

Asymptotic freedom is equivalent to the statement that the critical surface of any renormalization group transformation is at $g^2 = 0$. The location of a fixed point involves some relation among the c_j's. From the discussion in the previous paragraph, setting the bare gauge coupling to some small but nonzero value amounts to defining the bare theory slightly away from the critical surface. Under blocking transformations, the theory flows away from the critical surface, along the renormalized trajectory. This will happen regardless of the value of the c_j's. Now we see how lattice QCD can produce predictions. If some discretization of lattice QCD is in the same universality class as continuum QCD (or more precisely QCD with some continuum cutoff), they correspond to theories of the form of Eq. (4.31) with different c_i's. Their long distance behavior is given by the behavior of the theory along the same renormalized trajectory and will therefore be identical. The only thing special about the lattice theory is that we use it for expediency: we know how to do nonperturbative calculations with it.

4.2 Renormalization group equations

Suppose we consider a field theory with a cutoff, formulated initially in terms of bare fields and couplings, and we use it to compute Green's functions beyond the lowest order in perturbation theory. We may find that the Green's functions do not depend separately on the bare parameters and the cutoff. Instead, the Green's functions can be written in terms of products of functions of the appropriate kinematic variables (the particles' momenta, for example) times coefficients that are specific functions of the bare parameters and the cutoff. These coefficients are called "renormalized couplings." The overall normalization of a correlator can shift relative to what we would find in a tree-level calculation. We express this shift by considering correlators of "renormalized fields." All cutoff dependence disappears when we express our answer in terms of renormalized quantities. If that happens, our theory is said to be "renormalizable." Generally, this will occur only if the theory contains relevant and marginal operators. It is a remarkable property of renormalizable field theories, that one can make predictions with them without any knowledge of the cutoff: simply determine the renormalized coupling parameters, which are typically finite in number, from a finite set of experimental data, and then use them to make additional predictions.

If the bare theory had irrelevant operators, Green's functions would depend on the cutoff, containing extra terms that scale as inverse powers of the cutoff, so that as the cutoff momentum scale is taken to be large compared with the scale at which Green's functions are measured, the cutoff dependence would disappear. The concept of a renormalized coupling or field is still a useful one, as we will see.

Suppose we view our candidate theory in terms of its renormalized parameters. We need no longer care what the bare parameters are, nor what the scale of the cutoff is (as long as it is set to be a shorter distance scale than the scale at which we do experiments). But the theory we originally wrote down was the bare theory with a cutoff. Would it not be more natural just to compute with it? Not necessarily. If the theory is renormalizable, its renormalized fields and couplings are what characterize all Green's functions, so as soon as we require consistency with experiment, the bare parameters and the cutoff are not independent quantities.

The foregoing is the philosophy underlying renormalization and the renormalization group equations. In the following subsections we review renormalization in ϕ^4 field theory and then generalize to discuss the renormalization group equations themselves.

4.2.1 *Renormalization in scalar field theory*

We have see that it is useful to consider how the bare parameters and cutoff combine to give renormalized parameters in a model renormalizable field theory. For a concrete example, consider (yet again) the Lagrangian for a scalar field

$$\mathcal{L} = \frac{1}{2}(\partial_\mu \phi_0)^2 - \frac{1}{2} m_0 \phi_0^2 - \frac{\lambda_0}{4!}\phi_0^4 \tag{4.32}$$

and a momentum cutoff Λ. We assume perturbation theory is valid, and we want to consider various corrections to tree-level Green's functions. With one exception, we will restrict our calculation to the lowest nontrivial order in perturbation theory.

The simplest Green's function is the boson self energy, given by the

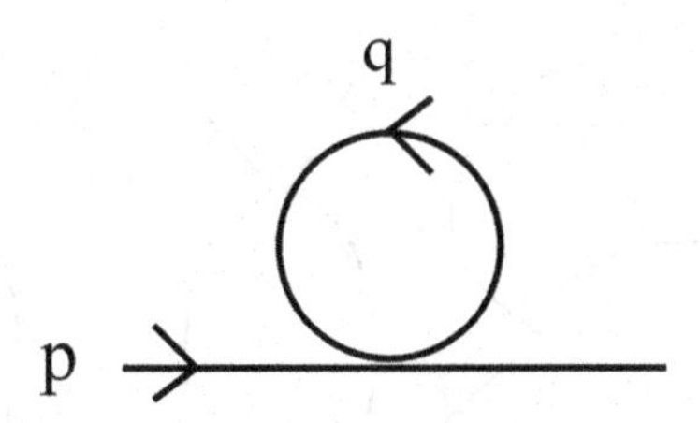

Fig. 4.4 The self energy graph.

Feynman graph of Fig. 4.4. It is

$$\Sigma(p) = -\frac{\lambda_0}{2}\int_0^\Lambda \frac{d^4q}{(2\pi)^4}\frac{1}{q^2+m_0^2}. \tag{4.33}$$

The integrand has no angular dependence, so

$$\Sigma = -\frac{\lambda_0}{2}\frac{\pi^2}{(2\pi)^4}\int_0^{\Lambda^2}\frac{x}{x+m_0^2}dx, \tag{4.34}$$

which for $\Lambda^2 \gg m_0^2$ simplifies to

$$\Sigma \to -\frac{\lambda_0}{32\pi^2}\left[\Lambda^2 - m_0^2 \ln\frac{\Lambda^2}{m_0^2}\right]. \tag{4.35}$$

How does the self energy Σ contribute to the boson propagator? We can sum up its contribution to all orders in λ_0,

$$G = G_0 + G_0\Sigma G_0 + G_0\Sigma G_0\Sigma G_0 + \cdots = G_0\frac{1}{1-\Sigma G_0}, \tag{4.36}$$

and if $G_0 = 1/(p^2+m_0^2)$,

$$G = \frac{1}{p^2+m_0^2-\Sigma}, \tag{4.37}$$

i.e., the particle has an effective mass

$$m^2 = m_0^2 - \Sigma = m_0^2 + \frac{\lambda_0}{32\pi^2}\left[\Lambda^2 - m_0^2\ln\frac{\Lambda^2}{m_0^2}\right]. \tag{4.38}$$

Next, let's look at the two-to-two scattering process shown in Fig. 4.5. We focus on the leftmost of the $M^{(2)}$ graphs, $M^{(2a)}$,

$$M^{(2a)} = \frac{\lambda_0^2}{2}\int_0^\Lambda \frac{d^4l}{(2\pi)^4}\frac{1}{[l^2+m_0^2][(p-l)^2+m_0^2]}, \tag{4.39}$$

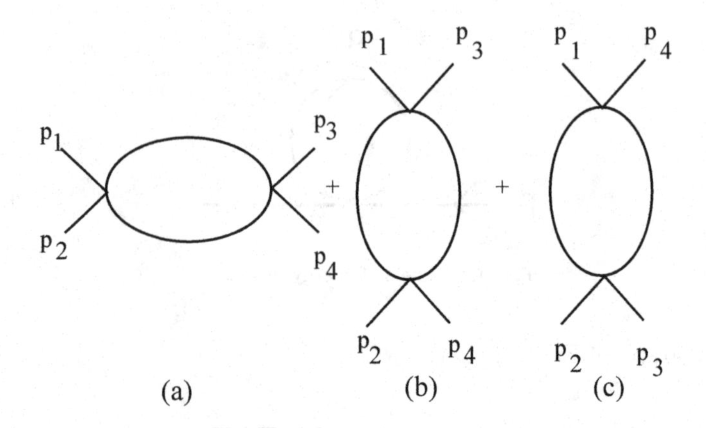

Fig. 4.5 One-loop scattering amplitudes in ϕ^4 field theory.

where $p = p_1 + p_2$. Introducing the Feynman parameter expression

$$\frac{1}{ab} = \int_0^1 dx \frac{1}{[ax + b(1-x)]^2} \tag{4.40}$$

we find

$$M^{(2a)} = \frac{\lambda_0^2}{2(2\pi)^4} \int d^4 l dx \frac{1}{[l^2 + m_0^2 - 2l \cdot px + p^2 x]^2}. \tag{4.41}$$

After a change of variables, $l' = l - px$, the denominator becomes $[l'^2 + m_0^2 + p^2 x(1-x)]^2$, the angular dependence disappears from the integrand, the integral over l' can be done, and we have

$$M^{(2a)} = \frac{\lambda_0^2}{32\pi^2} \int_0^1 dx \left[\ln\left(1 + \frac{\Lambda^2}{m_0^2 + p^2 x(1-x)}\right) - \frac{\Lambda^2}{\Lambda^2 + m_0^2 + p^2 x(1-x)} \right]. \tag{4.42}$$

If $\Lambda^2 \gg m_0^2$, we can drop the last term and the "1" in the first term. Let's also separate the p^2 dependence from the Λ dependence by introducing an arbitrary mass scale μ^2 (μ is called the "regularization point"). Then

$$\begin{aligned} M^{(2a)} &= \frac{\lambda_0^2}{32\pi^2} \int_0^1 dx \left[\ln \frac{\Lambda^2}{\mu^2} - \ln \frac{m_0^2 + p^2 x(1-x)}{\mu^2} \right] \\ &= \frac{\lambda_0^2}{32\pi^2} \left[\ln \frac{\Lambda^2}{\mu^2} - \hat{I}(p^2) \right]. \end{aligned} \tag{4.43}$$

Note that $\hat{I}(p^2)$ is finite as $\Lambda \to \infty$, but is also arbitrary (we must specify some value of μ). The other two diagrams are identical except the momenta vary. With the tree-level term included, the complete invariant amplitude up to second order in λ_0 is

$$M = -\lambda_0 + \frac{3\lambda_0^2}{32\pi^2} \ln \frac{\Lambda^2}{\mu^2} - \frac{\lambda_0^2}{32\pi^2}[\hat{I}(s) + \hat{I}(t) + \hat{I}(u)], \tag{4.44}$$

where $s = (p_1+p_2)^2$, $t = (p_1-p_3)^2$, and $u = (p_1-p_4)^2$ are the Mandelstam invariants.

The momentum-independent term

$$\lambda = \lambda_0 - \frac{3\lambda_0^2}{32\pi^2} \ln \frac{\Lambda^2}{\mu^2} \tag{4.45}$$

is interpreted as the renormalized coupling. Indeed, if we computed to higher order, we would always find the same combination of bare coupling and cutoff. In particular, we would be led to replace λ_0^2 with λ^2 in the one loop term above.

One of the popular original renormalization procedures added counterterms to the Lagrangian to cancel the divergences in the renormalization of coupling and mass. Since the divergent terms involving Λ are just constant shifts in the mass and coupling, the required counterterms are just a new mass and ϕ^4 term. No new operators are required. That was the original definition of renormalizability.

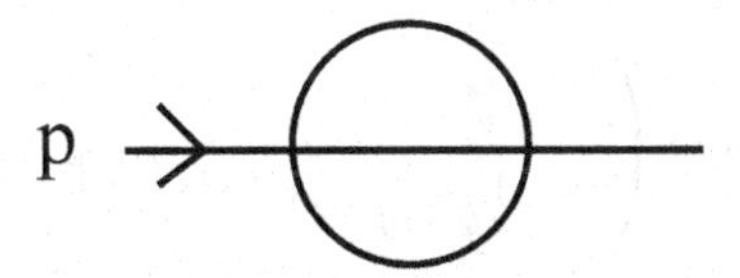

Fig. 4.6 Lowest-order contribution to field renormalization.

There is one other kind of Λ dependence in Green's functions, a contribution to $\Sigma(p)$ that is proportional to p^2. It comes from the graph of Fig. 4.6, and it happens that it is equal to

$$\left[\frac{\lambda_0^2}{24(16\pi)^2} \ln \frac{\Lambda^2}{\mu^2}\right] p^2. \tag{4.46}$$

If we recalculate the ϕ_0 propagator while including it, we find

$$\langle\phi_0(-p)\phi_0(p)\rangle = \frac{1}{\left(1-\frac{\lambda_0^2}{24(16\pi)^2}\ln\frac{\Lambda^2}{\mu^2}\right)p^2+m_0^2+\frac{\lambda_0}{32\pi^2}(\Lambda^2-m_0^2\ln\frac{\Lambda^2}{m_0^2})} \tag{4.47}$$

or

$$\langle\phi_0(-p)\phi_0(p)\rangle \equiv \frac{1}{Z_\phi}\frac{1}{p^2+m^2} \tag{4.48}$$

where $Z_\phi = 1-\frac{\lambda_0^2}{24(16\pi)^2}\ln\Lambda^2/\mu^2$ is also cutoff-dependent. We can split Z_ϕ into two terms, one for each ϕ_0

$$\phi = \frac{1}{\sqrt{Z_\phi}}\phi_0, \tag{4.49}$$

so

$$\langle\phi(-p)\phi(p)\rangle = \frac{1}{p^2+m^2}. \tag{4.50}$$

Thus, to this order in perturbation theory, all the cutoff dependence is confined to three places,

$$m^2 = m_0^2 - \Sigma = m_0^2 + \frac{\lambda_0}{32\pi^2}\left[\Lambda^2 - m_0^2\ln\frac{\Lambda^2}{m_0^2}\right], \tag{4.51}$$

$$\lambda = \lambda_0 - \frac{3\lambda_0^2}{32\pi^2}\ln\frac{\Lambda^2}{\mu^2}, \tag{4.52}$$

and

$$\phi = \left(\frac{1}{1-\frac{1}{2}\frac{\lambda_0^2}{24(16\pi)^2}\ln\frac{\Lambda^2}{\mu^2}}\right)\phi_0. \tag{4.53}$$

In higher order, the cutoff dependence continues to appear only in m^2, λ, and ϕ. We can interpret these relations as showing that the quantities m^2, λ, and ϕ (which we imagine being measured in experiments) have no cutoff dependence, and that Green's functions parameterized in terms of m, λ and ϕ are finite as the cutoff is made very large. In order to do this, we must think of the bare parameters m_0, λ_0, and ϕ_0 as implicit functions of the cutoff Λ; as Λ is varied, we must tune m_0, λ_0, and ϕ_0 so that m, λ and ϕ are unchanged. The regularization point μ is fixed when we (arbitrarily) choose some prescription to define the scattering amplitude. For example, from Eq. (4.44) we could fix the value of λ knowing the value of M at some fiducial choice of the external momenta.

A technical aside: the reader has presumably seen renormalization of ϕ^4 field theory before, perhaps in the context of dimensional regularization. UV divergences are regulated by continuation away from four dimensions. The regularization point μ then appears because away from four dimensions the coupling has a dimension. Its physics content, that it is an arbitrary point used to separate an expression into a cutoff-dependent part and a cutoff-independent part, is identical to what we have seen here.

4.3 Renormalization group equations for the scalar field

Let us generalize the discussion and consider a renormalized Green's function for a set of renormalized fields, a renormalized coupling, and a regularization point μ, (while still assuming that the mass is zero)

$$G^{(n)}(p_1, \ldots, p_n, \mu, g) = \langle \phi(p_1) \ldots \phi(p_n) \rangle. \tag{4.54}$$

It is also equal to a correlator of the bare fields

$$Z_\phi^{-n/2} \langle \phi_0(p_1) \ldots \phi_0(p_n) \rangle \equiv Z_\phi^{-n/2} G_0^{(n)}. \tag{4.55}$$

If we shift the regularization point $\mu \to \mu + \delta\mu$, and tune the renormalized couplings to compensate, $g \to g + \delta g$, the bare Green's function remains unchanged, and so

$$\frac{\partial G}{\partial \mu}\delta\mu + \frac{\partial G}{\partial g}\delta g + \frac{n}{2}\frac{1}{Z_\phi}\frac{\partial Z_\phi}{\partial \mu} G \delta\mu = 0. \tag{4.56}$$

Multiplying by $\mu/\delta\mu$, converting variations to partial derivatives, and defining

$$\mu \frac{\partial g}{\partial \mu} = \beta(g); \qquad \frac{1}{2}\frac{\mu}{Z_\phi}\frac{\partial Z_\phi}{\partial \mu} = \gamma_\phi, \tag{4.57}$$

we get

$$\left[\mu \frac{\partial}{\partial \mu} + \beta(g)\frac{\partial}{\partial g} + n\gamma_\phi\right] G^{(n)}(p_1, \ldots, p_n, g, \mu) = 0. \tag{4.58}$$

This is an example of a renormalization group equation. It tells us that the functions β and γ compensate the Green's function for variation in the renormalization scale.

To simplify the discussion, we assume that our fields are massless. If we assume that $G^{(n)}$ has an "engineering" dimension $(\text{energy})^{d_n}$, and if we scale all momenta by $p \to sp$, and $\mu \to s\mu$, simple dimensional analysis gives

$$\left[\mu \frac{\partial}{\partial \mu} + s\frac{\partial}{\partial s} - d_n\right] G^{(n)}(sp_1, \ldots, sp_n, g, \mu) = 0 \tag{4.59}$$

and, combining the two previous equations gives

$$\left[s\frac{\partial}{\partial s} - \beta(g)\frac{\partial}{\partial g} - n\gamma_\phi - d_n\right] G^{(n)}(sp_1, \ldots, sp_n, g, \mu) = 0. \tag{4.60}$$

This equation tells us the variation of the Green's function with respect to variation in the external momenta. If we could solve it, we would know $G^{(n)}(sp)$ in terms of $G^{(n)}(p)$. This requires knowledge of $\beta(g)$, and γ_ϕ. If we knew them, we could introduce a scale-dependent coupling constant

$$s\frac{\partial \overline{g}(s)}{\partial s} = \beta[\overline{g}(s)]; \qquad \overline{g}(1) = g \tag{4.61}$$

and then integrate Eq. (4.60) to yield

$$G^{(n)}(sp, g, \mu) = s^{d_n} G^{(n)}[p, \overline{g}(s), \mu] \exp\left(-n\int_1^s \frac{ds'}{s'}\gamma_\phi[\overline{g}(s')]\right). \tag{4.62}$$

In words, a change in the momentum scale $p \to sp$ is equivalent to

- multiplication (rescaling) by s^{d_n}
- a modified coupling constant that flows according to the beta function
- an extra factor that gives rise to a so-called "anomalous dimension"

The behavior of $G^{(n)}$ is very complicated, but there is a special case where it becomes simple. That is, if the running coupling reaches a value g^* such that further changes in momentum do not affect it. That happens when $\beta(g^*) = 0$, *i.e.*, at fixed points in the renormalization group flow of coupling constants. Then we can replace g by g^* in the last term of Eq. (4.62) and replace γ_ϕ by $\gamma_\phi^* = \gamma_\phi(g^*)$, which is independent of s'. Then

$$G^{(n)}(sp, g, \mu) = s^{d_n - n\gamma_\phi^*} G^{(n)}(p) \tag{4.63}$$

or

$$G^{(n)}(p, g, \mu) \sim p^{d_n - n\gamma_\phi^*}. \tag{4.64}$$

This is the scaling behavior of Eq. (4.30).

We might want to consider more complicated correlators. We can think of these correlators as not just being simple products of fields, but as correlators of operators (which are, of course themselves products of fields). We can regard these operators either as functions of the bare parameters, or of the renormalized ones. A renormalization constant would relate the bare and renormalized quantities, carrying all the cutoff and scale dependence:

$$O_0 = Z_O(\mu)O(\mu). \tag{4.65}$$

One can generalize the discussion of the previous paragraphs by considering mixed Green's functions

$$G^{(n,i)} = \langle \phi_1 \dots \phi_n O_1 \dots O_i \rangle = Z_\phi^{-n/2} Z_0^{-i} G_0^{(n,i)}. \tag{4.66}$$

This Green's function also obeys a renormalization group equation

$$\left[\mu \frac{\partial}{\partial \mu} + \beta(g) \frac{\partial}{\partial g} + n\gamma_\phi + i\gamma_O \right] G^{(n,i)} = 0, \tag{4.67}$$

where

$$\gamma_O = \frac{\mu}{Z_O} \frac{\partial Z_O}{\partial \mu}. \tag{4.68}$$

Finally, we have not discussed renormalization group equations for massive theories. We can think of a massive theory as a massless theory that happens also to contain an operator O with a coupling constant m that carries a dimension. We can think of Green's functions for the massive theory to be Green's functions of the form of Eq. (4.66), which obey the renormalization group equation Eq. (4.67). Just as dimensionless couplings have beta functions, $\beta(g) = \mu \partial g / \partial \mu$, so do dimensionful ones like the mass. The standard language for the analog of its beta function is the dimensionless quantity $\gamma_m = (\mu/m)\partial m / \partial \mu$. Green's functions are then written in terms of the running coupling constant $\bar{g}$, and a running mass $\bar{m}$. We will encounter running masses in Ch. 12 and the more generalized problem of dealing with complicated operators in the context of QCD and the Standard Model, in Ch. 16.

4.4 Effective field theories

Let us return to our original discussion of a field theory, in which we broke the fields into low momentum and high momentum components, and integrated out the high momentum components. The momentum scale where we made a distinction between the two components was Λ. We recall that

$$\int [d\phi_H][d\phi_L] \exp[-S(\phi_L, \phi_H)] = \int [d\phi_L] \exp[-S_\Lambda(\phi_L)] \qquad (4.69)$$

with

$$\exp[-S_\Lambda(\phi_L)] = \int [d\phi_H] \exp[-S(\phi_L, \phi_H)]. \qquad (4.70)$$

Now for some terminology. The action $S_\Lambda(\phi_L)$ is called a "low energy" or "Wilsonian" effective action. S_Λ is nonlocal in time scales and over distance scales of order $1/\Lambda$, but we are going to use it to describe physics on time scales much greater than $1/\Lambda$ or distances much greater than $1/\Lambda$. For such uses, it is local. We can, therefore, write it as a sum of local operators of the ϕ_L's, which we label as O_i,

$$S_\Lambda = \int d^D x \sum_i g_i O_i. \qquad (4.71)$$

The sum runs over all local operators allowed by the symmetries of the original action. That's an infinite number of terms! How can we organize them, so that we can actually perform calculations with the low energy effective action?

One approach is to use dimensional analysis. This amounts to assuming that the dimension of some variable is just its engineering dimension, or very close to it. This assumption will be valid if, in fact, the system is weakly interacting. In perturbation theory, anomalous dimensions are given by power series in the coupling constant, and they are small if the coupling is small. If the system is strongly interacting, the anomalous dimensions will typically be large (and difficult to compute), and we will lose control over the calculation. So let us work with the engineering dimension.

The action S_Λ is dimensionless. Then if O_i has an energy dimension E^{δ_i}, which we write as $[O_i] = E^{\delta_i}$, its coupling has dimension $[g_i] = E^{D-\delta_i}$. For example, in a scalar field theory, $S = \frac{1}{2}\int d^D x (\partial_\mu \phi)^2$, so $[\phi] = E^{-1+D/2}$, and if O_i is built of M ϕ's and N derivatives, then

$$\delta_i = M\left(-1 + \frac{D}{2}\right) + N. \qquad (4.72)$$

Let us define dimensionless couplings $\lambda_i = \Lambda^{\delta_i - D} g_i$. We will assume the λ_i's are numbers that are of order unity. Now for a process at scale E, we estimate the size of its term in the action to be

$$\int d^D x \, O_i \sim E^{\delta_i - D} \tag{4.73}$$

so the size of some process governed by this term has an energy dependence on the order of

$$\lambda_i \left(\frac{E}{\Lambda}\right)^{\delta_i - D} . \tag{4.74}$$

Now if $\delta_i > D$, this term becomes less and less important at small energy: it is "irrelevant." An operator whose scaling dimension is $\delta_i = D$ remains constant with energy and is called marginal or normalizable. And operators with $\delta_i < D$ are again relevant. Of course, all of this discussion is just repetition from the last section.

Typically, there are a finite number of relevant or marginal operators. We can see that from our scalar field theory example. We need

$$\delta_i = M\left(-1 + \frac{D}{2}\right) + N \leq D \tag{4.75}$$

for an operator not to be irrelevant. In $D = 4$ in a theory whose symmetry structure permits only even powers of ϕ, only $(M, N) = (2, 0)$ is relevant, and only $(M, N) = (2, 2)$ and $(4, 0)$ are marginal. So only three couplings characterize the low energy properties of these theories, and we can use an effective theory containing only three operators to make predictions in terms of these three couplings. From the point of view of the low energy effective theory, they are fundamental constants and they can only be determined from experiment. However, the low energy theory was derived from a short distance theory. Calculations with the short distance theory should be able predict their values, if we knew how to compute in that theory. (Of course, the parameters of the short distance theory, which are used to compute the values of the parameters in the low energy effective theory, must also ultimately be fixed by experiment, too.)

Why do we emphasize the free action in this particular analysis? It is purely a matter of expediency: like looking for a lost item only under a street light. If the theory is weakly coupled, we can hope to perform calculations of Green's functions, and the free action might reliably determine the size of a typical Green's function. The effective action contains free field terms, which are bilinear in the field variables, and interaction terms with higher powers of fields. We want to compare the sizes of these interaction terms with the free field part of the effective action. A convenient way to make this comparison is to define the scaling transformation such that the kinetic term in the action remains invariant. In the scalar case the kinetic term is $\frac{1}{2}(\partial_\mu \phi)^2$, with $M = N = 2$, $\delta = D$. Scale all energies and momenta by a factor $s < 1$, so lengths and times scale as $1/s$. The volume and derivative in the kinetic term scale as s^{2-D}, so to preserve the size of the kinetic term, fluctuations in ϕ must scale as as $s^{-1+D/2}$, exactly as we saw before. Then an operator O_i, with engineering dimension δ_i, scales with respect to the kinetic term as $s^{\delta_i - D}$. Again, whether the exponent of s is positive, zero, or negative determines whether the operator is relevant, marginal, or irrelevant.

Table 4.1 Examples of effective field theories

	high energy theory	scale	low energy theory
(1)	string theory	$M_{string} \sim 10^{18}$ GeV	field theory of gravity and matter
(2)	grand unified theory	$M_{GUT} \sim 10^{16}$ GeV	$SU(3) \times SU(2) \times U(1)$
(3)	Weinberg-Salam model	$M_W \sim 80$ GeV	Fermi theory
(4)	QCD	0.5-1 GeV	pions (and nucleons)
(5)	lattice field theory	(lattice spacing)$^{-1}$	continuum field theory

Some examples of high energy theories and their low energy effective field theories are shown in Table 4.1. In cases (2) and (3), both the high energy and low energy theories can be analyzed reliably in perturbation theory over their entire domains of applicability. In case (3) there are neither relevant nor marginal operators in the low energy theory (since the dimension of a fermion field $[\psi]$ is $3/2$, $[(\bar{\psi}\Gamma\psi)^2]$ has $\delta = 6$). In case (4) the fields on the right are not the fields on the left: at long distances, QCD becomes strongly interacting, and the only light modes are the Goldstone boson pions.

Presumably, no field theory we'll ever encounter gives a complete description of nature up to arbitrarily high energies. They are all effective field theories, valid up to some cutoff scale. If that is the case, is renormalization unimportant, because the cutoff renders all integrals finite? No, renormalization tells us that low energy physics depends on the short distance theory only through the relevant and marginal couplings, and possibly through some leading irrelevant couplings, if one can measure small enough effects.

So far, we have only considered weakly coupled theories, where scaling is governed by exponents that deviate from the engineering dimensions of operators by only a small amount. When there are interactions, the renormalization group still tells us how coupling constants run. This dependence is given by Eq. (4.62), which has the approximate solution near the fixed point $g = g^*$

$$G^{(n)}(sp, g, \mu) \approx s^{\delta_n - n\gamma_\phi(g^*)} = s^{\delta_n'}. \tag{4.76}$$

If $\gamma_\phi(g^*)$ is large, δ_n' can be very different from the engineering dimension δ_n. This happens in spin models in $D = 2$ or 3 dimensions. Whether an operator is relevant or marginal might be different from what a naive treatment might give.

If the dimensional coupling scales as $g = E^{D-\delta_i}$, then its renormalization group equation is

$$E\frac{\partial g}{\partial E} = (D - \delta_i)g + \dots. \tag{4.77}$$

Marginal operators, for which $D - \delta_i = 0$, can show important deviations from naive expectations. The renormalization group equation for their coupling generally looks like

$$E\frac{\partial g}{\partial E} = bg^2 + \mathcal{O}(g^3) \tag{4.78}$$

since the leading term In Eq. (4.77) vanishes. If $b > 0$, g drops as E drops, so the strength of the operator falls away at low energy: we speak of a "marginally irrelevant" operator. If $b < 0$, then g rises as E falls. This is a "marginally relevant operator." Integrating Eq. (4.78) gives

$$g(E) = \frac{g(\Lambda)}{1 + bg(\Lambda)\ln(\Lambda/E)} \tag{4.79}$$

and $g(E)$ becomes large at $E \sim \Lambda \exp[1/bg(\Lambda)]$. In this case, we generally don't know what will happen at low energy. In QCD confinement sets in.

Irrelevant or marginally irrelevant operators are useful because they tell us the range of validity of the theory at high energy. They grow as the energy scale rises. When they get to be of order unity something new has to happen. Relevant operators are dangerous, however. Think about the operator ϕ^2 in scalar field theory, appropriate for a description of a paramagnetic or ferromagnetic solid. In the effective action there is a term $(\lambda_{\phi^2}\Lambda^2)\phi^2$, where presumably $\lambda_{\phi^2} \sim 1$. It takes a special effort – some fine tuning in the short distance theory – to make λ_{ϕ^2} small. Otherwise, if $\lambda_{\phi^2} \sim 1$ the ϕ particle is heavy, and it should not appear in the low energy effective field theory at all. In the context of ferromagnetism, a ferromagnetic-paramagnetic transition occurs when λ_{ϕ^2} is tuned to vanish, since when $\lambda_{\phi^2} > 0$ the vacuum expectation value of ϕ can be zero, whereas if $\lambda_{\phi^2} < 0$ the internal symmetry is spontaneously broken.

If we were in the business of designing short distance completions of a theory (that is, inventing a more fundamental short distance theory whose low energy limit reproduces an effective field theory) we might adopt an aesthetic rule, that the low energy effective field theory should be "natural" in the sense that its masses are forbidden by symmetries. A natural effective theory could have gauge interactions, because gauge invariance forbids a mass. It could have fermions, if their masses are forbidden by chiral symmetry. And it could have scalars, if their masses are forbidden by supersymmetry. In this sense, the Standard Model itself is not natural, since the Higgs boson has a mass that is presumably not more than a few hundred GeV. There has to be new physics at an energy scale no more than an order of magnitude greater than its mass to prevent it from running off to the cutoff scale.

How is this discussion relevant to lattice QCD? Lattice QCD is the short distance theory, and we want it to reduce to ordinary continuum QCD at distances that are much greater than a lattice spacing, though still small enough that we understand what we are looking for (because as we run to even larger scales, continuum QCD itself becomes strongly interacting). In order to do that, the lattice theory must respect the symmetries of continuum QCD (since if it did not, its low energy theory also would not). This means that, at a minimum, the lattice theory should preserve local gauge invariance. If the lattice fermions possess a chiral symmetry, the low energy theory will inherit it. If they do not, it will be necessary to fine-tune some parameter in the lattice theory to recover it. This is in fact what happens when the fermions are discretized as Wilson-type fermions. And in contrast to the real-world case of beyond-standard-model physics, where we want irrelevant operators to tell us about the underlying short-distance theory, irrelevant operators in a lattice context are artificial: the short-distance theory is unphysical, and we do not want to see its effects.

On the lattice, if all quark masses are set to zero, the only dimensionful parameter is the lattice spacing a. Lattice Monte Carlo predictions are of dimensionless ratios of dimensionful quantities, or of dimensionless numbers made by combining a dimensionful prediction (like a mass) with an appropriate power of the cutoff (that is, a calculation produces the product $a \times m$). One can determine the lattice spacing by fixing one mass from experiment, and then one can go on to predict any other dimensionful quantity, or one can predict dimensionless ratios of dimensionful quantities. Imagine computing some masses at several values of the lattice spacing. (In practice, this could be done by picking several values of the bare parameters and calculating masses for each set of couplings.) If the lattice spacing is small enough, the typical ratio behaves as

$$[am_1(a)]/[am_2(a)] = m_1(0)/m_2(0) + \mathcal{O}(m_1 a) + \mathcal{O}[(m_1 a)^2] + \dots \quad (4.80)$$

(modulo powers of $\log(m_1 a)$, which could arise from operators whose coefficients are generated at higher order in the running coupling). The leading term does not depend on the value of the UV cutoff. That is our cutoff-independent prediction. Everything else is an artifact of the calculation. We say that a calculation "scales" if the $a-$dependent terms in Eq. (4.80) are zero or small enough that one can extrapolate to $a = 0$, and generically refer to all the $a-$dependent terms as "scale violations." Clearly our engineering goal is to formulate the lattice theory to minimize scale violations.

What are the implications for the behavior of the coupling constant as we take the lattice spacing to zero? Each dimensionless combination $am(a)$ can be expressed as a function of the bare coupling(s), $am = f(\{g(a)\})$. As $a \to 0$ we must tune the couplings so that

$$\lim_{a \to 0} \frac{1}{a} f(\{g(a)\}) \to \text{constant}. \quad (4.81)$$

Of course, that constant is just the mass. That is, the fixed "physical" quantities m determine how the couplings $\{g(a)\}$ vary with a. In units of a, however, the correlation lengths $1/ma$ must diverge as a vanishes. From the discussion of Section 4.1, we see that the couplings $\{g(a)\}$ must be tuned to approach a fixed point.

What does this mean for lattice QCD? QCD is a theory with a marginally relevant operator, parameterized by the usual gauge coupling. A particular lattice QCD action has, in addition, a set of irrelevant operators arising from discretization effects. Ignoring them, we focus on the running coupling constant, whose beta function for small g is given in perturbation theory by

$$\mu \frac{\partial g}{\partial \mu} = -\frac{\beta_0}{16\pi^2} g^3 - \frac{\beta_1}{(16\pi^2)^2} g^5 \quad (4.82)$$

with $\beta_0 = 11 - (2/3)N_f$ and $\beta_1 = 102 - (38/3)N_f$ for $SU(3)$. It vanishes at $g = 0$, which because of the sign of the leading term is an an ultraviolet attractive (or infrared repulsive) fixed point. The continuum limit of QCD is thus achieved by taking the bare coupling g to zero. When g^2 is small,

$$\frac{16\pi^2}{g(\mu)^2} = \beta_0 \ln \frac{\mu^2}{\Lambda^2} + \frac{\beta_1}{\beta_0} \ln \ln \frac{\mu^2}{\Lambda^2} + \ldots \tag{4.83}$$

in some regularization scheme associated with Λ (which we have not specified).

As we have seen, perturbative running is characterized by the Λ parameter. Instead of looking at mass ratios, we could demand that a ratio like m/Λ shows no cutoff dependence. Inverting the relation between g^2 and Λ, and replacing μ by $1/a$, we would ask that the lattice spacing vary with g as

$$a\Lambda = \left(\frac{16\pi^2}{\beta_0 g^2(a)} \right)^{\beta_1/(2\beta_0^2)} \exp\left(-\frac{8\pi^2}{\beta_0 g^2(a)} \right). \tag{4.84}$$

In the early days of lattice simulations people attempted to observe this behavior in a lattice simulation by determining the dependence of a dimensionful quantity on the bare coupling constant. For example the simulated mass of a hadron am in lattice units should decrease with decreasing bare coupling g so that $am/(a\Lambda)$ approaches a constant, with $a\Lambda$ given by Eq. (4.84). This behavior is called "asymptotic scaling." Please note that asymptotic scaling is not scaling. Scaling means that dimensionless ratios of physical observables do not depend on the cutoff. Asymptotic scaling involves perturbation theory and the definition of coupling constants. One can have scaling without asymptotic scaling.

In this chapter we make no claims for originality: everything we have said can be found (at greater length) in many books. Our favorite suggestions for further reading about the renormalization group are Ma (1976) and Wilson and Kogut (1974). We turn to Ramond (1981) and Peskin and Schroeder (1995) for field theory questions. The literature on effective field theories is also vast; for a good introduction, see Polchinski (1992). We will encounter the language of this chapter in many places to follow. Running coupling constants and masses will reappear when we discuss the lattice QCD predictions for Standard Model tests in Ch. 16. Effective field theories are used to describe heavy quarks on the lattice in Ch. 6, and to connect QCD to low energy phenomena in Ch. 14.

Chapter 5

Yang-Mills theory on the lattice

In Ch. 3 we saw how the pure $U(1)$ and $SU(3)$ gauge theories are formulated on the lattice. Here we examine the pure $SU(3)$ gauge theory in greater depth.

5.1 Gauge invariance on the lattice

As we have seen, we need two key ingredients to construct a lattice action for a gauge theory. First, the action must have the correct continuum limit. Second, it should be gauge invariant. All popular actions place the gauge variables on the links of the lattice. They are elements of the fundamental representation of $SU(3)$, related to the vector potential through

$$U_{x\mu} = \exp[igaA_{\mu}^{c}(x)\lambda_c], \tag{5.1}$$

where λ_c for $c = 1, \ldots, 8$ are the eight generators of $SU(3)$ and g is the gauge coupling constant. The link matrix is the path-dependent gauge connection that relates the color space at site x to the site $x+\hat{\mu}$. The more symmetric notation

$$U(x, x+\hat{\mu}) \equiv U_{x,\mu} \tag{5.2}$$

helps to emphasize its relationship to the two neighboring sites. On a lattice of finite dimensions $L_x \times L_y \times L_z \times L_t$, a periodic boundary condition in the time coordinate is required to simulate the quantum partition function. It identifies the time coordinate $x_0 = L_t$ with $x_0 = 0$. It is common also to require the same periodicity in the spatial directions.

A gauge transformation rotates the color space on each site through an

$SU(3)$ matrix W_x. The link matrix transforms according to

$$U_{x,\mu} \longrightarrow W_x U_{x,\mu} W^\dagger_{x+\hat{\mu}} \tag{5.3}$$

This transformation of the gauge link matrices correlates with transformations of the matter fields on the lattice sites. For example, a quark field q_x on each site would transform as

$$q_x \longrightarrow W_x q_x, \tag{5.4}$$

and the "gauge inner product"

$$q^\dagger_x U_{x,\mu} q_{x+\hat{\mu}} \tag{5.5}$$

is then invariant under the gauge transformation. In the reverse direction we must have

$$q^\dagger_{x+\hat{\mu}} U^\dagger_{x,\mu} q_x. \tag{5.6}$$

That is, the gauge connection is directional. The links $U_{x,\mu}$ are forward connections and the links $U^\dagger_{x,\mu}$ are backward connections. In the alternative notation we have

$$U(x+\hat{\mu}, x) \equiv U^\dagger_{x,\mu}. \tag{5.7}$$

Gauge invariant expressions can be constructed from ordered products of gauge links. For example, a product of two links in the same direction connects the color spaces on sites x and $x + 2\hat{\mu}$, leading to the gauge-invariant combination

$$q^\dagger_x U_{x,\mu} U_{x+\hat{\mu},\mu} q_{x+2\hat{\mu}}. \tag{5.8}$$

More generally, we can define a gauge-invariant inner product of matter fields on any pair of sites x and y connected by a continuous path $C = \{(x_i, \hat{\mu}_i)\}$ of links:

$$q^\dagger_x P \prod_C U(x, x + \hat{\mu}_i) q_y \tag{5.9}$$

The P indicates that the product is taken in the order of the path. If the path traverses a link backwards, we use the adjoint of the corresponding link matrix.

Gauge invariant expressions can also be constructed from the links themselves. For any closed path of links C we form

$$W_C = \mathrm{Tr}[P\prod_C U(x, x+\hat{\mu}_i)]. \tag{5.10}$$

The result does not depend on the starting and ending point, since a cyclic reordering of the factors preserves both the trace and the path closure.

Note that the path C may close by virtue of lattice periodicity in any periodic direction, a generalization of the Polyakov loop, discussed in Ch. 3. It is easy to show that the corresponding gauge loop trace is also invariant.

5.2 Yang-Mills actions

Gauge invariant actions are constructed from the trace of path products along closed loops. In Ch. 3 we introduced the simplest closed loop, namely the plaquette:

$$U_{x,\mu\nu} = \mathrm{Tr}\left(U_{x,\hat{\mu}}U_{x+\hat{\mu},\nu}U^{\dagger}_{x+\hat{\nu},\mu}U^{\dagger}_{x,\nu}\right), \tag{5.11}$$

and the corresponding Wilson action

$$S_{G-SU(3)} = \frac{1}{g^2}\sum_x\sum_{\mu\neq\nu}\left(3 - \mathrm{Re}U_{x,\mu\nu}\right). \tag{5.12}$$

The link matrix is related to the vector potential through $U_{x,\mu} = \exp(igaA^c_\mu\lambda^c/2)$. In the limit of zero lattice spacing we obtain the standard continuum Yang-Mills action

$$S_{G-SU(3)} = \int_0^\beta d\tau\int d\mathbf{x}\frac{1}{4}(F^c_{\mu\nu})^2 + \mathcal{O}(a^2), \tag{5.13}$$

$$F^c_{\mu\nu} = \partial_\mu A^c_\nu - \partial_\nu A^c_\mu - gf_{abc}A^a_\mu A^b_\nu. \tag{5.14}$$

Lattice artifacts contribute at $\mathcal{O}(a^2)$.

Other gauge invariant choices are possible. Any arbitrary closed loop can be used to construct the Yang-Mills action. As long as it is suitably averaged over orientations and translations in space and time to obtain the correct space-time symmetries, it must reduce to the correct continuum action. At $\mathcal{O}(a^2)$, however, such actions may differ. Consequently, linear combinations of closed loops can be designed so the leading order lattice corrections are canceled. This is the basis of the Symanzik improvement program that seeks to reduce lattice artifacts at nonzero lattice spacing (Symanzik, 1980, 1983a,b). For example, a suitable linear combination of the plaquette term, a 2×1 rectangle, and a six-link "chair" or "parallelogram" makes a good choice. We discuss these improvements in greater depth in Ch. 12.

5.3 Gauge fixing

The partition function is given by the gauge-invariant integral over link matrices:

$$Z(\beta) = \int \prod_{x,\mu} dU_{x,\mu} \exp(-S_{G-SU(3)}). \tag{5.15}$$

The action and measure are invariant under the gauge transformation (5.3). These are the the so-called "flat" directions in the integration. In the continuum limit the gauge degrees of freedom are noncompact and the flat directions are of infinite measure. Consequently, the continuum path integral requires Fadeev-Popov gauge fixing to obtain a finite result. Fortunately, lattice regulation permits the use of a compact group of gauge transformations, thereby rendering the flat directions finite. We may integrate freely over all gauge choices.

Sometimes it is nevertheless useful to fix the gauge. For example, we might construct a meson from a source operator involving a quark on one lattice site and an antiquark on another. Including the gauge connection as in (5.9) makes the source gauge invariant. Fixing a suitable gauge also does the job. In this section we describe a few possible gauge choices.

5.3.1 *Maximal tree gauge*

One way to fix the gauge is to transform as many gauge links as possible to the unit matrix. That amounts to a change of variable in which the gauge transformation matrices W_x themselves become integration variables and the same number of links are set to a unit matrix and eliminated. To illustrate, consider the set of four links forming a plaquette. For simplicity of notation, call them $U(1,2)$, $U(2,3)$, $U(3,4)$, $U(4,1)$. Setting $W_1 = 1$ and $W_2 = U(1,2)$ transforms the $1-2$ link to a unit matrix, but also replaces $U(2,3)$ with $U(1,2)U(2,3)$. Similarly, setting $W_3 = U(1,2)U(2,3)$ transforms the altered $2-3$ link to unity and replaces $U(3,4)$ with

$U(1,2)U(2,3)U(3,4)$ Finally, if we set $W_4 = U(1,2)U(2,3)U(3,4)$ the first three links are transformed to unity and the last link inherits the original link product, thus preserving the plaquette trace.

Since the plaquette trace is gauge invariant, a gauge transformation clearly can never set all four links to one. The same conclusion applies to any closed loop. Complete gauge fixing in this manner creates a set of unit matrices on a "maximal tree" of links. A maximal tree fills the lattice in the sense that all sites are traversed by the tree. It has no closed loops and establishes a unique path connecting any pair of lattice sites. In effect the maximal tree aligns the color spaces on each site so a gauge-invariant inner product (5.9) formed from a path of links in the tree reduces to an ordinary dot product. This gauge is unique up to a global gauge transformation, i.e. $W_x = \text{const.}$

Temporal gauge is a special class of the maximal tree gauge. In this gauge all possible time-like links are set to unity. Since a loop can close through time-periodicity by wrapping around from $\tau = L_t - 1$ to $\tau = 0$, it isn't possible to include all time-like links. For example, we could exclude from the tree a set of links on one time-slice of the lattice. From Ch. 3 we notice that the integration over those links serves to enforce Gauss' Law in that gauge, so they must not be fixed. To complete the maximal tree in temporal gauge one must also fix a set of space-like links.

5.3.2 *Landau and Coulomb gauge*

In continuum gauge theory Landau gauge is defined in terms of the vector potential by

$$\partial_\mu A_\mu^c(x) = 0 \tag{5.16}$$

for $\mu = 0, \ldots 3$ and Coulomb gauge by

$$\nabla_i A_i^c = 0 \tag{5.17}$$

for $i = 1, 2, 3$. On the lattice these gauge conditions are replaced by minimizing (over all gauge choices)

$$I_L = \sum_{x,\mu} \text{ReTr} U_{x,\mu} \tag{5.18}$$

($\mu = 0, 1, 2, 3$) for Lorentz gauge and

$$I_C = \sum_{x,i} \mathrm{ReTr} U_{x,i} \tag{5.19}$$

($i = 1, 2, 3$) for Coulomb gauge. To see how this works we take the continuum limit, using (5.1). For simplicity, we consider an abelian gauge field and examine the Coulomb gauge prescription. A gauge transformation (3.20) does the mapping

$$A_{xi} \to A'_{xi} = A_{xi} + \lambda_x - \lambda_{x+\hat{i}}. \tag{5.20}$$

We substitute this expression into I_C and set the partial derivatives with respect to each λ_x to zero. We obtain

$$\sum_{i=1}^{3} (A'_{x,i} - A'_{x-\hat{i},i}) = 0 \tag{5.21}$$

which is a finite difference representation of (5.17). Similar reasoning works for Landau gauge and for $SU(3)$.

With either Coulomb or Landau gauge there is a residual, discrete gauge ambiguity discovered by Gribov (1978), which has never caused your authors trouble.

5.4 Strong coupling

The $SU(N)$ Yang-Mills theory is asymptotically free with an ultraviolet fixed point at zero coupling. The continuum limit is reached as the coupling g vanishes (weak coupling). To study nonperturbative phenomena in this limit requires numerical simulation. In the opposite limit of strong coupling the theory is easily solved by hand, and, as we will see in this section, it exhibits confinement. Although it is still not proved, all the evidence we have is that if the internal symmetry group is $SU(N)$, confinement persists to weak coupling.

By contrast the pure $U(1)$ theory is not asymptotically free. At weak coupling it is a trivial theory of noninteracting photons. At strong coupling the compact theory we have been discussing exhibits confinement for much the same mathematical reasons as $SU(N)$. A singularity (phase transition) of the partition function separates the strong and weak coupling regimes.

5.4.1 *Wilson loop and confinement*

As an introduction to strong coupling methods, we show that the $U(1)$ gauge theory is confining in the strong coupling limit. For this purpose we return to the $U(1)$ Wilson loop observable Eq. (3.72), which we write here as $W(R,T)$ where R is the spatial separation of the static charges and T is the time interval between creation and annihilation of those charges. As we saw from (3.67) and (3.71), in the limit of low temperature and large T this observable measures the potential energy of separation of the two static charges:

$$\langle W(R,T)\rangle = \int [dU] W(R,T) \exp(-S_{G-U(1)})/Z \sim Ae^{-TV(R)}, \quad (5.22)$$

where A is a constant and we have written $[dU]$ to denote the Haar invariant integration over the $U(1)$ link variables (equivalently, integration over θ_{xT} and ω_{xT}). Here, the excitation energy $E_1 - E_0 = V(R)$ of (3.67) is the difference in energy of the system with a pair of static charges and without any background charges (the vacuum). That is just the potential energy of the static pair of charges.

Confinement in this model is signaled by an indefinite linear rise in the potential as the charges are separated. That is to say, for large R,

$$V(R) \sim \sigma R, \quad (5.23)$$

where σ is called the string tension. Substituting this expression into (5.22) gives

$$-\ln(\langle W(R,T)\rangle) \sim \sigma TR, \quad (5.24)$$

an "area law" for large T and R.

In a nonconfining theory, the potential energy should become a constant at large distance. In that case we could have

$$-\ln(\langle W(R,T)\rangle) \sim e_0(T+R), \quad (5.25)$$

a "perimeter law" for large T and R.

Now let us calculate the Wilson loop expectation value in the compact $U(1)$ theory in leading order in $1/g^2$ (strong coupling limit). The Wilson plaquette action is given in the two-dimensional theory (3.42). In higher dimensions one simply sums over plaquettes in all planes as we showed explicitly for $SU(3)$ in (3.55). Since the Wilson loop is in one space and one time dimension, as we will see, to leading order in $1/g^2$ we need consider only the plaquettes in the same space-time plane.

First, the partition function is trivially evaluated. Since the action is proportional to $1/g^2$, its exponential is just 1 to leading order. Thus with a suitable choice of integration measure, to leading order,

$$Z \sim 1. \tag{5.26}$$

(We use the same measure for numerator and denominator in (5.22), so any normalization factors cancel.) In the numerator, the same leading term in $\exp(-S_{G-U(1)})$ contributes nothing, because the integral over the phase factors in $W(R,T)$ vanishes.

To find the leading nonvanishing contribution, consider, first, the case $R = T = 1$, namely a plaquette. In this case the first order term in $1/g^2$ in $\exp(-S_{G-U(1)})$ contains, among many other terms, the same plaquette traversed in the opposite sense. Multiplying it by $W(R,T)$ cancels the phase factors, resulting in a nonvanishing contribution at order $1/g^2$. All other terms in the expansion to this order give vanishing integrals. Next, consider the 1×2 Wilson loop. In this case a single plaquette from the expansion does not cancel all the phase factors. But a term at order $1/g^4$ contains the product of two plaquettes that fill the 1×2 Wilson loop and cancel all phase factors. This is the only term with a nonvanishing integral to this order.

For arbitrary R and T we must go to order $1/g^{2RT}$ to obtain the leading nonvanishing contribution, which consists of a product of plaquettes tiling the Wilson loop. It is represented pictorially in Fig. 5.1. It contributes

$$\langle W(R,T) \rangle \sim \left(\frac{1}{2g^2} \right)^{RT} \tag{5.27}$$

for a symmetric lattice ($a = \Delta\tau$), leading to area-law confinement with a string tension

$$\sigma = \log(2g^2). \tag{5.28}$$

More generally to any order in $1/g^2$ any nonvanishing contribution to the Wilson loop expectation can be represented by a surface in a four-dimensional space of any shape, including handles, bounded by the Wilson loop and tiled by plaquettes coming from an expansion of the exponential of the action. Such strong coupling expansions have been studied in considerable detail and to very high order (Drouffe and Itzykson, 1978). The

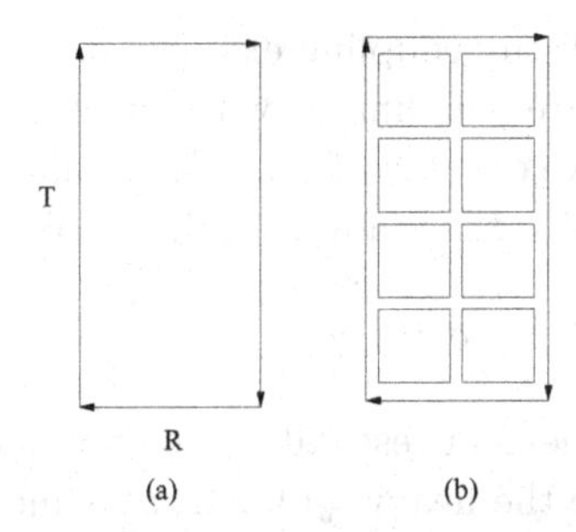

Fig. 5.1 An $R \times T$ Wilson loop (a) and the pattern of tiling that gives the leading order strong coupling contribution (b).

analogy with quantum fluctuations of a string is suggestive; such imagery makes a strong coupling expansion intuitively appealing, even if it goes in the opposite direction of the physical limit.

The same reasoning applies to $SU(N)$, except that we do not simply cancel phase factors. Instead we integrate using group orthogonality theorems

$$\int du\, U_{ij} = 0$$
$$\int dU\, U_{ij} U^*_{k\ell} = \delta_{ik}\delta_{j\ell}/N, \tag{5.29}$$

where we normalize the Haar measure to $\int dU = 1$. For $SU(3)$ we also have

$$\int dU\, U_{ij} U_{k\ell} U_{mn} = \epsilon_{ikm}\epsilon_{j\ell n}/3!, \tag{5.30}$$

where ϵ_{ikm} is the usual cross product symbol. For higher N the corresponding identity has N factors in the integrand, the generalization of the permutation symbol and $N!$. There are similar identities for integrands with still more $SU(N)$ factors.

These identities tell how to tile the Wilson loop to obtain nonvanishing contributions in the strong coupling expansion of the $SU(N)$ Wilson action. The first identity (5.29) requires that any link matrix be matched by at least one link matrix in the opposite sense. This leads to the same tiling that we required for the $U(1)$ theory. The second identity (5.30) introduces a new possibility: we can match a link matrix with two in the same direction and get a nonvanishing integral. Thus an alternative tiling replaces any plaquette of one orientation with a product of two plaquettes in the same space-time location, but of the opposite orientation. Such alternatives are higher order in the the strong coupling expansion.

To leading order we are left, finally, with the same leading order tiling of the Wilson loop that we obtained for $U(1)$, as shown in Fig. 5.1. Details of the calculation are left to the reader. Again we obtain confinement with

$$\sigma = \log(g^2 N). \tag{5.31}$$

Notice that this confinement test fails in the presence of light dynamical fermions. As we separate the heavy $Q\bar{Q}$ pair, at some point it will become energetically favorable to pop a light $\bar{q}q$ pair out of the vacuum, so that we are separating two color singlet mesons. The Wilson loop then shows a perimeter law behavior. With dynamical quarks, "confinement" simply means it isn't possible to isolate a quark at macroscopic distances.

5.4.2 *Glueball mass*

We continue our brief examination of the strong coupling limit by calculating an intuitively interesting result, namely the mass of a glueball in $SU(N)$ gauge theory.

To create a glueball we use gauge invariant interpolating operators. The simplest choice is the plaquette operator itself. Consider the real part of the space-space oriented plaquette. Viewed as an operator acting on the vacuum, as we have seen in Ch. 3, it serves to create a closed loop of color electric flux. If we sum over spatial rotations, we obtain an operator that belongs to the trivial representation of the cubic group. It contains the spin-parity $J^P = 0^+$ of a continuum theory among higher possible angular momenta. It is also charge-conjugation even. To see this, observe that charge conjugation reverses the direction of the color electric flux. So the charge conjugated plaquette operator is constructed by traversing the same links in the opposite direction. Such a reversed path is included in the sum over rotations. Let us denote this operator by $G_{0^{++}}$.

To compute the glueball mass we calculate the correlator (3.63) of a plaquette source at $\tau = \tau_1$ and a plaquette sink at $\tau = \tau_2$. Since the plaquette operator has a nonvanishing vacuum expectation value, we must subtract the "vacuum disconnected" contribution to isolate the glueball mass. That is, we want the connected correlator

$$C_{\text{conn}}(\tau_2 - \tau_1) = \langle 0| \mathcal{A}_{\tau_2}\mathcal{B}_{\tau_1} |0\rangle - \langle 0| \mathcal{A}_{\tau_2} |0\rangle \langle 0| \mathcal{B}_{\tau_1}] |0a\rangle . \tag{5.32}$$

The vacuum disconnected part (second term on the rhs above) in the present case is just

$$\langle 0| G_{0^{++}\tau_2} |0\rangle \langle 0| G_{0^{++}\tau_1} |0\rangle = \langle 0| G_{0^{++}} |0\rangle^2 . \tag{5.33}$$

In analogy with (3.67) we have

$$C_{\text{conn}}(\tau_2 - \tau_1) = \langle 0| G_{0^{++}} \left|0^{++}\right\rangle \left\langle 0^{++}\right| G_{0^{++}} |0\rangle \, e^{(-\tau_2+\tau_1)m_{0^{++}}} . \tag{5.34}$$

In functional integral language we want

$$\begin{aligned} C_{\text{conn}}(\tau_2 - \tau_1) = &\int [dU] G_{0^{++}\tau_2} G_{0^{++}\tau_1} \exp(-S_G)/Z \\ &- \left(\int [dU] G_{0^{++}} \exp(-S_G)/Z\right)^2 . \end{aligned} \tag{5.35}$$

We proceed to a calculation of this mass to leading order in the strong coupling expansion. This expectation value is analyzed in exactly the same way as the Wilson loop. We must find a tiling that bounds both the source and sink plaquettes. Topologically speaking there are two classes of surfaces, those that connect the source and sink and those that do not. The latter generate the vacuum disconnected part. The former contribute to the glueball mass. The leading order connected contribution forms a square cylinder as shown in Fig. 5.2. We obtain

$$C_{\text{conn}}(T) \sim k \left(\frac{1}{N^4 g^8}\right)^T \tag{5.36}$$

for constant k leading to

$$m_{0^{++}} \sim 4 \log(N g^2). \tag{5.37}$$

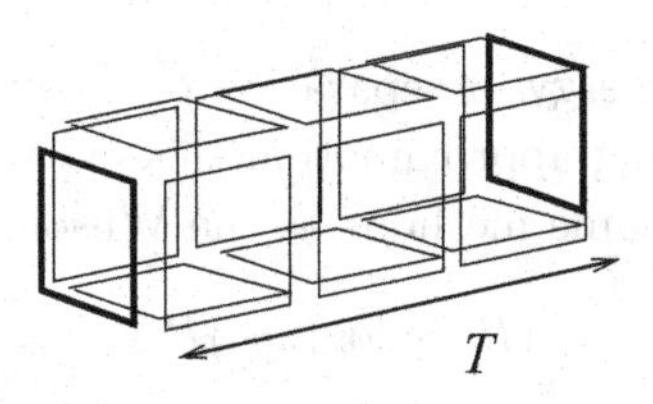

Fig. 5.2 A pair of plaquettes separated by a distance T and the pattern of tiling that gives the leading order strong coupling contribution to the connected correlator.

5.4.3 *Polyakov loop*

We conclude our discussion of the strong coupling limit of pure Yang-Mills theory by returning to the Polyakov loop, a loop that wraps around the lattice by virtue of periodicity. A single such loop has vanishing expectation value. A physical interpretation is tempting. As we recall from Ch. 3, the expectation value of a single Polyakov loop gives

$$\langle \mathcal{P}(x) \rangle = \exp(-\beta F_0) \tag{5.38}$$

where F_0 is the free energy of a single static quark. Confinement should require that it be infinite, forcing a zero expectation value. But it is zero for another reason.

To see this consider a particular symmetry of the functional integral. If we multiply all time-like links in one time slice by a member z of the center of the gauge group, the action and integration measure is unchanged. (By definition the center elements commute with all group elements.) In the case of $SU(3)$ these elements are just $\exp(2\pi ik/3)$ times the identity for $k = 0, 1, 2$, i.e. elements of the group $Z(3)$. Under the center transformation the Polyakov loop is multiplied by z. In the course of doing the path integration, these center-related configurations are visited symmetrically. The Polyakov loop expectation value averages to zero. However, this result is merely a test of our projection onto states that satisfy Gauss' Law! The integration over a set of time-like links includes a sum over the center transformation, so it enforces a requirement that all states be color neutral (triality zero). A single static charge in the fundamental representation fails this test, since there is no way to screen it in finite volume in a pure Yang-Mills theory.

On the other hand, a pair of Polyakov loops in opposite directions has zero triality. As we have seen in Ch. 3 its expectation value is

$$\langle \mathcal{P}(x)\mathcal{P}^*(x+R) \rangle = \exp[-\beta F(R)], \tag{5.39}$$

where $F(R)$ is the free energy of separation of a pair of static quarks. In the leading strong coupling approximation this expectation value is easily calculated following the same methods as the Wilson loop and gives

$$F(R) \sim \log(2g^2)R \tag{5.40}$$

at large R. At infinite separation we get a zero expectation value, the proper test of confinement. At high temperature, a phase transition is encountered, leading to a deconfined phase in which the free energy of separation is asymptotically constant. This result can be understood from the strong-coupling point of view by considering the entropy of fluctuations in the surface connecting the two Polyakov loops. As the temperature is increased at a fixed strong coupling, the rapidly increasing multiplicity of terms contributing at the same high order offsets the cost of going to high order. The flux tube connecting the static charges fluctuates freely, leading to deconfinement.

Chapter 6

Fermions on the lattice

In contrast to gauge fields, for which the replacement of the vector potential by the link is nearly the beginning and the end of the story, marrying light or massless fermions to the lattice has been a difficult problem of long standing. The kernel of the difficulty is chiral symmetry, and taking either the naive view of a chiral fermion (in terms of a helicity eigenstate) or a more sophisticated view (as a $(\frac{1}{2},0)$ representation of the Lorentz group), it is clear that it is the space-time lattice itself that is the source of the problem. Many approaches to lattice fermions have been developed over the years, none of which are without complications. At present, there are two directions to take: one can sacrifice full chiral symmetry on the lattice in return for simple discretizations of fermions that are easy to simulate, or keep full chiral symmetry at a considerably higher computational cost.

Heavy fermions are insensitive to chiral symmetry, but bringing them to the lattice is also not altogether straightforward. The problem now is that the product $m_q a$ can be much greater than unity. This product governs the scale of lattice artifacts. In that case the quark mass – but not other relevant quantities – is much greater than the scale of the UV cutoff, suggesting immediately that a solution of the problem can best be couched in the language of effective field theory through an expansion in inverse powers of the mass.

6.1 Naive fermions

Naive fermions are the simplest lattice implementation of fermions. They are rarely used in simulations, but it is interesting to begin with them, to illustrate the pitfalls associated with fermions on the lattice. The action for a continuum free fermion is

$$S = \int d^4x[\bar{\psi}(x)\gamma_\mu \partial_\mu \psi(x) + m\bar{\psi}(x)\psi(x)]. \tag{6.1}$$

As we have seen in Ch. 3, naive lattice fermions are constructed by replacing the derivatives by symmetric differences. We explicitly introduce the lattice spacing a in the denominator and write

$$S_L^{\text{naive}} = \bar{\psi} D^{\text{naive}} \psi = \sum_{n,\mu} \bar{\psi}_n \gamma_\mu \Delta_\mu \psi_n + m \sum_n \bar{\psi}_n \psi_n, \tag{6.2}$$

where the lattice derivative is

$$\Delta_\mu \psi_n = \frac{1}{2a}(\psi_{n+\hat{\mu}} - \psi_{n-\hat{\mu}}). \tag{6.3}$$

As we did in Ch. 3, gauge fields are introduced by using a link variable to connect neighboring sites. Then

$$S_L^{\text{naive}} \to \frac{1}{2a} \sum_{n,\mu} [\bar{\psi}_n \gamma_\mu U_\mu(n) \psi_{n+\hat{\mu}} - \bar{\psi}_n \gamma_\mu U_\mu(n-\hat{\mu})^\dagger \psi_{n-\hat{\mu}}] + m \sum_n \bar{\psi}_n \psi_n. \tag{6.4}$$

The free propagator is easy to construct:

$$\frac{1}{a} S(p) = (i\gamma_\mu \sin p_\mu a + ma)^{-1} = \frac{-i\gamma_\mu \sin p_\mu a + ma}{\sum_\mu \sin^2 p_\mu a + m^2 a^2}, \tag{6.5}$$

while in the weak coupling limit, with $U_\mu(x) = \exp[igaV_\mu(x)]$, the vertex is

$$S_i = \sum_{p,q,k} \bar{\psi}(p) g \gamma_\mu \cos\left[\frac{1}{2}(p+q)_\mu a\right] \psi(q) V_\mu(k) \tag{6.6}$$

with $p + k + q = 0$.

Now the lattice momentum components p_μ range from $-\pi/a$ to π/a. A continuum fermion with its propagator $(i\gamma_\mu p_\mu + m)^{-1}$ has a large contribution at small p from four modes that are bundled together into a single Dirac spinor. The lattice propagator has these modes too, at $p = (0, 0, 0, 0)$, but, as we have seen in Ch. 3, there are other degenerate ones, at $ap = (\pi, 0, 0, 0)$, $(0, \pi, 0, 0)$, ..., (π, π, π, π). As a goes to zero, the propagator is dominated by the places where the denominator is small, and there are sixteen of these, in all the corners of the Brillouin zone. Thus our action is a model for sixteen light fermions, not one. This is the famous "doubling problem."

Doubling is an exact symmetry of the naive fermion. Let us follow the pioneering analysis of Karsten and Smit (1981) and call a location where the propagator has a pole $\bar{p}$. There are 16 symmetry transformations $\psi'_n = T_n\psi_n$, ,$\bar{\psi}'_n = \bar{\psi}_n T_n^\dagger$ that interchange the $\bar{p}$'s: $T = 1, \gamma_\mu\gamma_5(-)^{x_\mu/a}, \ldots$. For example, $\gamma_1\gamma_5 e^{i\pi x_1/a}$ shifts p_1 to $p_1 + \pi a$. We can label the sixteen T's by where they take momenta near $(0,0,0,0)$ and call them $T(\bar{p})$. We further label the spin part of $T(\bar{p})$ as $s(\bar{p}) = 1, \gamma_\mu, \ldots$. Let us assume that the momentum p is close to some $\bar{p}$, $p = k + \bar{p}$, and write

$$\frac{1}{a}S(p) = \frac{-i\gamma_\mu \sin(k+\bar{p})_\mu a + ma}{\sum_\mu \sin^2(k+\bar{p})_\mu a + m^2a^2} \tag{6.7}$$

or

$$S(p) = s(\bar{p})\frac{m - i\gamma_\mu k_\mu}{m^2 + k^2}s(\bar{p})^{-1} + O(a). \tag{6.8}$$

To identify the particle content, rotate to Minkowski space, where $\gamma^0 = i\gamma_4$, $k^0 = ik_4$, and

$$S(p) = s(\bar{p})\frac{m - i\vec{\gamma}\cdot\vec{k} + i\gamma_0 k_0}{m^2 + \vec{k}^2 - k_0^2}s(\bar{p})^{-1} + O(a). \tag{6.9}$$

$S(p)$ has sixteen poles, corresponding to sixteen propagating Dirac particles. We can extract the wave functions of these particles from the residues of the poles: for $k_0 = \sqrt{\vec{k}^2 + m^2} > 0$, they are

$$s(\bar{p})[i\vec{\gamma}\cdot\vec{k} - i\gamma_0 k_0]s(\bar{p})^{-1} = \sum_\lambda [s(\bar{p})u_\lambda(\vec{k})][\bar{u}(\vec{k})s(\bar{p})^{-1}] \tag{6.10}$$

and for $k_0 = -\sqrt{\vec{k}^2 + m^2}$ they are

$$s(\bar{p})[i\vec{\gamma}\cdot\vec{k} - i\gamma_0 k_0]s(\bar{p})^{-1} = -\sum_\lambda [s(\bar{p})v_\lambda(-\vec{k})][\bar{v}(-\vec{k})s(\bar{p})^{-1}] \tag{6.11}$$

where u_λ and v_λ are the standard four-component free spinors. Thus the wave functions for the sixteen particles and antiparticles are $s(\bar{p})u(\vec{k})$, $s(\bar{p})v(\vec{k})$, $\bar{u}(\vec{k})s(\bar{p})^{-1}$, and $\bar{v}(\vec{k})s(\bar{p})^{-1}$.

Even if we restricted all initial and final state particles to be ones with $\bar{p} = 0$, this would not eliminate the other particles from contributing as intermediate states, since gauge bosons couple to them. Loops of these particles would be present in all processes. Furthermore, by glancing at the vertex, Eq. (6.6), we see that all these particles have the same vector charge, g, or in units of g, all fermions have $Q = 1$.

When the fermion mass is set to zero, the naive fermion action has a continuous global chiral symmetry $\psi \to e^{i\epsilon\gamma_5}\psi$, $\bar{\psi} \to \bar{\psi}e^{i\epsilon\gamma_5}$. Could we construct an action that is simultaneously chiral and undoubled? Such an action would have a propagator

$$S(p) = \left[\sum_\mu i\gamma_\mu P_\mu(p) \right]^{-1} \tag{6.12}$$

where $P_\mu(p)$ is some real function of p. For one Dirac fermion, we would expect that $P_\mu(p) = 0$ at $p_\mu = 0$. By periodicity, it must also cross the $P_\mu = 0$ axis at $p_\mu = 2\pi/a$. Then at some intermediate momentum, either P_μ must be discontinuous, or there must be another crossing of the $P_\mu = 0$ axis. The second case implies that the spectrum is doubled.

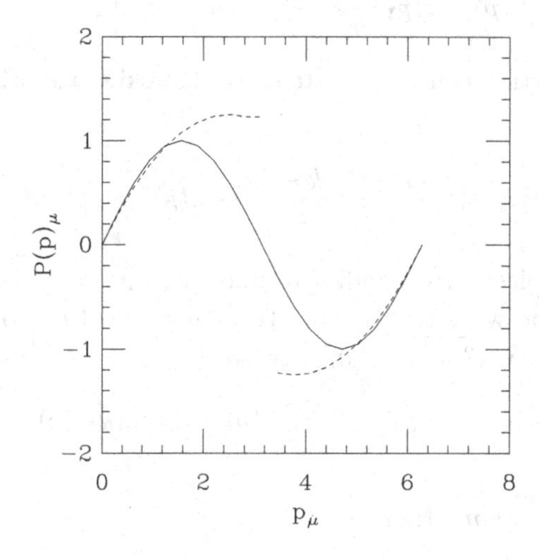

Fig. 6.1 Possible choices for the function $P(p)_\mu$.

In the first case, the corresponding coordinate space action would be nonlocal:

$$\Delta_\mu \psi_n = \sum_m f(m)(\psi_{n+m\hat{\mu}} - \psi_{n-m\hat{\mu}}) \tag{6.13}$$

where $f(m)$ would fall off as a power of m. For example, a linear $P(p)_\mu$ would give a derivative operator like $f(m) \sim (-)^m/m$. Such actions are unacceptable, since they do not correspond to local field theories. Thus, we seem to be stuck: if an action is chiral, it must be doubled.

One way to avoid doubling would be to alter the dispersion relation so that it has only one low energy solution. The other solutions are forced to

$E \sim 1/a$ and become very heavy as a is taken to zero. The simplest version of this solution, called a Wilson fermion, adds an irrelevant operator, a second-derivative-like term

$$S^W = r\bar{\psi}D_W\psi = -\frac{r}{2a}\sum_{n,\mu}\bar{\psi}_n(\psi_{n+\hat{\mu}} - 2\psi_n + \psi_{n-\hat{\mu}}) \simeq -\frac{ar}{2}\bar{\psi}D^2\psi \quad (6.14)$$

to S^{naive}. (We are obviously considering the free theory; gauge fields could be introduced straightforwardly as before.) The propagator will become

$$\frac{1}{a}S(p) = \frac{-i\gamma_\mu \sin p_\mu a + ma - r\sum_\mu(\cos p_\mu a - 1)}{\sum_\mu \sin^2 p_\mu a + [ma - r\sum_\mu(\cos p_\mu a - 1)]^2}. \quad (6.15)$$

It remains large at $p_\mu \simeq (0,0,0,0)$, but the "doubler modes" are lifted at any fixed nonzero r to masses that are order $1/a$, so $G(p)^{-1}$ has one four-component minimum. Unfortunately, the Wilson term is not chiral. Apparently, if an action is undoubled, chiral symmetry is explicitly broken.

Nielsen and Ninomiya (1981a,b,c) analyzed this problem in depth and proved a famous "no-go" theorem about doubling and chirality. In detail, the theorem assumes

- A quadratic fermion action $\bar{\psi}(x)iH(x-y)\psi(y)$, where H is Hermitian, has a Fourier transform $H(p)$ defined for all p in the Brillouin zone, and has a continuous first derivative everywhere in the Brillouin zone. $H(p)$ should behave as $\gamma_\mu p_\mu$ for small p_μ.
- A local conserved charge Q defined as $Q = \sum_x j_0(x)$, where j_0 is a function of the field variables $\psi(y)$ where y is close to x.
- Q is quantized.

The statement of the theorem is that, once these conditions hold, $H(p)$ has an equal number of left handed and right handed fermions for each eigenvalue of Q.

There is a folkloric version of the theorem that says that no lattice action can be undoubled, chiral, and have couplings that extend over a finite number of lattice spacings ("ultra-locality"). This is actually not what the theorem says. It is the quantization of the charge that governs whether the theorem is evaded or not. (This is the constraint modern four-dimensional chiral actions relax.)

Ultra-locality is a historical engineering constraint on lattice action design. What is needed for a proper field theoretic description is locality, meaning that the range of the action is restricted to be on the order of the size of the spatial cutoff. It is believed that having lattice couplings that fall off exponentially with distance (measured in units of the lattice spacing), *i.e.* $S = \sum_{x,r} \bar{\psi}(x) C(r) \psi(x+r)$ with $C(r) \simeq \exp(-r/\xi)$, $\xi \propto a$, corresponds to a local action in the continuum limit, but that slower fall off (power law, for example) does not.

The connection between chirality and doubling deepens when we think about the anomaly, for there is a theorem due to Adler that states that it is not possible to regularize a theory with a single fermion in a way consistent with chiral symmetry, such that the axial current is conserved. With a lattice regulator, either we must add fermions to cancel the anomaly (naive and staggered fermions), or explicitly break chiral symmetry (Wilson-type fermions), redefine what is meant by a chiral rotation (overlap fermions) or change the dimensionality of space-time (domain wall fermions).

With naive fermions, the anomaly is canceled among the doublers. To see that, imagine coupling the naive fermion to an axial vector field (or alternatively, couple the axial vector field to the chirally invariant part of a Wilson fermion action). This is easily done by introducing a new link field $U_\mu(x) = \exp[iga\gamma_5 J^5_\mu(x)]$. The interaction term is

$$S_i = \sum_{p,q,k} \bar{\psi}(p) g \gamma_\mu \gamma_5 \cos\left[\frac{1}{2}(p+q)_\mu a\right] \psi(q) J^5_\mu(k) \tag{6.16}$$

with $p + k + q = 0$. The vertices give the axial charges of fermions near some $\bar{p}$:

$$\begin{aligned} [\bar{u}(p+\bar{p}) s^{-1}(\bar{p})] \gamma_\mu \gamma_5 \cos(\frac{1}{2}(p+q+2\bar{p})_\mu a) [s(\bar{p}) v(q+\bar{p})] = \\ \bar{u}(p) [s^{-1}(\bar{p}) \gamma_\mu \gamma_5 s(\bar{p})] \cos(\bar{p}_\mu a) v(q) + \mathcal{O}(a). \end{aligned} \tag{6.17}$$

Near $\bar{p} = 0$, $s(p) = 1$, and the coupling $[s^{-1}(\bar{p}) \gamma_\mu \gamma_5 s(\bar{p})] = \gamma_\mu \gamma_5$, so the axial charge is $Q_5 = 1$. If $\bar{p}_\nu$ is near π/a then $s(\bar{p}) = \gamma_\nu \gamma_5$, the coupling is $-\gamma_\mu \gamma_5$, and $Q_5 = -1$. In all, eight of the doublers have $Q_5 = 1$ and eight have $Q_5 = -1$. (This is the charge quantization of the Nielsen-Ninomiya theorem.) The anomaly is canceled; the spectrum is doubled.

6.2 Wilson-type fermions

We have already introduced the undoubled but non-chiral Wilson fermion, whose free action appends the Wilson term $r\bar{\psi} D_W \psi$, Eq. (6.14), to S^{naive}. While the true Wilson action of classical antiquity is $\bar{\psi}[D^{\text{naive}} + D_W]\psi$, and in addition the parameter r is conventionally set to unity, there are actually two dimension-five operators that can be added to a naive fermion action. The Wilson term is just one of them. The other dimension-five term is a magnetic moment term

$$S_{SW} = \bar{\psi}(x)\sigma_{\mu\nu}F_{\mu\nu}\psi(x). \tag{6.18}$$

(Note our convention: $\sigma_{\mu\nu} = -(i/2)[\gamma_\mu, \gamma_\nu]$ or $\gamma_\mu\gamma_\nu = \delta_{\mu\nu} + i\sigma_{\mu\nu}$.) If both terms are included the action is called the "Sheikholeslami-Wohlert" (Sheikholeslami and Wohlert, 1985) or "clover" action. The latter name arises because the lattice version of $F_{\mu\nu}$ (we'll call it $C_{\mu\nu}$ below) is the sum of imaginary parts of the product of links around the paths shown in Fig. 6.2. These days, pure Wilson fermions are rarely used, having been replaced by the clover action (with various choices used to set the value of the clover term).

Specifically, the clover term is a sum of four oriented loops of links

$$iC_{\mu\nu}(x) = \frac{1}{8}(f_{\mu\nu} - f_{\mu\nu}^\dagger). \tag{6.19}$$

The $f_{\mu\nu}$ is the sum of loops in Fig. 6.2: the upper right corner is

$$U_\mu(x)U_\nu(x+\hat{\mu})U_\mu(x+\hat{\nu})^\dagger U_\nu(x)^\dagger. \tag{6.20}$$

In the Abelian limit

$$C_{\mu\nu} = \frac{1}{4}\sum_c A_\mu(c) \tag{6.21}$$

is the sum of the vector potentials from the links running counter-clockwise around the outside of the two-by-two plaquette. It is equal to $F_{\mu\nu}$ plus lattice artifacts. Of course, many other definitions are possible, though less often used.

In contrast to naive fermions, whose Dirac operator is antihermitian, the Wilson fermion action is neither hermitian nor antihermitian. However, it obeys a property called "γ_5-hermiticity," meaning that $D^\dagger = \gamma_5 D\gamma_5$. This relation implies that if the Dirac operator has a complex eigenvalue, $D\phi_i = \lambda_i\phi_i$, then λ_i^* is also an eigenvalue. (To see this, introduce factors of γ_5 to replace the eigenvalue equation by $\gamma_5 D\gamma_5\gamma_5\phi_i = \lambda_i\gamma_5\phi_i$ or, taking the Hermitian conjugate, $(\phi_i\gamma_5)D = \lambda_i^*(\phi_i\gamma_5)$. The equivalent secular equation for this equation, a left eigenvalue equation, is $\det(D - \lambda_i^* I) = 0$, which shows the pairing.) D can also have real eigenvalues, which in principle could be of either sign and any magnitude.

In the Wilson fermion literature it is conventional to talk about "hopping parameter" $\kappa = \frac{1}{2}(ma + 4r)^{-1}$ rather than the quark mass, and to rescale the fields $\psi \to \sqrt{2\kappa}\psi$. The action for an interacting clover fermion is conventionally written in terms of two parameters κ and c_{SW} as

$$
\begin{aligned}
S = &\sum_n \bar{\psi}_n \psi_n \\
&-\kappa \sum_{n\mu} \left[\bar{\psi}_n (r - \gamma_\mu) U_\mu(n) \psi_{n+\mu} + \bar{\psi}_n (r + \gamma_\mu) U_\mu^\dagger(n - \hat{\mu}) \psi_{n-\hat{\mu}}\right] \\
&- \frac{r\kappa c_{SW}}{2} \sum_{n\mu\nu} \bar{\psi}_n \sigma_{\mu\nu} C_{\mu\nu} \psi_n .
\end{aligned}
\tag{6.22}
$$

Of course, one could consider many variations of Wilson fermions: one could change the definition of the derivative, or of the Wilson term, or of the anomalous magnetic moment term. In the literature, these variations are often given their own names, but they share the common property that they are undoubled and nonchiral. We will continue to call any ultralocal lattice action that is undoubled and nonchiral a Wilson action.

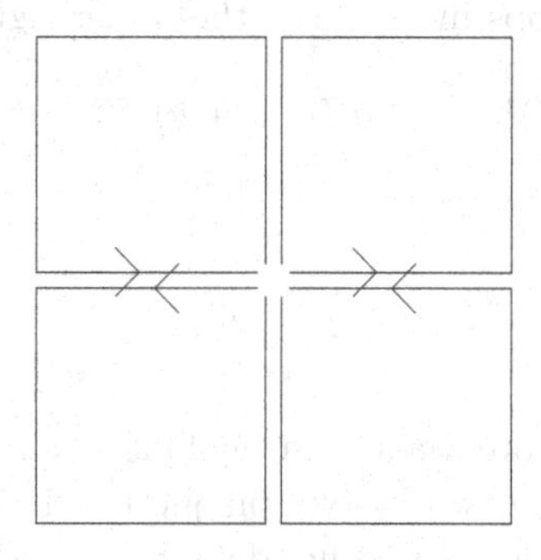

Fig. 6.2 The "clover term".

Why the Wilson term? Why the clover term? We will return to a fuller version of this story when we discuss "improvement," but let us give an abbreviated explanation. Lattice actions are merely bare actions, constructed as a sum of relevant, marginal, and irrelevant operators, each multiplied by its own coupling constant. If we neglect interactions, naive dimensional analysis tells us the relevance of any operator. We want to find a lattice analog of a continuum action $\bar{\psi}(\not{D} + m)\psi$, whose lattice artifacts are removed order by order in a or g^2, or both.

Tree level improvement is simplest: use the organizing principle of dimensionality to construct a set of fermionic operators, first in the continuum, and then on the lattice. The fermion field ψ has mass dimension 3/2, so the operator $O_0 = \bar{\psi}\psi$ is dimension 3, $O_1 = \bar{\psi}\not{D}\psi$ is dimension 4. There are two dimension-5 operators, $O_2 = \bar{\psi}\not{D}\not{D}\psi$ and $O_3 = (1/2)\bar{\psi}\sigma_{\mu\nu}F_{\mu\nu}\psi$, fifteen dimension-6 operators, and so on. Lattice operators, which we will label with a superscript L, can be written as continuum operators plus higher order artifacts: for example, the naive fermion derivative is

$$\Delta_\mu = \partial_\mu + a^2 \frac{1}{6}(\partial_\mu)^3 + \ldots. \tag{6.23}$$

The Wilson term and the clover term are particular linear combinations of O_2^L and O_3^L.

To remove all order a errors, we would need a lattice Lagrangian

$$\mathcal{L}^L = c_0 O_0^L + c_1 O_1^L + c_2 O_2^L + c_3 O_3^L. \tag{6.24}$$

Clearly, the choice $c_0 = m$, $c_1 = 1$, $c_2 = c_3 = 0$ reproduces the continuum action through $O(a)$. (In fact, this is the naive fermion action.) However, this action is still doubled.

How can we remove the doublers while avoiding introducing more lattice artifacts? A scheme for doing this systematically was introduced by Lüscher and Weisz (1985b) and is called "on shell improvement." The particular version of this program that is applied to the undoubling problem is to add an operator to $\mathcal{L}^L$ that does not alter the good low momentum behavior we just found, but does lift the doublers (which, after all, live at $p \sim 1/a$). We will describe this procedure in more detail in Ch. 10, but here is a synopsis:

One way not to alter the low energy behavior is to choose a combination of operators that could appear in a redefinition of the field variables, as it could be removed by inverting the redefinition. As an example, in the continuum, suppose that we shift the fields infinitesimally to

$$\psi \to e^{\epsilon a(\not{D}+m)}\psi; \qquad \bar{\psi} \to \bar{\psi} e^{\epsilon' a(\not{D}+m)}. \tag{6.25}$$

(This is called an "isospectral transformation.") Then the leading change in the Lagrangian is

$$\delta\mathcal{L} = a(\epsilon + \epsilon')\bar{\psi}(m\not{D} + \not{D}\not{D})\psi. \tag{6.26}$$

The $\not{D}$ term can be absorbed into O_1^L by rescaling the fields. The operator $\not{D}\not{D}$ is

$$\not{D}\not{D} = D^2 + \frac{1}{2}\sigma_{\mu\nu}F_{\mu\nu}. \tag{6.27}$$

The Jacobian does not contribute at $O(\epsilon)$ (Sheikholeslami and Wohlert, 1985). The $\not{D}\not{D}$ term is a "redundant operator", *i.e.*, if we had had it to start with, it could have been removed from the Lagrangian by undoing the field redefinition.

Now we pass to the lattice and write down some discretization of the continuum action, which will, of course have discretization artifacts proportional to some power of the lattice spacing. Without the isospectral transformation, we would have discretized $\bar{\psi}\not{D}\psi$, which would have given us the naive action. This action has $O(a^2)$ discretization errors, but it is doubled. Doing the isospectral transformation before we pass to the lattice guides us toward a discretization of the Lagrangian $\mathcal{L} = \bar{\psi}[\not{D} + \epsilon a \not{D}\not{D}]\psi$. Of course, the lattice analog of $\not{D}\not{D}$ is a particular linear combination of the Wilson and clover terms. Their relative weights are fixed from Eq. (6.27). To order (a^2), this fixes $r = c_{SW}$. However, the overall weight of the $\not{D}\not{D}$ term is free because it came from an isospectral transformation: regardless of the value of r, all lattice actions with the Wilson term and the simple clover term, but with $r = c_{SW}$, will have $O(a^2)$ lattice artifacts. Clearly, we want to avoid "near doubling," and so $r = 1$ is a convenient choice.

(Notice that while the transformation is called 'isospectral," we only consider its effect to leading order in the infinitesimal parameter ϵa. We have also not considered the effect of the field redefinition on currents. We will return to this discussion in Ch. 10.)

At the order we are working, there are only two new parameters in the action, and their ratio was fixed using Eq. (6.27). However, that will not be the case in higher order, so let us consider the other dimension-5 operator (the clover term) by itself. We can fix its value (as we would for the values of all operators which are not redundant in the sense of isospectral transformations) by performing a lattice calculation of some physical process and matching it to the desired continuum result. One way to do this is to turn on a quark mass and compute two quantities: the dispersion relation for very small momentum, and the energy of the quark in an external magnetic field. Written as an energy, the result of these two calculations is

$$E(\vec{p}, B) = m_1 + \frac{p^2}{2m_2} + \frac{\vec{\sigma} \cdot \vec{B}}{2m_B}. \tag{6.28}$$

If the bare fermion mass is m_0, then we can show (for $r = 1$)

$$am_1 = \log(1 + am_0)$$

$$(2m_2)^{-1} = \frac{2 + 4am_0 + (am_0)^2}{2m_0(1 + am_0)(2 + am_0)}$$
$$(2m_B)^{-1} = (2m_2)^{-1} - \frac{1 - c_{SW}}{2m_0(1 + m_0 a)}. \tag{6.29}$$

To make the magnetic moment agree with its continuum value, we need $m_B = m_2$ implying $c_{SW} = 1$ $(= r)$ at tree level. The difference between m_1 and m_2 just corresponds to an additive shift in the mass of a hadron, which does not contribute to mass differences. This will be described more in Sec. 6.6.

Wilson-type fermions contain explicit chiral-symmetry breaking terms. This can cause problems in simulations or in their analysis. The most obvious difficulty is that the zero bare quark mass limit is not respected by interactions; the quark mass is additively renormalized. The value of the bare quark mass m_q at which the pion mass vanishes is not known a priori before beginning a simulation. Simulations with Wilson type fermions have to map out m_π vs am_q by direct observation.

We can see how the Wilson term affects observables by recalling our discussion of Ward identities. This requires a digression into a discussion of conserved currents for naive and Wilson fermions. The conserved vector and axial currents for naive fermions are

$$J_\mu(n) = \frac{1}{2}[\bar{\psi}_n \gamma_\mu U_\mu(n)\psi_{n+\hat{\mu}} + \bar{\psi}_{n+\hat{\mu}} \gamma_\mu U_\mu(n)^\dagger \psi_n] \tag{6.30}$$

and

$$J^5_\mu(n) = \frac{1}{2}[\bar{\psi}_n \gamma_\mu \gamma_5 U_\mu(n)\psi_{n+\hat{\mu}} + \bar{\psi}_{n+\hat{\mu}} \gamma_\mu \gamma_5 U_\mu(n)^\dagger \psi_n]. \tag{6.31}$$

This is easily checked from the equation of motion: $\Delta_\mu J_\mu(n) = 0$ with the backward difference $\Delta_\mu \phi_n = \phi_n - \phi_{n-\hat{\mu}}$. The Wilson term spoils the conservation of both currents. It is possible to define a conserved vector current for Wilson fermions,

$$J^c_\mu(n) = \frac{1}{2}[\bar{\psi}_n(\gamma_\mu - r)U_\mu(n)\psi_{n-\hat{\mu}} + \bar{\psi}(n+\hat{\mu})(\gamma_\mu + r)U_\mu(n)^\dagger \psi_n] \tag{6.32}$$

but no conserved axial current exists. Because the Wilson action is not chiral the PCAC relation for on-shell matrix elements of the axial current includes an extra term

$$\langle\alpha|\Delta_\mu J^5_\mu|\beta\rangle = \langle\alpha|2m\bar{\psi}\gamma_5\psi + r\{\gamma_5, D_W\}|\beta\rangle. \tag{6.33}$$

The mass term is the ordinary chiral-symmetry violating term, but the second term represents the explicit chiral symmetry breaking of the Wilson term. If we neglect gauge interactions, D_W is an irrelevant operator, $D_W \sim a\nabla^2$, so it dies away in the naive continuum limit. In fact, if we evaluate it between two free quark states, it is

$$\langle p_2 | r\{\gamma_5, D_W\} | p_1 \rangle = ra(m_1^2 + m_2^2)\bar{u}(p_2)\gamma_5 u(p_1) + O(a^2). \tag{6.34}$$

At nonzero a, and in the interacting theory, the situation is a bit more complicated. Because the action is not chiral, nothing protects the quark mass from acquiring an additive shift, so that the value of m at which correlators show chiral behavior such as a conserved axial vector current is no longer at $m = 0$.

In perturbation theory the loop graphs that contribute to the additive mass renormalization have a $1/a$ divergence (regulated by the lattice). This $1/a$ factor combines with the intrinsic a dependence of the operator to give a mass shift $\delta m a = O(g^2)$. (We have reintroduced the lattice spacing to write a sensible dimensionless expression; recall that on the lattice, all dimensionful parameters scale with the appropriate dimension of a since it is the only scale. What is important for current algebra is the size of chiral symmetry violations compared with other scales.)

The situation we are describing was analyzed in detail by Bochicchio *et al.* (1985). They found that the Wilson term mixes with the pseudoscalar current and the divergence of the axial vector current,

$$\{\gamma_5, rD_W\} \rightarrow 2\delta M \bar{\psi}\gamma_5\psi - (Z_A - 1)\Delta_\mu J_\mu^5, \tag{6.35}$$

where δM and $Z_A - 1$ are both $O(g^2)$, so that Eq. (6.33) is replaced by

$$Z_A \langle \alpha | \Delta_\mu J_\mu^5 | \beta \rangle = \langle \alpha | 2(m + \delta M)\bar{\psi}\gamma_5\psi | \beta \rangle. \tag{6.36}$$

By fine-tuning m to be equal to $-\delta M$, one can return to the chiral point. Within perturbation theory this is the complete story: only one parameter needs to be tuned for $g^2 \neq 0$. If we think of the coupling as a scale dependent $g(a)$, then at sufficiently small a, it runs (logarithmically) to zero. If there is a regime where perturbation theory is applicable, then in that regime chiral symmetry breaking for Wilson fermions is like chiral symmetry breaking in the continuum.

A second serious problem with the use of a nonchiral action is that operators that would not mix if the action were chiral can now do so. This seriously compromises weak interaction matrix element calculations done with Wilson-type fermions. (See Ch. 16 for more discussion.)

Finally, if the action were chiral, the Dirac operator would have an eigenmode spectrum that would be strictly bounded: all of its real eigenvalues on a background gauge configuration would have the same sign, regardless of the values of the gauge field variables. Because the action is not chiral, no such constraint exists. This is a serious problem because of the way simulations are done: one picks a value of bare mass m and constructs a set of propagators, inverting $D + m$ on an ensemble of gauge configurations. It can happen that one configuration in the ensemble has an eigenmode whose eigenvalue is near $-m$. Then $D+m$ is non-invertible: the propagator blows up. This situation is called "encountering an exceptional configuration" in the literature. The reader can imagine that one or two exceptional configurations can seriously compromise the statistical quality of a data set, and indeed, they present a practical problem in using Wilson type fermions for simulations at small quark masses.

When $D+m$ has a zero eigenvalue, the fermionic determinant $\det(D+m)$ is zero, so such gauge configurations should have zero weight in the ensemble. To steer around them correctly puts strong demands on the simulation algorithm. Of course such configurations are a serious problem for the quenched approximation, which omits the fermionic determinant altogether from the integration measure. But even with the determinant properly included, there can be problems. There are arguments in the literature that full QCD with Wilson-like fermions at nonzero lattice spacing has unphysical phases in which QCD is realized differently from continuum expectations. That would preclude doing simulations relevant to phenomenology at those mass and lattice spacing parameters.

6.2.1 *Twisted-mass fermions*

A formulation of lattice fermions (Frezzotti *et al.*, 2000) that removes exceptional configurations is called "twisted mass QCD." This is a scheme for $N_f = 2$ flavors in which the lattice Dirac operator is expanded to become

$$D_{\text{twist}} = D + m + i\mu\gamma_5\tau_3. \tag{6.37}$$

The isospin generator τ_3 acts in flavor space. The extra term is called the "chirally twisted mass." It protects the Dirac operator against exceptional configurations for any finite μ, since $\det D_{\text{twist}} = \det((D+m)^\dagger(D+m)+\mu^2)$. Doublers are avoided when a Wilson-type operator is used for D.

One might think that twisted mass QCD reduces to ordinary QCD only at $\mu = 0$, but that is not so. If we take the continuum twisted-mass action

$$\mathcal{L} = \bar{\psi}(D + m + i\mu\gamma_5\tau_3)\psi \tag{6.38}$$

and perform a chiral rotation,

$$\begin{aligned} \psi_0 &= \exp\left(i\alpha\gamma_5\frac{\tau_3}{2}\right)\psi \equiv R(\alpha)\psi \\ \bar{\psi}_0 &= \bar{\psi}\exp\left(i\alpha\gamma_5\frac{\tau_3}{2}\right) \equiv \bar{\psi}R(\alpha), \end{aligned} \tag{6.39}$$

we observe that if $\tan\alpha = \mu/m$ the action becomes

$$\mathcal{L} = \bar{\psi}_0(D + M)\psi_0 \tag{6.40}$$

with $M = \sqrt{m^2 + \mu^2}$.

A simulation in twisted mass QCD begins by picking a lattice discretization of D, such as the Wilson or clover actions. [The improvement program for twisted mass QCD is practically identical to the program for any Wilson-type action (Frezzotti *et al.*, 2001).] One also chooses the parameter α. One then measures some Euclidean Green's functions

$$\langle O(\psi, \bar{\psi})\rangle_\alpha = Z^{-1}\int [d\psi d\bar{\psi} dU] O(\psi, \bar{\psi}) e^{-S}, \tag{6.41}$$

which are then related to the ordinary QCD Green's functions (of the fields ψ_0, $\bar{\psi}_0$) by considering the sum of twisted-mass correlators

$$\langle O(\psi_0, \bar{\psi}_0)\rangle = \langle O[R(\alpha)\psi, \bar{\psi}R(\alpha)]\rangle_\alpha. \tag{6.42}$$

At nonzero lattice spacing, the twist term breaks flavor and parity symmetries. These symmetries are restored in the continuum limit. Flavor symmetry breaking produces a mass splitting between the charged and neutral pions. At "maximal twist" ($\alpha = \pi/2$) symmetry is restored at a rate $O(a^2)$ (Frezzotti and Rossi, 2004). Maximal twist is emerging as the favorite way to implement twisted mass QCD because for many interesting operators, Eq. (6.42) simplifies for its special value of α.

Notice, finally, that, as is the case for all other parameters in Wilson-type actions, the twist angle is not given by the bare parameters: it is shifted by interactions. One way to determine this would be to compute the renormalized mass m_R in terms of the bare parameters:

$$m_R = Z_m[m(1 + b_m am) + \bar{b}_m a\mu^2] + O(a^2); \tag{6.43}$$

one would have to somehow (via perturbation theory or some additional nonperturbative calculation) find the coefficients Z_m, b_m and $\bar{b}_m$. Then $\alpha = \pi/2$ corresponds to $m = -\bar{b}_m a\mu^2$. For further details we refer the reader to the literature.

6.3 Staggered fermions

The sixteen-fold degenerate doublers of naive fermions can be condensed to a four-fold degeneracy by means of the local transformation $\psi_n \to \Omega_n \psi'_n$, $\bar{\psi}_n \to \bar{\psi}'_n \Omega^\dagger_n$ where

$$\Omega_n = \gamma_0^{n_0} \gamma_1^{n_1} \gamma_2^{n_2} \gamma_3^{n_3}. \tag{6.44}$$

There are sixteen different Ω's. Using

$$\Omega^\dagger_n \gamma_\mu \Omega_{n+\hat{\mu}} = (-1)^{n_0+n_1+\cdots+n_{\mu-1}} \equiv \alpha_\mu(n), \tag{6.45}$$

(and noting $\Omega^\dagger_n \Omega_n = 1$) we rewrite the action as

$$S = \frac{1}{2a} \sum_{n,\mu} \bar{\psi}'_n \alpha_\mu(n) \left[U_\mu(n) \psi'_{n+\hat{\mu}} - U_\mu(n-\hat{\mu})^\dagger \psi'_{n-\hat{\mu}} \right] + m \sum_n \bar{\psi}'_n \psi'_n. \tag{6.46}$$

Written in terms of ψ', the action is diagonal in spinor space. Although ψ' is a four-component spinor, all components interact independently and identically, so we can reduce the multiplicity of naive fermions by a factor of four simply by discarding all but one Dirac component of ψ'. The resulting one-component field χ_n is the "staggered fermion" field with the *one-component action*

$$\begin{aligned} S &= \frac{1}{a} \bar{\chi} M(U) \chi \\ &= \frac{1}{2a} \sum_{n,\mu} \bar{\chi}_n \alpha_\mu(n) \left[U_\mu(n) \chi_{n+\hat{\mu}} - U_\mu(n-\hat{\mu})^\dagger \chi_{n-\hat{\mu}} \right] + m \sum_n \bar{\chi}_n \chi_n. \end{aligned} \tag{6.47}$$

Because of the sign alternation in $\alpha_\mu(n)$, the natural unit cell for the staggered fermion field is the 2^4 hypercube. One may then think of the sixteen hypercube components of the field as four sets of four Dirac components. This residual "doubler" degree of freedom is called "taste:" a single staggered fermion corresponds to four tastes of continuum fermions.

In the older literature, taste was sometimes associated with physical flavors: up, down, etc. This was not a fruitful idea; it is not possible to break flavor symmetry by introducing different masses for the tastes. In contemporary practice one introduces a new staggered fermion species with its own mass for each new flavor. So each flavor carries four tastes. We defer until the end of this section dealing with the additional multiplicity. For simplicity for the most part we discuss a one-flavor, four-taste theory.

The one-component action (6.47) is invariant under a modified $U(1)$ chiral transformation. Notice that the kinetic term in the action connects only even with odd sites. The mass term connects even with even and odd with odd. So at zero mass the one-component action is invariant under the transformation

$$\chi \to \exp(i\Gamma_5\theta)\chi \quad \bar{\chi} \to \bar{\chi}\exp(i\Gamma_5\theta) \tag{6.48}$$

where Γ_5 is diagonal in the site and color index and at site n

$$\Gamma_{5n} = 1 \quad \text{even } n; \quad \Gamma_{5n} = -1 \quad \text{odd } n. \tag{6.49}$$

At any mass the fermion action (6.47) satisfies

$$M(U)^\dagger = \Gamma_5 M(U)\Gamma_5. \tag{6.50}$$

This remnant chiral symmetry is key to the practical utility of this action. It suppresses additive quark mass renormalization and produces a natural Goldstone boson. At zero quark mass M is antihermitian and thus has imaginary eigenvalues. The spectrum of $M^\dagger M$ is bounded from below by $(am)^2$, permitting simulation at quark masses considerably smaller than the Wilson actions. Thus staggered fermions are preferred over Wilson ones in situations in which the chiral properties of the fermions dominate the dynamics.

A convenient change of basis helps to expose the taste multiplicity (Gliozzi, 1982; Duncan *et al.*, 1982; Kluberg-Stern *et al.*, 1983). For simplicity we begin with free fermions. Label the sixteen sites of the hypercube with four-component vectors η with components $\eta_\mu = 0$ or 1. Then define a four-taste Dirac field through a unitary change of basis:

$$\psi_y^{\alpha a} = 1/8 \sum_\eta \Omega_\eta^{\alpha a} \chi_{2y/a+\eta}. \tag{6.51}$$

(Use the same Ω as before to get the same Dirac gamma matrices.) The field ψ has four Dirac spinor components α and four taste components a

and lives on the hypercube with origin at $2y$. This is called the *staggered fermion spin-taste basis.* The inverse transform is

$$\chi_{2y/a+\eta} = 2\mathrm{Tr}[\Omega_\eta^\dagger \psi_y]. \tag{6.52}$$

With a little algebra the free staggered fermion action can then be expressed in the spin-taste basis:

$$S = \sum_{y,\mu} b^4 \bar{\psi}_y \left[(\gamma_\mu \otimes I) \, \Delta_\mu + \frac{1}{2} b(\gamma_5 \otimes \gamma_\mu^* \gamma_5) \Box_\mu \right] \psi_y$$
$$+ mb^4 \sum_x \bar{\psi}_y I \otimes I \psi_y, \tag{6.53}$$

where the first and second block derivatives ($b = 2a$) are

$$\Delta_\mu \psi_y = \frac{1}{2b}[\psi_{y+b\hat{\mu}} - \psi_{y-b\hat{\mu}}] \tag{6.54}$$

and

$$\Box_\mu \psi_y = \frac{\psi_{y+b\hat{\mu}} + \psi_{y-b\hat{\mu}} - 2\psi_y}{b^2}. \tag{6.55}$$

The sum over y runs over all hypercubes of the blocked lattice. The tensor product notation is spin $\otimes$ taste. Taste symmetry is four-fold, so the use of Dirac matrices for its generators is natural. The irrelevant dimension-five term involving $(\gamma_5 \otimes \gamma_\mu^* \gamma_5)$ breaks taste symmetry at nonzero lattice spacing.

In the interacting theory the action in the spin-taste basis is a bit more complicated. Since the transformation from the one-component hypercube basis to the spin-taste basis collects fields from different lattice sites, to preserve gauge invariance in the interacting theory, the basis change must include the gauge connection:

$$\psi_y^{\alpha a} = 1/8 \sum_\eta \Omega_\eta^{\alpha a} W(2y, 2y+\eta) \chi_{2y/a+\eta}. \tag{6.56}$$

where $W(2y, 2y+\eta)$ is a product of gauge link matrices $U_\mu(n)$ connecting sites $2y$ and $2y+\eta$, or a suitable linear combination of such products [sometimes then projected to $SU(3)$]. The inverse transformation is then

$$\chi_{2y/a+\eta} = 2W^{-1}(2y, 2y+\eta)\mathrm{Tr}[\Omega_\eta^\dagger \psi_y]. \tag{6.57}$$

With interactions included the lattice action in the spin-taste basis does not have a simple expression. Expanded in powers of the lattice spacing, the action becomes

$$S = \sum_{y,\mu} b^4 \bar{\psi}_y [(\gamma_\mu \otimes I) D_\mu \psi_y + a S_{tb,1} + \mathcal{O}(a^2)$$
$$+ m b^4 \sum_x \bar{\psi}_y I \otimes I \psi_y]. \tag{6.58}$$

The zeroth order (in a) term is the continuum action with a four-fold taste degeneracy. The irrelevant first order taste breaking (and Lorentz symmetry breaking) contribution $S_{tb,1}$ contains dimension five fermion bilinears in $\gamma_5 \otimes \gamma_\mu \gamma_5 D_\mu^2$, $(\gamma_\mu - \gamma_\nu) \otimes 1 F_{\mu\nu}$, and $\gamma_5 \sigma_{\mu\nu} \otimes (\gamma_\mu + \gamma_\nu)\gamma_5 F_{\mu\nu}$. So even in the interacting theory only irrelevant operators break the taste symmetry; in the continuum limit taste-symmetry breaking is suppressed, and we obtain a theory with four degenerate tastes.

At nonzero lattice spacing most of the staggered fermion taste and spin rotations are replaced by shifts and rotations in the hypercube. The continuous symmetries of continuum QCD are broken to discrete symmetries. In particular, continuum taste symmetry is broken and taste multiplets are split. But at zero mass, one lattice symmetry transformation, a $U(1)_A$ symmetry, survives in continuum form:

$$\psi_y \to \exp(i\theta \gamma_5 \otimes \gamma_5)\psi_y \tag{6.59}$$
$$\bar{\psi}_y \to \bar{\psi}_y \exp(i\theta \gamma_5 \otimes \gamma_5).$$

This is just Eq. (6.48), reexpressed in the spin-taste basis. We see that the remnant continuous chiral transformation mixes taste as well as spin. It is easy to check that the taste-breaking terms in the interacting theory listed after Eq. (6.58) are invariant as they should be.

Following the contemporary practice of introducing a new staggered fermion species for each new flavor, we write the conserved currents in a world with two degenerate flavors of quarks, namely up and down. The fermion field is now a spinor in flavor space and we introduce the flavor Pauli spin matrix τ_i. There is a conserved vector current based on the spin-taste operator $\tau_i \gamma_\mu \otimes 1$, which is represented in the one-component basis by

$$J_{i,\mu}(n) = \frac{1}{4}\left[\bar{\chi}_{n+\hat{\mu}} \alpha_\mu(n) \tau_i U_\mu^\dagger(n) \chi_n + \bar{\chi}_n \alpha_\mu(n) U_\mu(n) \tau_i \chi_{n+\hat{\mu}}\right]. \tag{6.60}$$

At zero mass there is also a conserved axial vector current based on the taste mixing operator $\tau_i \gamma_\mu \gamma_5 \otimes \gamma_5$:

$$J^5_{i,\mu}(n) = \frac{1}{4}\left[\bar{\chi}_{n+\hat{\mu}}\alpha_\mu(n)(-)^n \tau_i U^\dagger_\mu(n)\chi_n + \bar{\chi}_n \alpha_\mu(n)(-)^n \tau_i U_\mu(n)\chi_{n+\hat{\mu}}\right]. \tag{6.61}$$

The spontaneous breaking of this symmetry produces an isovector of Goldstone pions with taste label γ_5.

A conventional isovector chiral transformation, on the other hand, would be generated by the taste singlet operator $\tau_i \gamma_5 \gamma_\mu \otimes 1$, and a isosinglet chiral transformation by $\gamma_5 \gamma_\mu \otimes 1$. In the one-component basis such currents connect $\bar{\chi}$ and χ at pairs of points separated by three links in the hypercube. They are not conserved at nonzero lattice spacing. Since the axial anomaly originates in the gluon sector, which carries no taste, it must be a taste singlet and couple to the pseudoscalar density generated by $\gamma_5 \otimes 1$. A consequence is that there is no exact connection between zero modes of the staggered fermion Dirac operator and the winding number of the gauge fields. These modes emerge only in the continuum limit. Their chirality is not precisely ± 1.

Let us consider the meson spectrum with degenerate up and down quarks. First we consider the continuum limit. At zero mass the action has a full $SU(8)_L \otimes SU(8)_R \otimes U(1)_V$ symmetry, which spontaneously breaks to $SU(8)_V \otimes U(1)_V$. Without taste we would be speaking only of $SU(2)$ and we would have the familiar isovector of three Goldstone pions. With tastes included, each of these pions comes in a taste multiplet of 16. The isoscalar pseudoscalar eta-prime is similarly multiplied by 16. It is modified by the gauge anomaly. As we have observed, the gauge anomaly is a taste singlet, so only the taste singlet eta-prime acquires a mass. The remaining taste-nonsinglet members survive as Goldstone bosons. They would be physical in an $SU(8)_V \otimes U(1)_V$ world, but they are unphysical in the $SU(2)_V \otimes U(1)_V$ world that is closer to our own. We must exclude them from consideration in constructing "physical" asymptotic states. We discuss this "valence projection" further in Ch. 14.

At nonzero lattice spacing but zero quark mass only an $SU(2)_V \otimes U(1)_V \otimes U(1)_A$ symmetry survives. Spontaneous breaking of this symmetry results in a single isovector Goldstone boson. The other 15 partners in their multiplets remain massive. Turning on the lattice spacing splits the taste multiplets. Because the surviving continuous $U(1)_A$ symmetry involves the taste generator γ_5, the lightest member of the multiplet (the one whose mass vanishes at zero quark mass, regardless of lattice spacing)

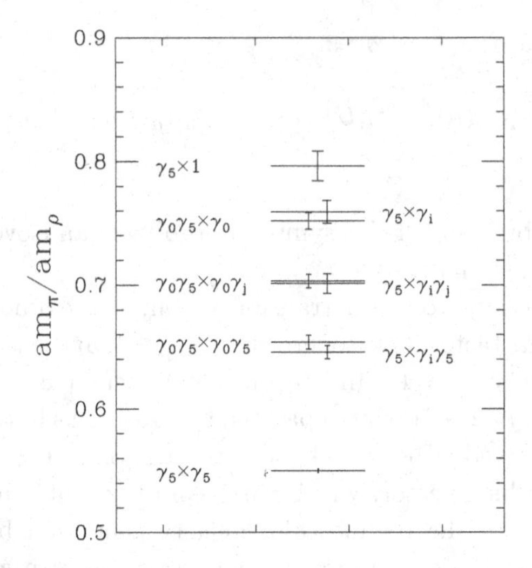

Fig. 6.3 An example of taste symmetry breaking in an improved staggered action. The notation $\Gamma_S \times \Gamma_T$ specifies the spin-taste construction of the pseudoscalar interpolating operator. (See Sec. 11.2.3.) Data are from Orginos *et al.* (1999). For an explanation of the splitting, see Lee and Sharpe (1999).

has taste component γ_5. An example of splitting is shown in Fig. 6.3. This pattern was first described by Lee and Sharpe (1999).

A glance at Eq. (6.58) suggests that taste breaking effects are $\mathcal{O}(a)$ (first order in lattice spacing). Appearances are misleading, however. In the meson spectrum, the splitting is expected to be $\mathcal{O}(a^2\alpha^2)$ for the following reason. First, in the free staggered action, the free quark dispersion relation must be identical to the naive free quark action, which has no taste splitting. Thus taste splitting occurs only as a result of the terms in the action involving gluon exchange – in particular gluons with momentum components close to the lattice cutoff. In a quark-antiquark meson, two such exchanges are required, one to change the tastes of both quark and antiquark and one to change them back again. Thus we require a factor $\mathcal{O}(\alpha^2)$. The further factor of $\mathcal{O}(a^2)$ is best understood from the point of view of an effective theory in which the gluonic degrees of freedom are integrated out. Then taste splitting is induced by dimension six four-fermion operators (Lee and Sharpe, 1999), which therefore require a dimensional factor of a^2.

To improve the accuracy of numerical simulations it is desirable to reduce and gain control of taste mixing. This is done with two strategies. First, the simulations are done with improved versions of staggered fermions, which we will return to in Ch. 10. Second, the analysis of low quark mass simulation data is done in the context of a low energy effective theory appropriate to staggered fermions, i.e. a chiral Lagrangian with $4N_f$ flavors of fermions (the "4" is for four tastes) including explicit taste-breaking terms. We will return to a detailed discussion of this methodology in Ch. 10.

We have postponed until now the problem of dealing with the unwanted taste degree of freedom. To introduce the solution that one finds in the literature, consider the following general situation. We imagine having a Lagrangian for continuum QCD with N_f flavors of quarks. The Dirac operator $M = D + m$ depends on the underlying gauge fields. We use it to construct operators whose expectation values can be analyzed to compute physical observables, which we write generically as

$$\langle O \rangle = \frac{1}{Z} \int [dU][d\bar{\psi}][d\psi] O(\psi_i \bar{\psi}_j) \exp(-S_F - S_G). \tag{6.62}$$

At this point we integrate out the fermions to write the observable as a functional integral over only the gauge fields,

$$\langle O \rangle = \frac{1}{Z} \int [dU] O(M_{ij}^{-1}) \det M \exp(-S_G). \tag{6.63}$$

Suppose we wish to pass to the lattice, using a discretization for the fermions that does not experience doubling. We could do this at any point in the procedure, from before Eq. (6.62) to after Eq. (6.63). However, for staggered fermions, we have to confront the problem of the taste multiplicity. This involves the two places the Dirac operator appears in Eq. (6.63), and the standard solution is rather different in either case.

First, M appears in the fermionic determinant $\det M$. To reduce the unwanted multiplicity of sea quark loops, simulation algorithms take the fourth root of the fermion determinant. To the extent the eigenvalues of the fermion matrix are nearly four-fold degenerate, the fourth-root approximation is effective. Since the splitting of the degeneracy is induced by an irrelevant operator, this approximation is thought to become exact in the continuum limit and produce the same results as an undoubled theory. At nonzero lattice spacing the approximation produces artifacts. We will see an example in Sec. 14.3. As of this writing the fourth root approximation is controversial. Simulations based on it have been very successful in reproducing experimental results, but there is no proof that the continuum limit is smooth.

Second, M appears in the operator $O(M_{ij}^{-1})$ involving valence quarks. From our previous discussion, the taste mixing terms in M are irrelevant operators, whose effects die away at least as $\mathcal{O}(a)$ as a is taken to zero. It is possible to construct operators that couple only to a single taste rather than to multiple members of a taste multiplet. Correlators of these single-flavor operators can mix with different tastes as intermediate states, but because this mixing is induced by irrelevant operators, it is simply a scaling violation. We expect universal behavior in the continuum limit when comparing results from staggered fermions with ones that are taste-symmetric. We will describe how to construct these single-flavor operators in Sec. 11.2.3.

The treatment of valence and sea quarks can be unified through the usual generating functional for Green's functions. For example for an operator $O(\psi_i\bar{\psi}_j)$ constructed from local mesonic combinations $\psi_i(x)\bar{\psi}_j(x)$ we have

$$Z(\sigma_{ij}) = \int [dU][d\bar{\psi}][d\psi] \exp(-S_G) \det[M + \sigma_{ij}(x)]^{1/4}. \tag{6.64}$$

so that $\langle O \rangle = O(\delta/\delta\sigma_{ij}) \log Z(\sigma_{ij})$. The sources σ_{ij} can be designed to couple only to selected tastes.

6.4 Lattice fermions with exact chiral symmetry

Far and away, the easiest path to a chiral fermion comes simply by modifying the definition of a chiral rotation from the usual $\delta\psi = i\epsilon\gamma_5\psi$, $\delta\bar{\psi} = i\epsilon\bar{\psi}\gamma_5$ to either

$$\delta\psi = i\epsilon\gamma_5 \left(1 - \frac{a}{2r_0}D\right)\psi; \qquad \delta\bar{\psi} = i\epsilon\bar{\psi}\left(1 - \frac{a}{2r_0}D\right)\gamma_5 \tag{6.65}$$

or

$$\delta\psi = i\epsilon\gamma_5 \left(1 - \frac{a}{r_0}D\right)\psi; \qquad \delta\bar{\psi} = i\epsilon\bar{\psi}\gamma_5. \tag{6.66}$$

The operator D should be local, and a is the lattice spacing. The naive $a \to 0$ limit of the new chiral rotation is just the usual one, so this is just the usual chiral rotation modified by the addition of an irrelevant operator. It is (nearly) ubiquitous to take D in the rotation to be the Dirac operator used in the action, so that the Lagrange density is $\mathcal{L} = \bar{\psi}D\psi$. Requiring $\delta\mathcal{L} = 0$ under either of the altered chiral rotations replaces the usual anticommutation relation for the Dirac operator by the new constraint

$$0 = \gamma_5 D + D\gamma_5 - \frac{a}{r_0} D\gamma_5 D. \tag{6.67}$$

This expression is called the "Ginsparg-Wilson relation," named after its inventors (Ginsparg and Wilson, 1982). They proposed it as a natural consequence of a renormalization group calculation, but we can take it as the consequence of an alteration of the definition of a chiral rotation. (Experts will recognize that we could have written the chiral rotation as $\delta\psi = i\epsilon\gamma_5(1 - aRD)\psi$ where R is some local operator. Then the Ginsparg-Wilson relation would have been $\{\gamma_5, D\} = aD\gamma_5 RD$. This extension has not received widespread practical use.)

The Ginsparg-Wilson relation immediately implies

$$\{\gamma_5, D^{-1}\} = \frac{a}{r_0} \tag{6.68}$$

and, using $D^\dagger = \gamma_5 D \gamma_5$,

$$D + D^\dagger = \frac{a}{r_0} D^\dagger D. \tag{6.69}$$

Fermions obeying the Ginsparg-Wilson relation have a number of interesting properties. First, their eigenvalues are confined to a circle of radius r_0/a in the complex plane, centered at the point $(r_0/a, 0)$. This is easy to see if we write the eigenvalue equation as $D\phi = \lambda\phi$ with $\lambda = x + iy$. The Ginsparg-Wilson relation via Eq. (6.69) says that $2x = (a/r_0)(x^2 + y^2)$ or $(x - r_0/a)^2 + y^2 = (r_0/a)^2$. Second, if an eigenmode of D is chiral, $\gamma_5\phi = \pm\phi$, then λ is real and equal to either 0 or $2r_0/a$:

$$(\gamma_5 D + D\gamma_5 - \frac{a}{r_0} D\gamma_5 D)\phi = (\pm 2\lambda\phi \mp \frac{a}{r_0}\lambda^2)\phi \tag{6.70}$$

so $\lambda[2 - (a/r_0)\lambda] = 0$.

As a crutch to further understand the spectrum of D, let us introduce the hermitian Dirac operator $H = \gamma_5 D$. Eq. (6.69) becomes

$$\gamma_5 H + H\gamma_5 = \frac{a}{r_0} H^2. \tag{6.71}$$

If an eigenstate $|\psi\rangle$ of H is also an eigenstate of γ_5, from Eq. (6.71), its H eigenvalue must be $(0, \pm 2r_0/a)$. Then it is an eigenstate of all four operators H, H^2, D, and γ_5. Otherwise, $|\psi\rangle$ and $\gamma_5 |\psi\rangle$ form a two-dimensional subspace that is invariant under the action of H and γ_5 (and consequently D), and on which H has eigenvalues $\pm\epsilon$. This is easiest to see by beginning with an eigenvector of H, which obeys $H|\psi\rangle = \epsilon|\psi\rangle$ for $\epsilon \neq 0$ or $\pm 2r_0/a$, and constructing the orthogonal state

$$|\phi\rangle = N[\gamma_5|\psi\rangle - |\psi\rangle\langle\psi|\gamma_5|\psi\rangle], \tag{6.72}$$

where N is a normalization factor, $1/N^2 = 1 - |\langle\psi|\gamma_5|\psi\rangle|^2$. (To prevent $1/N^2 = 0$, the construction requires that $|\psi\rangle$ not be an eigenstate of γ_5.) Computing $H|\phi\rangle$ and using the Ginsparg-Wilson relation, Eq. (6.71), we get

$$H|\phi\rangle = -\epsilon|\phi\rangle + |\psi\rangle N\epsilon\left(\epsilon\frac{a}{r_0} - 2\langle\psi|\gamma_5|\psi\rangle\right). \tag{6.73}$$

So we see that the subspace is invariant under H. It follows that $|\phi\rangle$ is an eigenstate of H with eigenvalue $-\epsilon$ and $\langle\psi|\gamma_5|\psi\rangle = \epsilon/(2r_0/a)$. Finally, we see that H^2 is just ϵ^2 on this subspace and it commutes with γ_5, so we can diagonalize H^2 and γ_5 simultaneously.

From these considerations, when we switch to the simultaneous eigenbasis of H^2 and γ_5, it is easy to show that H is a direct sum of two by two blocks, each of which is

$$H = \frac{1}{2r_0/a}\begin{pmatrix} \epsilon^2 & \epsilon\sqrt{(2r_0/a)^2 - \epsilon^2} \\ \epsilon\sqrt{(2r_0/a)^2 - \epsilon^2} & -\epsilon^2 \end{pmatrix}. \tag{6.74}$$

[On this basis $\gamma_5 = \mathrm{diag}(+1, -1)$]. In addition H may have simple diagonal terms for $\epsilon = 0$ or $2r_0/a$.

The operator $D = \gamma_5 H$ reduces, of course, to block-diagonal form on the same two- or one-dimensional eigenspaces of H^2. Where the eigenvalues of H are $\pm\epsilon$, the eigenvalues of D come in complex conjugate pairs,

$$\lambda = \frac{\epsilon^2 \pm i\epsilon\sqrt{(2r_0/a)^2 - \epsilon^2}}{2r_0/a}, \tag{6.75}$$

or they are real: $\epsilon = 0$ corresponds to $\lambda = 0$ and $\epsilon = \pm 2r_0/a$ corresponds to $\lambda = 2r_0/a$. Where eigenmodes of H with nonzero eigenvalue $\pm\epsilon$ have matrix elements $\langle\phi|\gamma_5|\phi\rangle = \pm\epsilon/(2r_0/a)$, from Eq. (6.67) it is easy to show that nonzero eigenvalue eigenmodes of D have $\langle\psi|\gamma_5|\psi\rangle = 0$.

Since our Hilbert space of Dirac fields is complete and finite dimensional we must have $\mathrm{Tr}\,\gamma_5 = 0$. If we take the trace on the eigenbasis of D, we find that the only nonzero contributions come from the eigenspaces 0 and $2r_0/a$. In the eigenspace 0, $\mathrm{Tr}\,\gamma_5$ counts the difference $n_+ - n_-$ between zero eigenmodes of positive and negative chirality. In the eigenspace $2r_0/a$ it counts the corresponding difference $n'_+ - n'_-$. For the net trace to vanish, we must have $(n_+ - n_-) + (n'_+ - n'_-) = 0$. Similar reasoning shows that $\mathrm{Tr}\ \gamma_5 D = (2r_0/a)(n'_+ - n'_-)$. This leads us to the identity

$$Q = \frac{a}{2r_0}\mathrm{Tr}\ \gamma_5 D = -\mathrm{Tr}\ \gamma_5 \left(1 - \frac{a}{2r_0}D\right) = n_- - n_+. \tag{6.76}$$

This is the first part of the index theorem, which equates the topological charge (winding number) to Q, the difference between zero eigenmodes of positive and negative chirality, gauge configuration by configuration. To complete the theorem we must relate Q to the topological charge. For a smooth background gauge configuration, showing that $\mathrm{Tr}\ \gamma_5 D$ is proportional to the integral over $\epsilon^{\alpha\beta\mu\nu}F^a_{\alpha\beta}F^a_{\mu\nu}$ involves manipulations similar to the continuum calculation (Peskin and Schroeder, 1995).

It is convenient (and useful, to make contact with five-dimensional formulations, as we will see in the next section) to define the massive overlap operator for fermion mass m as

$$D(m) = \left(1 - \frac{m}{2r_0/a}\right) D + m. \tag{6.77}$$

The "shifted propagator" $\hat{D}(m)^{-1}$ is often used in calculations:

$$\left(1 - \frac{m}{2r_0/a}\right) \hat{D}(m)^{-1} = D(m)^{-1} - \frac{1}{2(r_0/a)}. \tag{6.78}$$

It obeys

$$\{\gamma_5, \hat{D}(m)^{-1}\} = 2m\hat{D}(m)^{-1}\gamma_5\hat{D}(m)^{-1}. \tag{6.79}$$

One could not ask for anything more chiral. $\hat{D}$, the inverse shifted propagator, is generally not local, but its anti-commutator is quite familiar: $\{\gamma_5, \hat{D}\} = 2m\gamma_5$.

Because the eigenvalues of D lie on a shifted circle, one can write D as

$$D = \frac{r_0}{a}[1 + V] \tag{6.80}$$

where V is a unitary operator. In practice, V is chosen to be a function of some ordinary, non-chiral, undoubled massless lattice Dirac operator d and a mass term equal to $-r_0/a$. The universally-used (perhaps unique?) choice is the overlap action of Neuberger (1998a,b)

$$D = \frac{r_0}{a}\left[1 + \frac{d - r_0/a}{\sqrt{(d - r_0/a)^\dagger(d - r_0/a)}}\right] \tag{6.81}$$

or, introducing the hermitian operator $h(m) = \gamma_5(d+m)$,

$$D = \frac{r_0}{a}[1 + \gamma_5 \epsilon[h(-r_0/a)]] \tag{6.82}$$

where $\epsilon(h)$ is the matrix step function, $\epsilon(h) = h/\sqrt{h^2}$. The choice of coefficients is made so that if d is "small" compared to r_0/a, $D \simeq d$.

Of course, with such a complicated expression, D cannot be ultra-local. If this action is to be well behaved, it must be local. Hernandez *et al.* (1999) have proved that $D(x-y)$ is local, falling off like $\exp[-C(x-y)/a]$ if the gauge connections are sufficiently smooth. As the gauge fields become rougher, at large lattice spacing, it is believed that this constraint is no longer satisfied. Golterman and Shamir (2003) have argued, and presented numerical evidence, that when the underlying gauge configuration becomes too rough, eigenmodes of the kernel action $d(-r_0/a)$ cease to be localized; their eigenmodes spread in space, overlap, and mix into a band. This may cause the overlap operator to become nonlocal. As a practical comment, if one is doing a calculation with overlap fermions, it is reasonably straightforward to check for locality by computing the norm of a vector

$$\psi(x) = D(x-y)\delta_{y,0} \tag{6.83}$$

and checking that it dies exponentially with x.

Only a cursory glance at Eq. (6.81) reveals the computational difficulty associated with the overlap. Because D is not ultra-local, its matrix representation is not sparse. We defer a discussion on practical methodology for computing the overlap operator to Sec. 8.6, and merely remark that the most useful approach to $\epsilon(h)$ involves an approximation

$$\epsilon(h) \to \epsilon_N(h) = h \sum_{j=1}^{N} \frac{b_j}{h^2 + c_j}. \tag{6.84}$$

Thus an application of the overlap operator to a trial vector involves finding inverses of the operator $h^2 + c_j$.

An interesting way to see that the Ginsparg-Wilson relation gives a lattice action with full chiral symmetry (other than simply noting that $\delta\mathcal{L} = 0$) is to consider a multi-fermion Green's function. As we saw in our discussion of Wilson fermions, in a non-chiral theory the Ward identities pick up extra terms involving $\bar\psi\{\gamma_5, D\}\psi$,

$$\langle \delta O(x,y)\rangle = \langle \delta O\rangle_{cons} + \langle J(x)\bar\psi(y)\{\gamma_5, D\}\psi(y)\rangle. \tag{6.85}$$

It is usually the case that $J(x) = \bar{\psi}(x)\Gamma\psi(x)$ for some Dirac matrix Γ. The Ginsparg-Wilson relation replaces the anti-commutator by $\bar{\psi}(x)D\gamma_5 D\psi(x)$ and so the extra term in the Ward identity is

$$\langle J(x)\bar{\psi}(y)\{\gamma_5, D\}\psi(y)\rangle = \langle \bar{\psi}(x)\Gamma\psi(x)\bar{\psi}(y)D\gamma_5 D\psi(y)\rangle. \tag{6.86}$$

However, $\psi(x)\bar{\psi}(y) = D(x,y)^{-1}$ and $D(x,y)^{-1}D(y) = \delta(x-y)$: the extra term is just a contact term. If the correlator is not evaluated at zero separation, it will give results consistent with chiral symmetry. Thus all the usual continuum chiral symmetry properties are present in Ginsparg-Wilson fermions, even at nonzero lattice spacing.

6.5 Exact chiral symmetry from five dimensions

One could just terminate a discussion of chiral lattice fermions with the Ginsparg-Wilson relation, but it is a worthwhile exercise to consider another ultraviolet completion of four-dimensional interacting fermions: we imagine that the world is really five-dimensional, and that a four-dimensional chiral fermion arises from structure in the fifth dimension. We will show that the fermions in the low-energy effective field theory at a particular location in the space obey the Ginsparg-Wilson relation. One could either work directly with them, or simulate the original five-dimensional theory on a lattice. The second alternative is called a "domain wall fermion."

6.5.1 *Five dimensions in the continuum*

The description of the five-dimensional world becomes much more transparent if we preserve continuum language. [A nice place to begin reading is Kaplan and Schmaltz (1996).] After we have outlined the main ideas, we can return to the lattice and describe the discrete version of these fermions. The lattice versions of these ideas originated with Kaplan (1992) and the method which is most widely used was developed by Shamir (1993).

Consider first free fermions in a five-dimensional Euclidean world: the usual coordinates $\mu = 1$ to 4, labeled by x_μ, and a fifth dimension labeled by s. We imagine a free Dirac operator

$$D_5 = D_4 + \gamma_5\partial_5 - M(s) \tag{6.87}$$

where $D_4 = \gamma_\mu\partial_\mu$, $\partial_5 = \partial/\partial s$, and the "mass parameter" is assumed to vary with s, interpolating between $-M$ and M as shown in Fig. 6.4a.

We can assume a solution of the Dirac equation $D_5\chi = 0$ where $\chi(x,s) = \exp(ipx)u(s)$, $p = (iE, \vec{p})$ in Euclidean space, where $p^2 = -E^2 + \vec{p}^2 = -m^2$. If $D_5\chi = 0$, then

$$[\gamma_5\partial_5 - M(s)]u = -i \not{p} u. \tag{6.88}$$

Squaring the equation in the usual fashion (multiplying on the left by $-(\gamma_5\partial_5 - M)$ and on the right by $i \not{p}$) we find

$$[-\partial_5^2 + V(s)]u = m^2 u. \tag{6.89}$$

Here $V(s) = M^2 + \gamma_5\partial_5 M$.

$V(s)$ is shown in Fig. 6.4b. The spectrum has three kinds of modes: If $m^2 > M^2$, the spectrum is continuous. If $m^2 < M^2$ there can be discrete levels, but if the only scale in the problem is M, and m and M are the same order of magnitude, these eigenvalues are also $O(M)$. Finally (and most important), there can be massless modes

$$[-\gamma_5\partial_5 + M(s)]u(s) = 0, \quad \gamma_\mu p_\mu u = 0 \tag{6.90}$$

with wave functions

$$u = \exp\left(\pm \int_0^s ds' \, M(s')\right) v \tag{6.91}$$

with $\gamma_\mu p_\mu v = 0$ and v chiral: $P_\pm v = v$; $P_\pm = (1 \pm \gamma_5)/2$. If the potential is as shown in Fig. 6.4, only the negative chirality mode is normalizable.

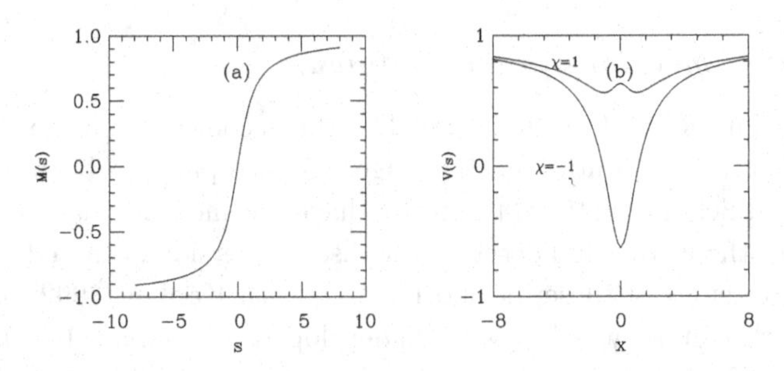

Fig. 6.4 (a) A typical potential $M(s)$ that supports a zero mode. (b) The potential in the two chiral sectors; the negative chirality zero mode is bound deep in the well.

This is a general situation, independent of any specific details about $M(s)$. Because $V(s)$ commutes with γ_5, its eigenstates come in chiral pairs – unless one of the wave functions is non-normalizable. In our example, that happens for the positive chirality zero mode. If $M(s)$ had additional zeros, there would be additional zero modes, corresponding to additional chiral fermions if the wave function of one chirality was non-normalizable.

Since the situation does not depend on a particular form for M, let us replace it by a wall with Dirichlet boundary conditions at $s = 0$. We restrict ourselves to $s > 0$ and take $M(s) = M$ in that domain. For this simplified case the left-handed zero mode is pinned to the $s = 0$ surface and its wave function falls away like $\exp(-Ms)$. Now let us find the four dimensional effective field theory of the massless mode. We can do this by finding the propagator $G(x, s_1; y, s_2)$ between two sites in the five-dimensional manifold and restrict our attention to the case $s_1 = s_2 = 0$ (since that is where the massless mode is confined). The Green's function obeys

$$(D_4 + \gamma_5 \partial_5 - M)G(x, s_1; y, s_2) = \delta(x - y)\delta(s_1 - s_2) \tag{6.92}$$

with

$$P_+ G(x, s_1; y; s_2) = 0 \tag{6.93}$$

at $s_1 = 0$. One can show that

$$G(x, s_1 = 0; y, s_2 = 0) = 2MP_- S(x, y)P_+, \tag{6.94}$$

where $S(x, y) = D(x, y)^{-1}$ if $x \neq y$,

$$D = M[1 - \gamma_5 \frac{Q}{\sqrt{Q^2}}], \tag{6.95}$$

and $Q = \gamma_5(M - D_4)$. That is, D obeys the Ginsparg-Wilson relation.

We reproduce the derivation of this result given by Lüscher (2000). It proceeds through the intermediary of the operator $\mathcal{G}(s_1, s_2)$ acting on four-dimensional Dirac fields $\psi(x)$ via $\mathcal{G}(s_1, s_2) = \int d^4y\, G(x, s_1; y, s_2)\psi(y)$. $\mathcal{G}$ obeys the equation

$$D_5 \mathcal{G}(s_1, s_2) = \delta(s_1 - s_2) \tag{6.96}$$

with the chiral constraint

$$P_+ \mathcal{G}(s_1 = 0, s_2) = 0. \tag{6.97}$$

Since D_4 does not depend on the fifth dimension,

$$\mathcal{G}(s_1, s_2) = [e^{(s_1 - s_2)Q}\theta(s_1 - s_2) + C]\gamma_5 \tag{6.98}$$

before we impose boundary conditions on our solution. Our first constraint is that we want $\mathcal{G}(s_1, s_2)$ to vanish as $|s_1 - s_2| \to \infty$. We can force that by introducing positive and negative eigenmode projection operators for Q, $\hat{P}_\pm = \frac{1}{2}(1 \pm Q/\sqrt{Q^2})$, writing the identity operator as $\delta(x-y) = \hat{P}_+ + \hat{P}_-$, and shuffling terms in Eq. (6.98) so that the exponential falls away appropriately. The identity $2MP_+\hat{P}_- = P_+D$ can be employed to show that the operator

$$\mathcal{G}(s_1,s_2) = \{e^{(s_1-s_2)Q}[\theta(s_1-s_2)\hat{P}_- - \theta(s_2-s_1)\hat{P}_+] + e^{s_1Q}\hat{P}_-\frac{2M}{D}\hat{P}_+e^{-s_2Q}\}\gamma_5 \tag{6.99}$$

is a solution of the equation of motion that also solves all boundary conditions. Equation (6.94) follows, improbably, by writing $\epsilon = Q/\sqrt{Q^2}$ with $\epsilon^2 = 1$, $MD^{-1} = \sum_n (\gamma_5\epsilon)^n$, shifting some sums, and inserting or removing contact terms as needed.

6.5.2 *Five dimensions on the lattice*

Everything we have described in the previous subsection immediately connects to our earlier lattice-based discussion. We first introduce a lattice spacing in the four dimensions of low-energy interest, while holding the fifth dimension continuous. The gauge fields in the five dimensional world are assumed merely to have no variation in the fifth dimension, so that the action continues to be a sum of a four-dimensional and a five-dimensional piece. Then, replacing M by r_0/a, we have

$$D_5 = d(U)_4^{\mathrm{latt}} + \gamma_5\partial_5 - r_0/a. \tag{6.100}$$

The overlap action is once again obtained as the four dimensional action at $x_5 = 0$.

If the action obeys the Ginsparg-Wilson relation, the transformation

$$\psi \to \psi + \epsilon\gamma_5\left(1 - \frac{D}{M}\right)\psi; \qquad \bar{\psi} \to \bar{\psi} + \epsilon\bar{\psi}\gamma_5 \tag{6.101}$$

leaves the action invariant. This transformation is certainly asymmetric, but in Euclidean space, ψ and $\bar{\psi}$ are integrated over separately, and a chiral rotation is just a change of integration variables.

Now define

$$\hat{\gamma}_5 = \gamma_5\left(1 - \frac{D}{M}\right). \tag{6.102}$$

It happens that $\hat{\gamma}_5^\dagger = \hat{\gamma}_5$ and $\hat{\gamma}_5^2 = 1$. We can define fermion chiral projectors to be $\hat{P}_\pm = \frac{1}{2}(1 \pm \hat{\gamma}_5)$ and anti-fermion projectors to be the usual $P_\pm$. Since

$$D = P_+ D \hat{P}_- + P_- D \hat{P}_+, \tag{6.103}$$

we could study a lattice-regulated theory consisting only of left-handed fermions by imposing the restrictions on our fields that $\hat{P}_- \psi = \psi$ and $\bar{\psi} P_+ = \bar{\psi}$.

We can also directly expose the chiral fermion in a five dimensional context. Write the five-dimensional field as $\Psi(x, s)$ and the action as

$$S = \int d^4 x ds\, \bar{\Psi}(x, s) D_5 \Psi(x, s). \tag{6.104}$$

Separate variables in Ψ to write

$$\begin{aligned} \Psi &= \sum_n [l_n(s) P_- + r_n(s) P_+] \psi_n(x) \\ \bar{\Psi} &= \sum_n \bar{\psi}_n(x) [l_n(s)^* P_+ + r_n(s)^* P_-]. \end{aligned} \tag{6.105}$$

The equations of motion separate into chiral sectors with

$$(-\partial_5 - M) l_n = -r_n \not{D}_4 \psi_n; \qquad (\partial_5 - M) r_n = -l_n \not{D}_4 \psi_n \tag{6.106}$$

Chiral zero modes (for which $\not{D}_4 \psi = 0$) again have exponentially growing or falling solutions in the fifth dimension. There are also the usual set of Kaluza-Klein modes (fermion modes with eigenvalues $O(M)$). For them, write $D_4^2 \psi_n = \omega_n^2 \psi_n$. If we replace the field variables in the action with the mode expansion of Eqs. (6.105), we can integrate over s to construct a four-dimensional action

$$S = \int d^4 x \left[\sum_{l=0} \bar{\psi}_l \not{D}_4 P_- \psi_l + \sum_{r=0} \bar{\psi}_r \not{D}_4 P_+ \psi_r + \sum_{n \text{ nonzero}} \bar{\psi}_n (\not{D}_4 + \omega_n) \psi_n \right]. \tag{6.107}$$

The three sums run, respectively, over the sets of negative chirality zero modes, positive chirality zero modes, and over all nonzero modes, for which chiralities are paired.

We can now describe the domain wall fermion of the lattice literature, as introduced by Shamir (1993). We take a five-dimensional space $0 < s < L$, with Dirichlet boundary conditions at both ends. There will be one zero mode of each chirality, localized on a boundary. Their wave functions are $A \exp(-Ms)$ or $A \exp[-M(L-s)]$, where $A^2 = 2M/[1-\exp(-2ML)]$ is the normalization factor. We must bundle the two chiral modes into a single

Dirac spinor. Shamir does this by orbifolding the fifth dimension. More simply said, he adds an extra term $mP_-\delta(s)\delta(s'-L) + mP_+\delta(s-L)\delta(s')$ to $D(x,s;x',s')$. After repeating the mode expansion, we replace the sum of positive and negative chirality massless modes by an action for a single Dirac particle

$$S = \int d^4x\, [\bar{\psi}(x)(\not{D}_4 + \mu)\psi(x)] \tag{6.108}$$

where the Dirac mass is

$$\mu = m\frac{2M}{1-\exp(-2ML)} \simeq 2mM. \tag{6.109}$$

As a final technical point, the would-be chiral modes dominate long distance physics, but it might be advantageous to suppress the Kaluza-Klein modes by hand. This can be done by adding a set of massive scalar fields (called "Pauli-Villars regulator fields" by Furman and Shamir (1995) after the familiar continuum regulators). If we take a discretized fifth dimension with sites labeled $s = 1$ to $2N$ (and include the orbifold mass m), a possible lattice partition function would be gotten by defining the regulator fields on every other site and giving them an action that has the same UV structure as the fermions:

$$Z = \int \prod_{x,\mu} dU_\mu(x) \prod_{s=1}^{2} N d\bar{\psi}_{x,s} d\psi_{x,s} \prod_{s=1}^{2N} d\phi^\dagger_{x,s} d\phi_{x,s} e^{-S}, \tag{6.110}$$

where

$$S = \sum_{x,s=1}^{2N} \bar{\psi}_{x,s} D_5(m)\psi_{x,s} + \sum_{x,s=1}^{N} \phi^\dagger_{x,s} D_5^\dagger(m=1) D_5(m=1)\phi_{x,s}. \tag{6.111}$$

Obviously, the boson site label runs over only every other fermion site. The complete functional determinant is then

$$\frac{\det D_5(m;2N)}{\det D_5^\dagger(m=1,N) D_5(m=1,N)}. \tag{6.112}$$

On the lattice the four dimensional term $\not{D}_4$ is replaced by any Wilson-type operator (to exclude doubling from the start). The Wilson term part of such an action is lumped with the M parameter of the five dimensional action. This prevents the exact separation of variables between the fifth dimension and the other four. [Relevant formulas are given by Shamir (1993).] The discretization of a finite-length fifth dimension affects the chirality of the action. A sketchy derivation of a lattice action on a finite volume [adapted from Edwards and Heller (2001)] illustrates these points. With L_5 sites in the fifth dimension and a lattice spacing in the fifth dimension of a_5, the action is

$$S = \sum_{i=1}^{L_5} \bar{\Psi}_i[(a_5 D_4(-M) + 1)\Psi_i - P_-\Psi_{i+1} - P_+\Psi_{i-1}] \tag{6.113}$$

with $P_-\Psi_{L_5+1} = mP_-\Psi_1$ and $P_+\Psi_0 = mP_+\Psi_{L_5}$. If one redefines $H_4 = \gamma_5 D_4$, sets $Q_\pm = a_5 H_4 P_\pm \pm 1$, and writes $\bar{\Psi}_i = (\mathcal{P}\chi)_i$, and $\bar{\psi}_i = \bar{\chi}_i Q_-^{-1}\gamma_5$, with $\mathcal{P}_{jk} = P_-\delta_{jk} + P_+\delta_{j-1,k}$ for $j = 1, 2, \ldots, L_5 - 1$, or $P_-\delta_{L_5,k} + P_+\delta_{1,k}$ if $j = L_5$, the action becomes

$$\begin{aligned} S = -\{&\bar{\chi}_1[(P_+ mP_+)\chi_1 - T^{-1}\chi_2] \\ &+ \sum_{i=2}^{L_5-1} \bar{\chi}_i[\chi_i - T^{-1}\chi_{i+1}] \\ &+ \bar{\chi}_{L_5}[\chi_{L_5} - T^{-1}(P_+ mP_-)\chi_1]\} \end{aligned} \tag{6.114}$$

where $T^{-1} = -Q_-^{-1}Q_+$. One can now integrate out the fields on successive sites, beginning with χ_{L_5} and $\bar{\chi}_{L_5}$ and going down to χ_1 and $\bar{\chi}_1$. The fermionic determinant is $\det \Delta_F$ where

$$\Delta_F = -[(T^{-L_5} + 1)\gamma_5]\left\{\frac{1+m}{2} + \frac{1-m}{2}\gamma_5 \frac{T^{-L_5} - 1}{T^{-L_5} + 1}\right\}. \tag{6.115}$$

The term in the square brackets is canceled by the determinant of the Pauli-Villars regulators, leaving the term in curly brackets,

$$\Delta_F = \frac{1+m}{2} + \frac{1-m}{2}\gamma_5 Q_{L_5}. \tag{6.116}$$

If we write

$$T^{-1} = \frac{1 + a_5 H_T}{1 - a_5 H_T} \tag{6.117}$$

(a convenient notation; actually, $H_T = H_4(2 + a_5\gamma_5 H_4)^{-1}$), then

$$Q_{L_5} = \frac{(1 + a_5 H_T)^{L_5} - (1 - a_5 H_T)^{L_5}}{(1 + a_5 H_T)^{L_5} + (1 - a_5 H_T)^{L_5}} \tag{6.118}$$

or

$$Q_{L_5} = \tanh\left(-\frac{L_5}{2}\log|T|\right). \tag{6.119}$$

These expressions were first derived by Neuberger. For finite L_5, Q_{L_5} is an approximation to the matrix step function $\epsilon(x)$ of the overlap operator [defined in Eq. (6.82)], of the form of Eq. (6.84). As L_5 grows, it becomes increasingly accurate.

This result is only one of many possible five-dimensional operators, which reduce to overlap-like four dimensional operators for large L_5. (For a recent discussion of several variant actions, see Edwards *et al.* (2005).) For any finite L_5 none of these domain wall actions are exactly chiral, just as no finite-N truncation of Eq. (6.84) would be chiral either. This means that these actions are members of the class of Wilson actions, and would show typical behavior of a Wilson action with regard to chiral symmetry. In particular, these actions would show an additive mass renormalization, which the domain wall literature refers to as a "residual mass," a shift away from zero for the value of the bare quark mass at which the pion becomes massless. As with ordinary Wilson-type fermions, this has to be determined via simulation. These actions will also fail to realize the index theorem precisely; they will not have chiral zero modes.

As a general rule, all finite-L_5 domain wall actions have much better chiral behavior than the generic Wilson or clover actions we discussed in Sec. 6.2. But, of course, some choices of lattice action will be better than others. Generally, the better the approximation to a step function, the better is the chiral behavior observed in a simulation. Doing domain wall simulations then requires confronting an engineering question, namely, for any action, how big must L_5 be for one's desired calculation?

6.6 Heavy quarks

The charm quark has a mass of about 1.5 GeV, and the bottom quark is at about 4.5 GeV. Only a very brave simulator would venture to a lattice spacing below 1/20 femtometer, so at best $M_c a > 0.4$ and $M_b a > 1.2$, but they are often very much greater. Unless they are carefully designed, lattice actions develop rather severe discretization effects once the product $M_Q a$ becomes large, so a direct attack on the properties of heavy quark systems is likely to encounter problems. Fortunately, one does not have to make such a direct attack, and a few minutes' reflection about effective field theories tells why; when the quark mass is the same size as the cutoff, and when one is interested in physical processes at scales far below the cutoff, relativity contributes little physics. Just as we organize a low energy effective action as a sum of terms of increasingly higher dimension, multiplied by a coupling constant that scales with a power of the cutoff, so we can construct an effective action for the heavy quark, whose terms are organized as powers of a small parameter like the inverse of the heavy quark mass or its velocity.

Heavy quarks (the charm or bottom quark) appear in two contexts: in bound states of a single heavy quark and one or more light hadrons, such as the D or B meson, or in bound states in which all the constituents are massive, such as the J/ψ or Υ systems. The physics of the two systems is slightly different. In systems with a single heavy quark, the center of mass coordinate is centered on the heavy quark: in the center of mass frame the heavy quark does not move. The light constituents contribute an amount $\Lambda \simeq$ a few hundred MeV to the energy of the system. These systems are described by Heavy Quark Effective Theory (HQET), whose expansion parameter is $1/M_Q$. The ratio Λ/M_Q would then be about 1/5 for charmed systems, 1/15 for bottom hadrons.

Quarkonia ($Q\bar{Q}$ states) are different. Think about them as you would positronium. In a Coulomb potential the kinetic energy $p^2/(2M)$ and potential energy $-\alpha_s/r$ balance, and the uncertainty principle says that $\langle p\rangle\langle r\rangle \simeq 1$, so $\langle p\rangle \simeq \alpha_s M_Q$ or $v \simeq \alpha_s$. The binding energy is roughly $\alpha_s^2 M_Q \sim M_Q v^2$. The small expansion parameter that characterizes the effective field theory is v; this expansion is called nonrelativistic QCD (NRQCD). In typical potential models, $v^2 \simeq 0.1$ for $b\bar{b}$ systems and 0.3 for $c\bar{c}$. Clearly, the expansion will work much better for bottom than for charm.

We discuss first the two effective field theories, and then return to consider the use of relativistic actions for heavy quarks.

6.6.1 *Heavy quark effective theory*

Let's return to Minkowski space and derive the heavy quark expansion. We begin with the Dirac Lagrangian

$$\mathcal{L} = \bar{\psi}[i\gamma^\mu D_\mu - M_Q]\psi. \tag{6.120}$$

Projecting out quark $\psi_+ = (1+\gamma_0)/2\psi$ and antiquark $\psi_- = (1-\gamma_0)/2\psi$ fields, we obtain the zeroth order Lagrangian describing a heavy quark at rest:

$$\mathcal{L}_0 = \psi_+^\dagger(iD_0 - M_Q)\psi_+, \tag{6.121}$$

and the propagator is just a straight Wilson line (P is the path-ordering operator)

$$S(x, x_0; x', x'_0) = -i\theta(x_0 - x'_0)\delta^3(\vec{x} - \vec{x}')e^{-iM_Q(x_0 - x'_0)} \times P \exp\left[-ig\int_{x'_0}^{x_0} dy\, A^0_\alpha(x, y)t^\alpha\right]. \tag{6.122}$$

To go further, we uncouple ψ_+ and ψ_- by writing

$$\begin{aligned}(iD_0 - M_Q)\psi_+ &= \vec{\sigma}\cdot\vec{D}\psi_- \\ (iD_0 + M_Q)\psi_- &= -\vec{\sigma}\cdot\vec{D}\psi_+\end{aligned} \tag{6.123}$$

or

$$(iD_0 - M_Q)\psi_+ = -(\vec{\sigma}\cdot\vec{D})\frac{1}{iD_0 + M_Q}(\vec{\sigma}\cdot\vec{D})\psi_+. \tag{6.124}$$

Expanding

$$\frac{1}{iD_0 + M_Q} = \frac{1}{2M_Q}\left[1 - \frac{iD_0 - M_Q}{2M_Q} + \left(\frac{iD_0 - M_Q}{2M_Q}\right)^2 + \dots\right], \tag{6.125}$$

and defining color electric and magnetic fields through the covariant derivatives $g\vec{E} = -i[D_0, \vec{D}]$ and $g\epsilon_{ijk}B_i = i[D_j, D_k]$, we transform the Dirac equation into

$$\begin{aligned}(iD_0 - M_Q)\psi_+ = -\Bigg[&\frac{(\vec{\sigma}\cdot\vec{D})^2}{2M_Q} + \frac{g}{8M_Q^2}[\vec{\sigma}\cdot\vec{D}, \vec{\sigma}\cdot\vec{E}] \\ &- \frac{(iD_0 - M_Q)}{8M_Q^2}(\vec{\sigma}\cdot\vec{D})^2 - (\vec{\sigma}\cdot\vec{D})^2\frac{(iD_0 - M_Q)}{8M_Q^2} + \dots\Bigg]\psi_+.\end{aligned} \tag{6.126}$$

Next, using the equations of motion to order $1/M_Q$ to eliminate time derivatives on the right hand side, we have

$$\begin{aligned}(-iD_0 + M_Q)\psi_+ = \Bigg[&\frac{D^2}{2M_Q} + g\frac{\vec{\sigma}\cdot\vec{B}}{2M_Q} \\ &+ \frac{g}{8M_Q^2}\{[D_i, E_i] + i(\vec{D}\times\vec{E} - \vec{E}\times\vec{D})\} + \dots\Bigg]\psi_+.\end{aligned} \tag{6.127}$$

The mass dependence of the leading term can be eliminated by redefining $\psi_+ \to \exp(-iM_Q t)\psi_+$. We have arrived at the Lagrangian for heavy quarks

$$\mathcal{L}_{\text{heavy}} = \psi_+^\dagger \left[iD_0 + \frac{D^2}{2M_Q} + g\frac{\vec{\sigma}\cdot\vec{B}}{2M_Q} \right. \\ \left. + \frac{g}{8M_Q^2}\{[D_i, E_i] + i(\vec{D}\times\vec{E} - \vec{E}\times\vec{D})\} \right] \psi_+ + O\left(\frac{1}{M_Q^3}\right). \tag{6.128}$$

For antiquarks, substitute $\psi_+ \to \psi_-^\dagger$, $\psi_+^\dagger \to \psi_-$, and $t^\alpha \to t^{\alpha *}$.

This is an effective action. One keeps as many terms as are needed to produce $1/M_Q^n$ accuracy. Keeping only the first term (iD_0) is called the "static limit;" it has very interesting symmetry properties (such as an $SU(2)$ associated with the quark's spin) and its own extensive literature and applications.

6.6.2 *Nonrelativistic QCD*

Before passing to the lattice, let us introduce nonrelativistic QCD, in the context of $\bar{Q}Q$ bound states. Here the expansion parameter is the quark's velocity. To organize the expansion, we have to think first about the sizes of fields and operators.

We start with the number operator:

$$\int d^3x\, \psi^\dagger \psi = 1. \tag{6.129}$$

From the uncertainty principle, the size of the bound state is $\langle x \rangle \sim 1/\langle p \rangle$ and so $\psi^\dagger\psi \simeq \langle p \rangle^3$ or $|\psi| \sim (M_Q v)^{3/2}$.

The kinetic energy is

$$KE \equiv Mv^2 = \int d^3x\, \psi^\dagger \frac{D^2}{2M_Q}\psi, \tag{6.130}$$

so of course the spatial derivative is $|D| \sim M_Q v$. The time derivative has a size $|D_0| \sim M_Q v^2$, because $[iD_0 + D^2/(2M_Q)]\psi = 0$.

We need color fields as well. If we work in Coulomb gauge, which is a reasonable choice for nonrelativistic bound states, the leading-order Schrödinger equation becomes

$$\left(i\partial_0 - gA_0 + \frac{\nabla^2}{2M_Q} \right)\psi = 0, \tag{6.131}$$

and so $|gA_0| \sim M_Q v^2$. Electric fields scale as gA_0/x, so they are $|gE| \sim M_Q^2 v^3$. The vector potential is a solution of the wave equation, so $(\partial_t^2 - \nabla^2) g\vec{A} = gJ = g^2 \psi^\dagger \vec{\nabla}\psi / M_Q$. Therefore, $|\vec{A}| \sim M_Q v^3$. Finally, $g\vec{B} = g\nabla \times A$ and so $|gB| \sim M_Q^2 v^4$. As expected, electric fields are bigger than magnetic fields.

We now construct the Lagrangian. We begin, of course, with the non-relativistic Schrödinger equation

$$\mathcal{L}^0_{\text{NRQCD}} = \psi_+^\dagger \left[iD_0 + \frac{D^2}{2M_Q} \right] \psi_+ \tag{6.132}$$

which we saw in the last section. We now want to add to it. To do this, we proceed a bit differently. We write down all interaction terms that are consistent with the symmetries we desire to include: gauge invariance, parity, rotational invariance, and so on. We assign an arbitrary coupling constant to each term, which we will have to determine by matching to a more fundamental theory (a relativistic fermion). Of course, we group terms by their power of velocity.

In order v^2,

$$\begin{aligned}
\delta\mathcal{L}^2 = {} & c_1 \frac{1}{M_Q^3} \psi_+^\dagger D^4 \psi_+ \\
& + c_2 \frac{g}{M_Q^2} \psi_+^\dagger (\vec{D}\cdot\vec{E} - \vec{E}\cdot\vec{D}) \psi_+ \\
& + c_3 \frac{ig}{M_Q^2} \psi_+^\dagger \vec{\sigma}\cdot(\vec{D}\times\vec{E} - \vec{E}\times\vec{D}) \psi_+ \\
& + c_4 \frac{g}{M_Q^3} \psi_+^\dagger \vec{\sigma}\cdot\vec{B}\psi_+ .
\end{aligned} \tag{6.133}$$

We have used the equations of motion to trade iD_0 for $D^2/(2M)$, for a reason that we will explain, below. At order v^4 there are spin-dependent terms

$$\begin{aligned}
\delta\mathcal{L}^4 = {} & f_1 \frac{g}{M_Q^3} \psi_+^\dagger \{D^2, \vec{\sigma}\cdot\vec{B}\} \psi_+ \\
& + f_2 \frac{ig}{M_Q^4} \psi_+^\dagger \{D^2, \vec{\sigma}\cdot(\vec{D}\times\vec{E} - \vec{E}\times\vec{D})\} \psi_+ \\
& + f_3 \frac{ig}{M_Q^3} \psi_+^\dagger \vec{\sigma}\cdot(\vec{E}\times\vec{E}) \psi_+
\end{aligned} \tag{6.134}$$

that are needed to get accurate spin splittings, and the list can go on, depending on the desired accuracy of the calculation. For a discussion of more operators, we refer the reader to the extremely well written paper by Lepage *et al.* (1992), where all of this was first described (and which we are merely following).

The c_i and f_i coefficients are determined by performing a calculation in relativistic QCD and in the effective field theory and forcing the result of the latter to agree with the former in their overlapping regions of validity. Here it is for momenta much smaller than M_Q. For example, because the first relativistic correction to the energy is $p^4/(8M_Q^3)$, $c_1 = 1/8$. Next, one can scatter a quark from a static electric field. The amplitude the underlying theory is

$$T_E(p,q) = \bar{u}(\vec{q})\gamma_0 g A_0(\vec{q}-\vec{p})u(\vec{p}). \tag{6.135}$$

Looking up the usual formula for a spinor wave function, we find

$$T_E(p,q) = \psi_+^\dagger[S(p,q) + i\vec{\sigma}\cdot(\vec{q}\times\vec{p})V(q,p)]gA_0\psi_+, \tag{6.136}$$

where $S(p,q) = [1 - (\vec{p}-\vec{q})^2/(8M_Q)^2]$. In this expression, the "1" matches the lowest-order Schrödinger calculation, and the second term can be matched from the effective Lagrangian if $c_2 = 1/8$. The c_3 part of $\delta\mathcal{L}^2$ gives the spin-dependent piece, if $c_3 = 1/8$, and scattering off a vector potential is matched by $c_4 = 1/2$. Of course, one could push the calculations to higher order in α and get more accurate matching coefficients.

If the reader will glance back at Eq. (6.128), she will notice that in NRQCD language, HQET is a theory with $c_4 = 1/2$, and $c_2 = c_3 = 1/8$, so the order $1/M_Q$ and $1/M_Q^2$ terms agree with the $O(v^2)$ expansion. However, other expansion parameters are different, and the terms that need to be bundled together to work to some desired order are also different. Notice, for example, that the v^4 terms in NRQCD are a mix of $1/M_Q^3$ and $1/M_Q^4$ terms.

Now we return to Euclidean space, and then to the lattice. The translation dictionary from Minkowski (M) to Euclidean (E) variables is that $x^0_{(M)} = -ix^4_{(E)}$, $\partial^0_{(M)} = i\partial^4_{(E)}$, $D_0^{(M)} = iD_4^{(E)}$, $A^0_{(M)} = -iA^4_{(E)}$ and $E^i_{(M)} = -iE^i_{(E)}$. We can keep our formulas for the actions with a few sign changes.

On the lattice, spatial derivatives are made covariant in the usual way, inserting link variables. Pretty much any discretization could be used, and since it is possible to get very high statistics in nonrelativistic simulations, the reader might want to discretize an a way that remains accurate all the way to larger lattice momenta. The electric and magnetic fields are components of $F_{\mu\nu}$ and they can be discretized using some variation on an appropriately-oriented clover term.

The time derivative is very special. Suppose we think of any of these lattice actions as a discretization of

$$\mathcal{L} = \psi(x,t)^\dagger D_4 \psi(x,t) + \psi^\dagger(x,t) H \psi(x,t), \tag{6.137}$$

and suppose that H involves only spatial derivatives. The simplest discretization of the time derivative gives

$$\mathcal{L} = \psi(x,t)^\dagger [U_t(x)\psi(x,t+1) - \psi(x,t) + aH\psi(x,t)]/a, \tag{6.138}$$

and the Green's function obeys the equation

$$U_t(x,t)G(x,t+1;0,0) - (\delta_{x,x'} - aH)G(x',t;0,0) = \delta_{x,0}\delta_{t,0}, \tag{6.139}$$

whose solution is

$$G(x,t+1;0,0) = U_t^\dagger(x,t)[\delta_{x,x'} - aH(x,x')]G(x',t;0,0). \tag{6.140}$$

This is an initial value problem. The complete Green's function can be computed with one pass through the lattice. Any more complicated derivative (say, including next-nearest neighbors) turns the initial value problem into a discretized partial differential equation, which can be solved only by an iterative procedure taking many passes through the lattice. Nobody ever does this.

Equation (6.140) is not precisely what the experts use. If the solution is to be stable, $|1 - aH|$ should be smaller than unity. In the free-field limit, $aH = \sum_i (2\sin p_i a/2)^2/(2M_Q a)$ and so one must have $M_Q a > 3$. To increase stability, slice up the time interval. If we call H_0 the Schrödinger term and δH everything else, Eq. (6.140) is replaced by

$$\begin{aligned} G(x,t+1;0,0) = {} & \left(1 - \frac{aH_0}{2n}\right)^n \left(1 - \frac{a\delta H}{2}\right) U_t^\dagger(x,t) \left(1 - \frac{a\delta H}{2}\right) \\ & \times \left(1 - \frac{aH_0}{2n}\right)^n G(x',t;0,0). \end{aligned} \tag{6.141}$$

It is not easy to take the continuum limit of a HQET or NRQCD calculation. These are effective field theories and need aM_Q large to make sense. However, to check results, one can compare simulations for a range of aM_Q's and check for stability. To go all the way to $M_Q a = 0$, we have to return to actions of relativistic quarks.

6.6.3 *Heavy, relativistic quarks*

Typical discretizations of light quarks develop lattice artifacts as am_q becomes order unity. A classic example is given by the simple Wilson or clover action's dispersion relation, from Eq. (6.29). There are two cures: One is to devise lattice actions for relativistic fermions that work well at all quark masses. This is easy to talk about, but hard to do, so the other alternative is to think about interpreting the results from simpler actions in a way that reduces lattice artifacts. The second choice is called the "Fermilab program" after the home of (two-thirds of) its inventors (El-Khadra *et al.*, 1997).

The reader should pause to re-read the discussion of the clover action from the point of view of on-shell improvement: absent any special discussion (about chiral symmetry, for example), one can write a lattice action as a sum of operators times coefficients, which we will now take to be functions of the bare mass in addition to the coupling g^2.

$$S = \sum_n c_n(m_0 a, g^2) S_n. \tag{6.142}$$

Furthermore, we do not enforce any space-time symmetry on the couplings: we are interested in low spatial momentum, not high energy. This of course allows the number of operators to increase: the dimension-4 operator $\bar{\psi} \not{D} \psi$ is replaced by the two operators $\bar{\psi}\gamma_4 D_4 \psi$ and $\bar{\psi}\vec{\gamma} \cdot \vec{D}\psi$. One then identifies redundant directions by performing an isospectral transformation (this is where doublers would be removed), and determines the other couplings by some physical improvement criterion. Through dimension-5 there are eight operators. Three are redundant, one is fixed by the bare mass, one by wave function normalization, one is the space-time asymmetry parameter, and the last two are the coefficients of $\vec{\Sigma} \cdot B$ and $\vec{\alpha} \cdot E$ interactions. It happens that the nonrelativistic limit of any space-time symmetric action takes the form

$$H' = \bar{\Psi} \left\{ M_1 + \gamma_0 A_0 - \frac{D^2}{2M_2} - \frac{i\vec{\Sigma} \cdot \vec{B}}{2M_B} - \gamma_0 \frac{[\vec{\gamma} \cdot \vec{D}, \vec{\gamma} \cdot \vec{E}]}{8M_E^2} + \dots \right\} \Psi, \tag{6.143}$$

where M_1, M_2, and M_B are all functions of $m_0 a$. The $\bar{\Psi}\, M_1 \Psi$ term does not contribute anything interesting to the dynamics; it (or rather, $M_1 - M_2$) is just an overall energy shift. One could obtain the correct $O(1/M_Q)$ corrections to HQET if one could tune $M_Q = M_2 = M_B$. The first tuning can be done by choosing am_0 to produce a desired M_2, either analytically or by simply fitting the dispersion relation. The second tuning can be done by varying the coefficient of $\vec{\Sigma} \cdot B$. (This is automatic at tree level for the clover action; recall Eq. (6.29), but it might have to be set by hand for other discretizations.) If the last term were needed, one could introduce a bare-mass-dependent $\vec{\alpha} \cdot E$ coupling.

Another lattice artifact that appears at large am_0 is a drift in the residue of the propagator. The continuum behavior $1/(2m)(1+\gamma_0)/2$ is generally modified by lattice artifacts into $Z(am_0)/(2m)(1+\gamma_0)/2$. One can restore the canonical behavior for the massive field by rescaling the lattice field by $1/\sqrt{Z(am_0)}$. The Wilson action is again the best-known example. While we earlier said that the connection between "continuum-like" field normalization and "κ-like" normalization was $\psi_c = \sqrt{2\kappa}\psi_\kappa$, the tree level connection is actually $\psi_c = \sqrt{1-6\kappa}\psi_\kappa$. In the massless limit of free field theory, $\kappa = \kappa_c = 1/8$, they agree, of course, but in the large mass limit κ vanishes and the large mass limit of $\sqrt{2\kappa}\psi$ is bad. A full nonperturbative determination of $Z(am_0)$ would presumably be needed to fix this unphysical result.

Chapter 7

Numerical methods for bosons

7.1 Importance sampling

A general functional integral for the partition function for a bosonic field $\phi(x)$ has the form

$$Z = \int [d\phi] \exp[-S(\phi)] \tag{7.1}$$

where $[d\phi] = \prod_n d\phi_n$ is an integration over the values of the field on each lattice site n. Any physical observable $\mathcal{O}$ can be expressed as a function of the field ϕ. Its expectation value is given by

$$\langle \mathcal{O} \rangle = \int [d\phi]\, \mathcal{O}(\phi) \exp[-S(\phi)] \Big/ \int [d\phi] \exp[-S(\phi)], \tag{7.2}$$

which is just the average of the observable with respect to the measure $[d\phi] \exp[-S(\phi)]$.

The massive multidimensional integration presents a numerical challenge. It is accomplished through a variety of importance sampling methods, designed to select a sufficiently complete subset of points $\{\phi^{(k)}, k = 1, \ldots, N\}$ in the integration domain such that the ratio of integrals is well estimated on this subset. These sample points are called "field configurations." One configuration specifies the value of the field on all lattice points. There are two broad classes of importance sampling methods: Monte Carlo and molecular dynamics.

7.1.1 *Monte Carlo methods*

The Monte Carlo method generates a sequence of random field configurations $\phi^{(k)}$ with a probability distribution tailored to the measure:

$$P(\phi) \propto \exp[-S(\phi)]. \tag{7.3}$$

An obvious precondition is that $\exp[-S(\phi)]$ be positive. Then the expectation value of the observable is just the simple average of the observable over the sample configurations:

$$\langle \mathcal{O} \rangle = \frac{1}{N} \sum_{k=1}^{N} \mathcal{O}(\phi^{(k)}) \tag{7.4}$$

in the limit $N \to \infty$. There are two Monte Carlo schemes in common use, namely, the method of Metropolis *et al.* (1953) and the heat bath method. Both methods start from an initial field configuration and approach the desired probability distribution through a sequence of prescribed random updates or transitions in the field values. Under suitable conditions the sequence of configurations $\phi^{(0)}, \phi^{(1)}, \ldots$ generated in this way becomes a "Markov chain." In analogy with statistical mechanics, the limiting probability distribution is called the "equilibrium distribution" and the process of reaching it is called "equilibration". The expectation value of the observable is then evaluated after the distribution is first suitably equilibrated.

The random transition takes ϕ to ϕ' according to a probability $R(\phi' \leftarrow \phi)$. Thus we may speak of a sequence of probability distributions $P^{(k)}(\phi)$ related by the Langevin evolution equation

$$P^{(k+1)}(\phi') = \int [d\phi]\, P^{(k)}(\phi) R(\phi' \leftarrow \phi). \tag{7.5}$$

In the transition the action changes from $S(\phi)$ to $S(\phi') = S(\phi) + \Delta S$. If the transition probability satisfies the "detailed balance" condition,

$$R(\phi' \leftarrow \phi)/R(\phi \leftarrow \phi') = \exp[-S(\phi')]/\exp[-S(\phi)] \tag{7.6}$$

then the distribution (7.3) is a fixed point of the equation. (The detailed balance condition is sufficient, but not necessary.) Whether, in fact, any starting point and any given transition scheme eventually leads to a unique end point, namely, the equilibrium distribution, is a deep question. In practice the transition methods are tailored to be sufficiently thorough that the desired result is achieved.

7.1.1.1 *Metropolis et al. method*

The standard method of Metropolis *et al.* (1953) makes a proposed change $\phi \to \phi'$ with probability $Q(\phi' \leftarrow \phi)$, such that the reverse transition has the same probability:

$$Q(\phi' \leftarrow \phi) = Q(\phi \leftarrow \phi'). \tag{7.7}$$

It then chooses a random number λ distributed uniformly on $[0, 1]$ and accepts the proposed change under the following conditions:

(1) If $S(\phi') < S(\phi)$, always accept the change.
(2) Otherwise, accept the change if $\exp(-\Delta S) > \lambda$.

This is called the "accept/reject" condition. If the change is rejected, another attempt is made. Clearly, this method generates a transition probability that satisfies detailed balance (7.6).

7.1.1.2 *Heat bath method*

The heat bath method visits each of the lattice sites and, while keeping the values at all other sites fixed, replaces the value of the field at the target site by a new value according to a probability distribution proportional to $\exp[-S(\phi)]$. The name comes from an analogy with a classical statistical ensemble with a hamiltonian and temperature defined so that $S(\phi) = H/kT$. Consider the dependence of the action on the single site coordinate ϕ_n with the field fixed at all other lattice sites,

$$H = h(\phi_n) + \epsilon_0. \tag{7.8}$$

The local objective is to achieve a thermal distribution $\exp[-h(\phi_n)/kT]$ of the local variable ϕ_n. For example, for the following simple discretization of free 4D scalar field theory, the lattice action couples only nearest neighbors

$$S = H/kT = -\sum_{n,\mu} \phi_{n+\mu}\phi_n + (8 + m^2) \sum_n \phi_n^2/2. \tag{7.9}$$

The dependence on the field at a single site ϕ_n is given by

$$h(\phi_n)/kT = -\phi_n \sigma_n + (8 + m^2)\phi_n^2/2, \tag{7.10}$$

where the environment is specified by

$$\sigma_n = \sum_\mu (\phi_{n+\mu} + \phi_{n-\mu}), \tag{7.11}$$

which is taken to be constant in this step. The distribution $\exp[-h(\phi_n)/kT]$ is simply Gaussian, so selecting a completely new ϕ_n in this case is as trivial as the theory itself.

The heat bath transition probability is independent of the starting configuration ϕ

$$R(\phi' \leftarrow \phi) \propto \exp[-S(\phi')] \tag{7.12}$$

and the condition (7.6) is clearly satisfied.

The method is repeated for each lattice site. Clearly the more local the action and the simpler the functional form of $h(\phi_n)$, the more efficient the updating process.

7.1.2 *Molecular dynamics method*

The molecular dynamics (MD) method uses classical dynamics and the ergodic hypothesis to obtain the desired statistical distribution (Callaway and Rahman, 1982, 1983). It introduces a fictitious momentum p_n conjugate to ϕ_n at each lattice site n and considers the hamiltonian

$$H(p, \phi) = \sum_n \frac{p_n^2}{2} + S(\phi). \tag{7.13}$$

This hamiltonian defines classical evolution in "molecular dynamics time" τ:

$$\dot{\phi}_n = p_n \quad \dot{p}_n = -\partial S / \partial \phi_n, \tag{7.14}$$

where the dot denotes the τ derivative. Starting from an initial choice $[p(\tau = 0), \phi(\tau = 0)]$ these equations define a trajectory $[p(\tau), \phi(\tau)]$ through phase space. The set of all such trajectories is area-preserving. The corresponding classical partition function is

$$Z = \int [dp] \, [d\phi] \exp[-H(p, \phi)]. \tag{7.15}$$

The Gaussian integral over momenta is trivially done, giving a result proportional to the path integral expression for the quantum partition function (7.1). So the quantum partition function can be evaluated using classical molecular dynamics. According to the ergodic hypothesis, the probability of visiting a point ϕ along the classical trajectory is proportional to

$\exp[-S(\phi)]$. Expectation values of observables are then computed by simply averaging over the MD "trajectory":

$$\langle \mathcal{O} \rangle = \frac{1}{T} \int_{\tau_0}^{\tau_0+T} d\tau \, \mathcal{O}[\phi(\tau)]. \tag{7.16}$$

As with other importance sampling methods, before averaging, one allows the system to evolve to a sufficiently large time τ_0 that thermal equilibrium is established.

7.1.3 *Refreshed molecular dynamics*

We are principally concerned that ϕ reaches the more probable equilibrium region quickly and then traverses the phase space efficiently. A measure of this efficiency is the rate of decay of the autocorrelation coefficient for any observable $\mathcal{O}$:

$$C(\tau) = N \int_0^\infty d\tau' \, \delta\mathcal{O}(\tau' + \tau)\delta\mathcal{O}[\tau'], \tag{7.17}$$

where $\delta\mathcal{O}(\tau) = \mathcal{O}[\phi(\tau)] - \langle \mathcal{O} \rangle$ and the normalization N is fixed by requiring $C(0) = 1$. The slower the autocorrelation coefficient decays, the longer it is necessary to follow the MD evolution to get a suitably small variance in the expected value of the observable.

"Refreshing" was proposed by Duane and Kogut (1985, 1986) as a method for speeding the decorrelation. In this method one starts with a random choice for the canonical momenta, evolves the system for a fixed molecular dynamics time, selects a completely new set of canonical momenta (the refreshing step) and without changing the coordinates ϕ starts a new trajectory. The process is then repeated. The random momenta are selected from the Gaussian distribution that appears in the classical partition function. This method also has the advantage of helping to realize ergodicity.

In practice refreshing is usually found to reduce autocorrelations. Why might that be? One can imagine that the classical trajectories include cycles that bring some of the coordinates back to their initial values, instead of visiting other parts of the phase space. Giving the system a thermal kick helps push it into a new region of phase space.

7.1.4 *Hybrid Monte Carlo*

Constructing the molecular dynamics trajectory requires a numerical integration of a system of first order differential equations. The numerical integration advances along the trajectory through a series of discrete steps in molecular dynamics time. To get a faithful sampling of the Feynman path integral the trajectory should be followed accurately. To do so requires making the step size small or the integration method more complicated. Either way the computational cost grows. Hybrid methods attempt to evade this problem.

Hybrid methods combine both the molecular dynamics and the Metropolis *et al.* Monte Carlo method. The molecular dynamics trajectory starts with a configuration ϕ and ends with ϕ'. This process can be viewed as defining a Metropolis *et al* transition function $Q(\phi' \leftarrow \phi)$. At zero integration step size the molecular dynamics evolution is time-reversal invariant, so Eq. (7.7) is satisfied. With a careful choice of the integration algorithm the evolution can also be time-reversal invariant for a discrete step size. One then proceeds with the usual accept/reject decision based on the change in the action $S(\phi)$. If the proposal is rejected, the configuration is restored to ϕ, the momenta are refreshed, and a new trajectory is followed.

Viewed as merely a definition of a Metropolis *et al.* transition function, the molecular dynamics evolution can be done with a coarse time step. However crude the integration method, it is essential that the stepping algorithm be time reversal invariant to high precision so that the reversibility condition Eq. (7.7) is satisfied. The leapfrog algorithm (see below) is a popular choice.

Whenever they are admissible, hybrid methods are the most popular and effective, since the accept/reject decision builds in an accuracy that is absent in plain molecular dynamics. Furthermore, the molecular dynamics trajectory provides a natural choice for the Monte Carlo transition function. For these reasons hybrid Monte Carlo methods are called "exact" methods.

7.1.5 *Leapfrog algorithm and improvements*

The leapfrog algorithm is a convenient second order integration method that gives the time reversal invariance needed for the Metropolis *et al.* transition function. It alternates coordinate and momentum updates. We illustrate it for a scalar field theory. We start by splitting the MD hamiltonian into kinetic and potential contributions as follows:

$$H = K + U \tag{7.18}$$

where $K = \sum_n p_n^2/2$ and $U = S(\phi)$. In analogy with quantum mechanical time evolution, the MD evolution over a time step ϵ can be realized through a classical time evolution operator if we use Poisson brackets in place of commutators,

$$T(\epsilon) = \exp(H\epsilon). \tag{7.19}$$

For a small time step ϵ we can approximate the operator by

$$T(\epsilon) \approx T_K(\epsilon/2)T_U(\epsilon)T_K(\epsilon/2) \equiv \exp(K\epsilon/2)\exp(U\epsilon)\exp(K\epsilon/2) + \mathcal{O}(\epsilon^3). \tag{7.20}$$

where $T_U(\epsilon)$ is the updating

$$\begin{aligned} p &\rightarrow p - \epsilon\frac{\partial U}{\partial \phi} \\ \phi &\rightarrow \phi \end{aligned} \tag{7.21}$$

and $T_K(\epsilon)$ is the updating

$$\begin{aligned} p &\rightarrow p \\ \phi &\rightarrow \phi + \epsilon p. \end{aligned} \tag{7.22}$$

Eq. (7.20) is clearly symmetric under time reversal. The operators T_K evolve only the coordinate and T_U evolves only the momentum. So at the expense of a local stepping error of $\mathcal{O}(\epsilon^3)$ we can accomplish the integration by evolving only the coordinate for a half step, followed by the momentum for a full step, followed by the coordinate for a half step.

If we repeat the process, two successive coordinate half steps combine to make a full step, and, except for the first and last iterations, we end up alternating full coordinate steps with full momentum steps. Since our goal is to integrate over a fixed time interval t we require t/ϵ steps, leading to a global error of $\mathcal{O}(t\epsilon^2)$; by definition, it is a "second order" method.

An improved leapfrog method was proposed by Sexton and Weingarten (1992), who showed that the coefficient of the error term is reduced if one changes the sequence to

$$T(\epsilon) \approx T_K(\epsilon/6)T_U(\epsilon/2)T_K(2\epsilon/3)T_U(\epsilon/2)T_K(\epsilon/6). \tag{7.23}$$

Although this is still a second order method, because the error is reduced, the step size can be lengthened somewhat.

7.2 Special methods for the Yang-Mills action

7.2.1 *Heat bath*

As we have seen in Ch. 5, the $SU(N)$ lattice Yang-Mills field is commonly represented by associating an $SU(N)$ matrix $U_{n\mu}$ with each lattice link. The heat bath method updates each link one by one, replacing it with a random new link with a probability determined by the contribution of that link to the full Yang-Mills action $S_{G,\mathrm{YM}}$. For the Wilson plaquette action, that contribution can be written as

$$S_{n\mu}(U_{n\mu}) = -\frac{\beta}{N}\,\mathrm{ReTr}U_{n\mu}V_{n\mu}, \tag{7.24}$$

where $V_{n\mu}$ is the sum of the "staples" connected to the link $U_{n\mu}$, *i.e.*, the product of the other three links in each plaquette containing $U_{n\mu}$. The new link is generated with probability

$$P_{n\mu}(U) \propto \exp[-S_{n\mu}(U)]. \tag{7.25}$$

The updating process for $SU(2)$ is simplest. We write the quaternion decomposition of an $SU(2)$ matrix U as

$$U = u_0 + i\vec{\sigma}\cdot\vec{u}, \tag{7.26}$$

where σ_j are the Pauli matrices and u_ν is a real four vector. (For a unitary matrix we also have $\sum u_\nu^2 = 1$.) Let v_ν be the corresponding components of $V_{n\mu}$. They are real because V is a sum of unitary matrices. The task reduces to one of generating a new unit four-vector u'_ν from the probability distribution

$$P(U) \propto \exp[\beta(u_0 v_0 - \vec{u}\cdot\vec{v}]. \tag{7.27}$$

Algorithms due to Creutz (1980) and Kennedy and Pendleton (1985) generate this distribution.

For $N > 2$ there is the method of Cabibbo and Marinari (1982). Select a diagonal $SU(2)$ subgroup of $SU(N)$. If we put the selected rows and columns first, we have the block decomposition:

$$U_{n\mu}V_{n\mu} = \left(\begin{array}{c|c} w & a \\ \hline a' & b \end{array}\right) \tag{7.28}$$

To preserve unitarity the update is accomplished by left multiplying $U_{n\mu}$ by a "hit" matrix

$$H = \left(\begin{array}{c|c} h & 0 \\ \hline 0 & 1 \end{array}\right). \tag{7.29}$$

After the hit the single-link action becomes

$$S_{n\mu}(HU_{n\mu}) = -\frac{\beta}{N}\,\mathrm{Re}[\mathrm{Tr}(hw) + \mathrm{Tr}b]. \tag{7.30}$$

In analogy with the $SU(2)$ heat bath, we select a random h according to the distribution

$$P(h) \propto \exp[\beta/N\,\mathrm{ReTr}(hw)]. \tag{7.31}$$

This process is then repeated for different $SU(2)$ subgroups to improve thermalization.

7.2.2 *Overrelaxed updates*

Overrelaxation methods seek to accelerate the decorrelation of the gauge configurations by introducing further shuffling of the gauge fields. The name comes from an analogy with a standard numerical method for improving the iterative solution of linear systems (Brown and Woch, 1987; Creutz, 1987).

The method is simplest for an $SU(2)$ gauge action. Each site is visited one by one. The action for a single gauge link $U_{n\mu}$ is $S_{n\mu}(U_{n\mu})$ where

$$S_{n\mu}(U) = -\beta/2\,\mathrm{Tr}(UV_{n\mu}). \tag{7.32}$$

The action is minimized when $U = U_0$, where

$$U_0 = V^{\dagger}_{m,\mu}/|V_{m,\mu}|. \tag{7.33}$$

and $|V|$ denotes $\det V$. Now consider the Metropolis *et al* move

$$U \to U' = U_0 U^{\dagger} U_0. \tag{7.34}$$

Repeating the move takes us directly back to U, so we satisfy the reversibility condition. Moreover, the action is unchanged:

$$S_{n\mu}(U') = S_{n\mu}(U), \tag{7.35}$$

so we accept the move with probability one.

In effect the move takes U to a "diametrically opposite" side of the minimum U_0 in the group manifold. Thus it helps generate a rather different gauge configuration without changing the action.

The generalization to $SU(N)$ follows closely the method of Cabibbo and Marinari (1982) discussed for the heat bath algorithm above. Instead of selecting a random h for the hit matrix, however, we take

$$h = (w^\dagger/|w|)^2. \tag{7.36}$$

The new contribution to the action becomes

$$S_{n\mu}(HU_{n\mu}) = -\beta/N \,\mathrm{Re}[\mathrm{Tr}(w^\dagger) + \mathrm{Tr}b]. \tag{7.37}$$

But since $\mathrm{Tr}w^\dagger = \mathrm{Tr}w$, the action is unchanged. This process is repeated for different $SU(2)$ subgroups to further the decorrelation.

The overrelaxation moves do not change the action. Thus if they are the only updates they constitute a "microcanonical" method. To explore the grand canonical phase space one alternates overrelaxation steps with heat bath steps or hybrid Monte Carlo trajectories.

7.2.3 *Molecular dynamics*

To set up the molecular dynamics (MD) equations one needs a momentum $H_{n\mu}$ for each link matrix $U_{n\mu} = \exp(iA_{n\mu})$. For gauge field molecular dynamics, it is convenient to treat the gauge links $U_{n\mu}$ as the coordinates, rather than the vector potential $A_{n\mu}$. Following Gottlieb *et al.* (1987), we start with a postulated equation of motion for the link matrices and a natural form for the MD hamiltonian. Then we complete Hamilton's equations by requiring energy conservation.

To preserve unitarity we require that $H_{n\mu}$ be traceless and hermitian and write the equation of motion for $U_{n\mu}$ as

$$\dot{U}_{n\mu} = iH_{n\mu}U_{n\mu} \tag{7.38}$$

and the MD hamiltonian as

$$H_{G,\mathrm{YM}} = \frac{1}{2}\sum_{n\mu} \mathrm{Tr}H_{n\mu}^2 + S_{G,\mathrm{YM}}(U). \tag{7.39}$$

The gauge momentum can be expanded in terms of the traceless hermitian generators T^a of the $SU(N)$ Lie algebra as

$$H_{n\mu} = \sum_a h^a_{n\mu} T^a. \tag{7.40}$$

Stochastic momenta are then obtained by selecting unit Gaussian random values for $h^a_{n\mu}$.

To obtain the MD equations, we require, first, that hamiltonian $H_{G,\mathrm{YM}}$ be a constant of the motion. So

$$\dot{H}_{G,\mathrm{YM}} = \sum_{n\mu} \mathrm{Tr}(\dot{H}_{n\mu} H_{n\mu}) + \dot{S}_{G,\mathrm{YM}}(U) = 0. \tag{7.41}$$

To be specific let us evaluate the time derivative of the Wilson plaquette action. In the notation of Eq. (7.24) it is

$$\dot{S}_{G,\mathrm{YM}}(U) = -\frac{\beta}{2N} \sum_{n\mu} \mathrm{Tr}[\dot{U}_{n\mu} V_{n\mu} + V^\dagger_{n\mu} \dot{U}^\dagger_{n\mu}], \tag{7.42}$$

where $V_{n\mu}$ is the formal partial derivative, $\partial S_{G,\mathrm{YM}}/\partial U_{n,\mu}$, *i.e.*, in differentiating, we also fix $U^\dagger_{n\mu}$. Putting together Eqs. (7.38), (7.41) and (7.42) gives

$$\sum_{n\mu} \mathrm{Tr}[H_{n\mu}(\dot{H}_{n\mu} + iF_{n\mu})] = 0, \tag{7.43}$$

where

$$F_{n\mu} = -\frac{\beta}{2N} \mathrm{Tr}(U_{n\mu} V_{n\mu} - V^\dagger_{n\mu} U^\dagger_{n\mu}). \tag{7.44}$$

The expression in parentheses in Eq. (7.43) is hermitian, so for the equation to hold for all traceless hermitian $H_{n\mu}$ we must have $\dot{H}_{n\mu} + iF_{n\mu} = cI$, *i.e.* a constant times the identity. For $H_{n\mu}$ to remain traceless, we must have $Nc = i\mathrm{Tr}\, F_{n\mu}$ so

$$i\dot{H}_{n\mu} = -\left.\frac{\beta}{N} \mathrm{Tr}(U_{n\mu} V_{n\mu})\right|_{\mathrm{TA}}, \tag{7.45}$$

where TA denotes the traceless antihermitian part: $W_{TA} = (W - W^\dagger)/2 - \mathrm{Tr}(W - W^\dagger)/(2N)$.

As a mnemonic, we can write this result in the form of Hamilton's equation

$$i\dot{H}_{n\mu} = \frac{\partial S_{G,\mathrm{YM}}}{\partial i A_{n\mu}} \tag{7.46}$$

where the formal derivative with respect to the vector potential $iA_{n,\mu} = \log U_{n\mu}$ means

$$\frac{\partial S_{G,\mathrm{YM}}}{i\partial A_{n\mu}} = -\frac{\beta}{N}\mathrm{Tr}\left(U_{n\mu}\frac{\partial S_{G,\mathrm{YM}}}{\partial U_{n,\mu}} \right)\Bigg|_{\mathrm{TA}} . \tag{7.47}$$

and the formal derivative $\partial S_{G,\mathrm{YM}}/\partial U_{n,\mu}$ means $V_{n\mu}$. The result is easily generalized to other gauge actions.

Equations (7.38) and (7.45) are the basis for gauge field molecular dynamics.

Chapter 8

Numerical methods for fermions

Calculations with dynamical fermions are expensive, so a great deal of effort has been invested in developing efficient algorithms. In this chapter we begin by reviewing strategies for treating the fermion determinant. We will see why essentially all large-scale simulations with dynamical fermions use some variation of a molecular dynamics update. We will discuss the standard family of algorithms first, and then briefly discuss improvements and special strategies for the chiral overlap fermion. At the heart of every method is the need to solve a large sparse linear system efficiently. So we conclude with a discussion of efficient sparse solvers and eigensolvers.

8.1 Taming the fermion determinant: the Φ algorithm

Let us begin with full QCD with a single flavor of fermion. Its partition function is

$$Z = \int [dU][d\bar{\psi}][d\psi] \exp[-S_G(U) - \bar{\psi}M(U)\psi] \tag{8.1}$$

where $M = D+m$. After integration over the fermion Grassmann variables, the partition function becomes

$$Z = \int [dU] \exp[-S_G(U)] \det M(U). \tag{8.2}$$

The determinant is nonlocal, so computing its change under a change in the gauge field is very expensive. One way to make the problem tractable is to simulate the determinant by introducing a color-triplet, scalar "pseudofermion" field Φ. This is done via the formal identity

$$\det M(U) = \int [d\Phi^* d\Phi] \exp[-\Phi^* M^{-1} \Phi]. \tag{8.3}$$

We can expand Φ in terms of eigenmodes ψ_j of M and the corresponding eigenvalues λ_j

$$\Phi^* M^{-1} \Phi = \sum_j \Phi^* \psi_j \frac{1}{\lambda_j} \psi_j^* \Phi. \tag{8.4}$$

This exposes a series of problems, all arising from the fact that the eigenvalues of lattice Dirac operators are complex and their real parts may not be positive-definite. Whatever the global properties of the eigenvalues, as long as the overlap of Φ onto the eigenstates is not carefully chosen, individual terms in the exponential can be complex or carry a net negative sign. Then the exponential in Eq. (8.3) cannot be interpreted as a conventional probability measure.

Fortunately, the determinant of the Dirac operator for any reasonable lattice fermion is real, although it may not be positive. Reasonable actions obey a property called "γ_5-hermiticity," meaning that $D^\dagger = \gamma_5 D \gamma_5$. This relation implies that if the Dirac operator has a complex eigenvalue, $D\phi_i = \lambda_i \phi_i$, then λ_i^* is also an eigenvalue.

For staggered fermions the Dirac operator is anti-hermitian, so all eigenvalues are imaginary, and γ_5-hermiticity means that they are paired or zero. We could then rewrite the determinant, $\det M = \prod_i (i\lambda_i + m)$ as

$$\det M = m^\nu \prod_{\text{pairs}} (\lambda_j^2 + m^2). \tag{8.5}$$

(for ν zero eigenvalues). If we could find an operator whose spectrum consisted of only the real factors above, we could use it in the pseudofermion exponential.

Nonchiral actions, such as the Wilson action, are typically neither hermitian nor anti-hermitian, and have eigenvalues that are paired complex-conjugates or unpaired and real: $\det M = \prod_{pairs}(|\lambda_j|^2 + m^2) \prod_r (\lambda_r + m)$. The absence of chiral symmetry means that the sign of the real eigenvalues is not protected, so the determinant can have either sign. Therefore, even an operator with real eigenvalues that reproduces the determinant would be unacceptable.

The eigenvalues of the chiral overlap operator lie on a circle of radius r_0 in the complex plane with nonnegative real parts. Like the Wilson operator, they are either paired complex conjugates or unpaired and real, but the determinant is nonnegative.

In practice, it is generally too difficult to find a "paired" operator that sums only over the paired eigenmodes and that can be constructed without knowledge of the eigenvectors. The solution involves an explicit doubling with corrections to come later. Because of γ_5–hermiticity, $\det D = \det D^\dagger$ and the determinant for two flavors is $\det D^2 = \det D^\dagger D$, just what we want. Our pseudofermion action has become

$$\begin{aligned} Z &= \int [dU d\Phi^* d\Phi] \exp[-S_G(U) - \Phi^*(M^\dagger M)^{-1}\Phi] \\ &= \int [dU] \exp[-S_G(U)] \det[M^\dagger(U)M(U)]. \end{aligned} \tag{8.6}$$

For the Wilson fermion and overlap action a doubling resulting in degenerate up and down quarks could be an acceptable approximation for some calculations. For calculations involving a different number of other flavors it is not. For the staggered fermion action the doubled action represents two flavors of staggered fermions. However, the combination $M^\dagger(U)M(U)$ decouples even and odd sites, so restricting the pseudofermion field Φ to even (or odd) lattice sites removes the additional doubling. The four-fold taste degeneracy remains and must be dealt with separately.

It is straightforward to generate a random pseudofermion field according to the distribution in Eq. (8.6). If the Gaussian random field R is distributed according to $\exp(-R^\dagger R)$, then

$$\Phi = M^\dagger(U)R. \tag{8.7}$$

is distributed according to the weight in (8.6).

The Φ algorithm of Gottlieb *et al.* (1987) combines a heat bath method for the pseudofermion field with a modified version of the refreshed molecular dynamics method for the gauge field, described in Ch. 7. One selects a random gauge momentum and a random pseudofermion field Φ. Then, keeping the Φ field fixed, one integrates the gauge field along a short molecular dynamics trajectory. The gauge momentum and Φ field are then refreshed and the process is repeated. In terms of the mnemonic Eq. (7.46) the molecular dynamics force term is the formal derivative of the effective action with respect to the vector potential $iA_{\mu n}$;

$$iF_{\mu n} = \frac{\partial}{i\partial A_{\mu n}} S_{\text{eff}} \tag{8.8}$$

where

$$S_{\text{eff}} = S_G(U) + \Phi^*[M(U)^\dagger M(U)]^{-1}\Phi. \tag{8.9}$$

Because the fermion matrix depends on the gauge field, the pseudofermion term contributes the "fermion force",

$$iF_{F,\mu n} = X^* \frac{\partial}{i\partial A_{\mu n}}[M(U)^\dagger M(U)]X, \tag{8.10}$$

where $X = (M^\dagger M)^{-1}\Phi$. Computing the derivative is straightforward but tedious. We refer the reader to the literature for details. To get X we must solve the large sparse linear system

$$(M^\dagger M)X = \Phi. \tag{8.11}$$

Since M depends on the gauge field, we must solve this linear system at each step in the gauge field molecular dynamics integration. For this reason, including fermions in the path integration adds considerably to the computational cost.

The "Φ" algorithm is easily converted to an "exact" hybrid Monte Carlo method, just as we did for the gauge field algorithm in Ch. 7. At the end of a gauge-field trajectory one makes an accept/reject decision, based on the change in the effective action S_{eff}.

All molecular dynamics methods suffer to some degree when the fermion matrix M has a small eigenvalue. Small eigenvalues appear at small quark mass and notably for gauge configurations with nontrivial topological winding number. Good chiral actions respect the index theorem: as the absolute value of the winding number increases, a new zero mode of the Dirac kinetic energy operator D appears. A small mass term in $M = D + m$ displaces the zero only slightly. The fermion determinant is then small; the fermion force, which involves inverting the fermion matrix, is large; and it changes rapidly when a zero mode appears. Larger forces require a smaller MD step size for accuracy or they lead to a lower acceptance rate. They tend to prevent changes in topological charge, so may impede an efficient exploration of the phase space.

How much of a problem is presented by topology change depends on how well the algorithm respects chirality. The classic Wilson fermion action has poor chiral properties and develops unwanted zero eigenvalues at not very small quark masses. Gauge configurations producing them are sometimes called "exceptional configuration". Their prevalence prevents efficient simulation with this action at moderately small quark masses.

Staggered fermion actions are protected from this defect by their remnant chiral symmetry, but at nonzero lattice spacing, there isn't enough symmetry to respect the index theorem precisely. The would be zero modes are only approximately zero. The large-force problem may worsen as the lattice spacing is reduced. With chiral fermion actions, such as domain wall and overlap, the problem is exquisitely apparent. When the topology changes the overlap operator changes discontinuously and the fermion force is momentarily infinite. We will revisit this question below when we discuss special methods for those actions.

8.2 Taming the fermion determinant: the R algorithm

The "R" algorithm works directly with the determinant, rather than reconstructing it from a pseudofermion field. In so doing it provides a framework in which one can hope to simulate a physical number of quark species.

Using the matrix identity $\ln(\det A) = \operatorname{Tr}\ln A$, we absorb the fermion determinant in Eq. (8.6) into an effective gauge action as

$$Z = \int [dU] \exp[-S_{\text{eff}}(U)], \tag{8.12}$$

where

$$S_{\text{eff}} = S_G(U) - \operatorname{Tr}\ln[M^\dagger(U)M(U)]. \tag{8.13}$$

The "R" algorithm of Gottlieb *et al.* (1987) builds a refreshed gauge-field molecular dynamics from this effective action. In this case the fermion force term involves

$$\begin{aligned} iF_{\mu n} &= \frac{\partial}{i\partial A_{\mu n}} \operatorname{Tr}\ln[M^\dagger(U)M(U)] \\ &= \operatorname{Tr}\left\{ \frac{1}{M^\dagger(U)M(U)} \frac{\partial}{i\partial A_{\mu n}} [M^\dagger(U)M(U)] \right\}. \end{aligned} \tag{8.14}$$

To evaluate this expression exactly, we need to know all the matrix elements of the inverse matrix $1/[M^\dagger(U)M(U)]$. Because the inverse is a dense matrix, that computation is prohibitively expensive. Instead, one estimates the inverse by a stochastic method. Let R be the Gaussian random field we introduced for the Φ algorithm. Averaged over this distribution, the product of two components at sites m and n and colors a and b satisfies

$$\overline{R^*_{an} R_{bm}} = \delta_{ab}\delta_{mn}, \tag{8.15}$$

where the overline denotes averaging over random R. Let

$$X = M^{-1}R = [M^\dagger(U)M(U)]^{-1}M^\dagger R. \tag{8.16}$$

Then

$$\overline{X_{an}X^*_{bm}} = [M^\dagger(U)M(U)]^{-1}_{an,bm}, \tag{8.17}$$

and

$$iF_{\mu n} = \overline{X^\dagger \frac{\partial}{i\partial A_{\mu n}}[M^\dagger(U)M(U)]X}. \tag{8.18}$$

The stochastic evaluation of the inverse matrix has brought us very close to the Φ algorithm. For example, note that if we define $\Phi = M^\dagger R$ in Eq. (8.16), the expression for the stochastic field X is exactly the same! But if we take seriously our intention to obtain a good stochastic approximation to the fermion force, we should refresh the vector X at least once in each gauge field molecular dynamics step, whereas the Φ algorithm permits holding the Φ field constant over the entire gauge field trajectory. The conventional R algorithm therefore refreshes at each step. This is the key difference between the two algorithms. We refer the reader to the literature for further details.

Because the effective action must be evaluated stochastically in the R algorithm, it is not possible to use it in an accept/reject decision. Therefore, it remains an "inexact" method. The molecular dynamics step size must be small enough that step-size errors are negligible, or results must be extrapolated to zero step size.

8.3 The fourth root approximation

The Φ algorithm has serious drawbacks. With Wilson fermions all quark species are doubled. A doubling of the strange quark is not an acceptable approximation to reality. Worse, with staggered fermions all quark species come in four degenerate tastes. An approximate solution takes the appropriate fractional power of the determinant. For staggered fermions it is

$$Z = \int [dU] \exp[-S_G(U)] \prod_f \det[M^\dagger(U)M(U)]^{N_f/4}, \tag{8.19}$$

where the product is over fermion flavors f with mass degeneracy N_f. For degenerate up and down quarks plus a strange quark $N_{ud} = 2$ and $N_s = 1$.

For Wilson fermions we would use $N_f/2$ instead.

Taking a fractional power in this way is not rigorously justified in field theory, so it is understandably controversial. In the continuum limit, it is expected that the eigenvalues of the fermion matrix are suitably degenerate, so the root is well defined. When they are not degenerate, the root produces artifacts. The unanswered question is whether the continuum limit of the modified theory is smooth enough that all artifacts vanish with the lattice spacing. So far we have neither proof that the approximation is harmless, nor any indications that it is harmful. Despite the controversy, some of the (apparently) most successful and precise results in contemporary lattice gauge theory are based on this approximation.

We modify the R algorithm to simulate the fractional power by rewriting the effective action as

$$S_{\text{eff}} = S_G(U) - \sum_f \frac{N_f}{4} \operatorname{Tr} \ln[M_f^\dagger(U) M_f(U)]. \tag{8.20}$$

In perturbation theory the trace in the fermion contribution to the effective action generates single sea quark loops. Multiple sea quark loops arise from reexponentiating those terms in the path integration. So the fourth-root reduces the contribution of each sea quark loop by a factor of four.

With these changes the R algorithm proceeds as before. The gauge field evolution is approximated by a series of small discrete time steps ϵ. With each step one generates a random Φ field based on the current gauge field, computes the fermion force, and updates the gauge momenta and field. To get a result at the end of a trajectory of fixed length that is good to second order in the step size, we use a second order method, such as the leapfrog algorithm. In the standard implementation, in one cycle it moves the gauge field by a half step, updates the gauge momenta by a full step there, then moves the gauge field by another half step.

To preserve second order accuracy in the presence of stochastic pseudofermion fields requires care. With the Φ algorithm the random pseudofermion field is generated at the beginning of the trajectory and remains a constant background throughout. No change in the leapfrog method is required. With the R algorithm the stochastic field depends on the instantaneous value of the gauge field. It becomes necessary to do one extra gauge-field update per cycle (Gottlieb *et al.*, 1987). For a given flavor, one begins by moving the gauge field through a time interval $\epsilon(1 - N_f/4)/2$. The stochastic field for that flavor is generated there. (With $N_f = 4$, the initial step is zero and the stochastic field is generated at the beginning of the interval.) The initial step is repeated to produce the stochastic fields for other flavors. Then the gauge field is moved to the leapfrog halfway point, where the standard algorithm resumes. The stepping asymmetry for $N_f < 4$ makes the method all the more unsuitable for a hybrid Monte Carlo accept/reject decision.

8.4 An exact algorithm for the fourth root: rational hybrid Monte Carlo

Another way to get a fractional power of the determinant is to start with a fractional power of the fermion matrix itself:

$$Z = \int [dU d\Phi^* d\Phi] \exp[-S_G(U) - \Phi^*(M^\dagger M)^{-N_f/4}\Phi] \tag{8.21}$$

$$= \int [dU] \exp[-S_G(U)] \det[M^\dagger(U)M(U)]^{N_f/4}. \tag{8.22}$$

There it can be approximated by a rational function (Clark and Kennedy, 2004a,b). A diagonal rational function has the partial fraction representation

$$(M^\dagger M)^{-N_f/4} \approx r(M^\dagger M) = a_0 + \sum_{n=1}^{N} \frac{a_n}{M^\dagger M + b_n}, \tag{8.23}$$

where a_n and b_n are constants. A Φ algorithm can then be constructed in much the same way as in Sec. 8.1. The principal advantage of this approach is that the effective action is computed deterministically, so it can be used in an accept/reject decision. Thus simulation results would not be subject to possible systematic errors due to a nonzero integration time step. Such a hybrid method is called "rational hybrid Monte Carlo."

The added complexity increases the computational cost. Instead of solving a single sparse linear system, a separate system must be solved for each term in the expansion of the rational function. Multishift solvers help here. The fermion force computation is similarly modified. The advantage of the method is that, despite the considerably larger cost per integration step, one can take a larger step size than accuracy would allow with the R algorithm, thereby gaining an advantage (to say nothing of the peace of mind that comes from using an exact algorithm).

8.5 Refinements

As the quark mass is decreased the condition number of the fermion matrix increases and the step size in the molecular dynamics evolution must be decreased to preserve accuracy of the R algorithm and maintain the acceptance rate of the Φ algorithm. Thus the computational expense grows. A variety of refinements attempt to reduce the cost by allowing a larger step size at the same accuracy or acceptance rate.

8.5.1 *Sexton-Weingarten scheme*

In Ch. 7 we discussed the leapfrog and improved leapfrog algorithms. The Sexton and Weingarten (1992) scheme makes a further improvement by splitting out the cheaper gauge dynamics from the more expensive fermion dynamics and giving it smaller steps. A regrouping of terms in the Hamiltonian gives

$$H = H_{YM} + S_F, \tag{8.24}$$

where $H_{YM} = K + S_G$ is the Hamiltonian for pure gauge molecular dynamics. Then, in analogy with the improvement to the leapfrog algorithm, for this split we make the approximation

$$T(\epsilon) \approx T_F(\epsilon/6)[T_{YM}(\epsilon/2m)]^m T_F(2\epsilon/3)[T_{YM}(\epsilon/2m)]^m T_F(\epsilon/6), \tag{8.25}$$

where $T_F(\epsilon) = \exp(S_F\epsilon)$ and $T_{YM}(\epsilon) = \exp(H_{YM}\epsilon)$. Note that the time step for the Yang-Mills evolution is ϵ/m, whereas it is ϵ for the fermion force term. The Yang-Mills dynamics is, in turn, approximated in the same manner:

$$T_{YM}(\epsilon) = T_K(\epsilon/6)T_G(\epsilon/2)T_K(2\epsilon/3)T_G(\epsilon/2)T_K(\epsilon/6). \tag{8.26}$$

where $T_G(\epsilon) = \exp(S_G\epsilon)$. Reducing the error from the gauge force permits a slightly bigger step for the fermion force. Of course one reaches a point of diminishing returns when the residual error from the fermion force dominates the overall error.

8.5.2 *Hasenbusch method*

Here is a preconditioning method that works directly with the fermion determinant and introduces an extra pseudofermion field, as follows (Hasenbusch, 2001; Hasenbusch and Jansen, 2003):

$$Z = \int [dU d\Phi_1^* d\Phi_1 d\Phi_2^* d\Phi_2] \exp\{-S_G(U) - \Phi_1^*(W^\dagger W)^{-1}\Phi_1 \\ - \Phi_2^*[(W^{-1}M)^\dagger (W^{-1}M)]^{-1}\Phi_2\}. \tag{8.27}$$

If W is chosen successfully, the condition number of both kernels is smaller than their product. An extra inversion is required, since the heat bath updating of Φ_2 starts from a Gaussian random R_2 and sets $\Phi_2 = W^{-1}MR_2$. The additional expense in calculating the fermion force is small, provided W is reasonably sparse.

An example of a possible preconditioner is one constructed from the fermion matrix itself: $W = M + m_0$, effectively increasing the mass of the quark.

The Hasenbusch scheme can be combined with the Sexton-Weingarten scheme for further improvement.

8.5.3 *Schwarz alternating method*

A good preconditioning strategy reduces the quark forces, whose effects are expensive to calculate, thereby permitting a larger step size at the same accuracy. The Lüscher implementation of the Schwarz alternating method attempts just that (Lüscher, 2003, 2004). It begins by partitioning the lattice into a set of regular hypercubic domains Λ of arbitrary size in such a way that they can be colored alternately white and black in chessboard style. The set of white sites is denoted Ω and black, Ω^*. The fermion matrix is then block LU decomposed:

$$M = \begin{pmatrix} D_{\Omega^*} & D_{\partial\Omega^*} \\ D_{\partial\Omega} & D_\Omega \end{pmatrix} = \begin{pmatrix} D_{\Omega^*} & 0 \\ D_{\partial\Omega} & 1 \end{pmatrix} \begin{pmatrix} 1 & D_{\Omega^*}^{-1} D_{\partial\Omega^*} \\ 0 & D_\Omega - D_{\partial\Omega} D_{\Omega^*}^{-1} D_{\partial\Omega^*} \end{pmatrix}. \tag{8.28}$$

The off-diagonal blocks contain terms in the fermion matrix that connect sites in adjacent hypercubic domains. The fermion determinant is then

$$\det M = \det D_\Omega \det D_{\Omega^*} \det[1 - D_\Omega^{-1} D_{\partial\Omega} D_{\Omega^*}^{-1} D_{\partial\Omega^*}]. \tag{8.29}$$

The first two factors are just the determinant of

$$D_1 \equiv \begin{pmatrix} D_{\Omega^*} & 0 \\ 0 & D_\Omega \end{pmatrix}. \tag{8.30}$$

Considered by itself, D_1 describes fermion dynamics decoupled with respect to the domains Λ. The fermion force on the gauge field is found by inverting the comparatively small matrices restricted to their domains. The gauge field dynamics in the domain also depends on the gauge force, which couples gauge links across domain boundaries. So it is necessary to fix a set of links at the surface to their values at the most recent update. They then define Dirichlet boundary conditions for the gauge field evolution within each domain.

The last factor is the determinant of

$$D_2 \equiv 1 - D_{\Omega}^{-1} D_{\partial\Omega} D_{\Omega^*}^{-1} D_{\partial\Omega^*}, \tag{8.31}$$

which couples the blocks and costs more to evaluate. But some economies are possible here. Inserting a projection P_{Ω^*} onto the boundary sites in the domain of $D_{\partial\Omega^*}$ as in

$$D_2' \equiv 1 - P_{\Omega^*} D_{\Omega}^{-1} D_{\partial\Omega} D_{\Omega^*}^{-1} D_{\partial\Omega^*} \tag{8.32}$$

does not change the determinant, and leads to a tolerably sparse matrix.

The product of the two determinants is simulated using a separate pseudofermion field for D_1 and D_2. It then helps to use the split time scale of the Sexton-Weingarten method, stepping the less expensive D_1 dynamics faster than the more expensive D_2.

Because the boundary gauge links need updating and, further, to avoid possible artifacts associated with the positioning of the domains, after each updating cycle, the domains are shifted by a random space-time displacement.

8.6 Special considerations for overlap fermions

Computational difficulties for overlap fermions arise even before one begins to think of using them in simulations. First one must evaluate the overlap operator itself. Recall that the overlap Dirac operator is

$$D = r_0\{1 + \gamma_5 \epsilon[h(-r_0)]\}, \tag{8.33}$$

where $h(-r_0) = \gamma_5(d - r_0)$ is the hermitian Dirac operator and $\epsilon(x) = x/\sqrt{x^2}$ is the matrix step function. The costly bottleneck in overlap simulations is, of course, the application of the step function to a trial vector.

One can approximate $\epsilon(h)$ either as a polynomial in h or as a rational function. A polynomial approximation would be

$$\epsilon(h) = h s_N(h^2) = h \sum_{n=0}^{N} c_n h^{2n}, \tag{8.34}$$

where $s_N(x)$ is an approximation to $1/\sqrt{x}$. This can be implemented as a series expansion in Chebychev polynomials $T_n(x)$,

$$s_N(x) = \sum_n c_n T_n(x). \tag{8.35}$$

In our experience diagonal rational functions are more efficient. One writes

$$\begin{aligned} \epsilon_N(h) &= h \frac{\sum a_n h^{2n}}{\sum b_n h^{2n}} \\ &= h \left[c_0 + \sum_{j=1}^{N} \frac{c_j}{h^2 + d_j} \right]. \end{aligned} \tag{8.36}$$

The sum of inverses can be computed using a multishift conjugate gradient algorithm described below.

For real x any good approximation $\epsilon_N(x)$ has a range $0 < x_{\text{min}} < x < x_{\text{max}}$ where it works reasonably well, but outside that range, it fails. (These end points can be adjusted by a multiplicative change of scale, but the lower end point is never zero.) In order for the approximation to succeed for the matrix function $\epsilon(h^2)$, we require that the good interval contain all the eigenvalues of h^2 from smallest λ^2_{min} to largest λ^2_{max}. The largest eigenvalue is bounded by the lattice cutoff, so in a typical simulation it is very stable; the smallest presents a problem. Indeed, in a dynamical simulation, λ_{min} crosses the origin when the topology of the gauge configuration changes, so one always encounters arbitrarily small values of λ.

The cure to this problem is to isolate the small eigenmodes and include their contributions exactly. Arrange the eigensolutions $h(-r_0)|j\rangle = \lambda_j |j\rangle$ in order from smallest λ_j^2 to largest. Then apply $\epsilon(h)$ to a vector $|\psi\rangle$ as follows:

$$\epsilon(h)|\psi\rangle = \epsilon_N(h) \left(|\psi\rangle - \sum_{j=1}^{J} |j\rangle\langle j|\psi\rangle \right) + \sum_{j=1}^{J} |j\rangle \epsilon(\lambda_j) \langle j|\psi\rangle. \tag{8.37}$$

Now $\epsilon_N(h^2)$ has to be a good approximation only for the range λ_J^2 to λ^2_{max}.

Two examples of ϵ_N's are the polar expression and the Zolotarev approximation. The polar expression [referring to Eq. (8.36)] has $c_0 = 0$ and

$$c_k = \frac{1}{N\cos^2\left[\pi(k-\frac{1}{2})/2N\right]}; \qquad d_k = \tan^2\left[\pi(k-\frac{1}{2})/2N\right] \tag{8.38}$$

It is actually the power series expansion of the approximate step function used by the finite-fifth-dimension domain wall fermion. It is not robust enough to use in most simulations.

The Zolotarev formula represents the state of the art (as of 2006). Referring readers to the mathematical literature for its derivation, we present a cookbook recipe (since we have encountered misprints in the literature). The coefficients are defined in terms of the elliptic integrals cel, cn, sn, and dn, which we list due to the plethora of conventions,

$$\mathrm{cel}(k,p,a,b) = \int_0^\infty \frac{a+bx^2}{(1+px^2)\sqrt{(1+x^2)(1+k^2x^2)}}dx$$
$$y = \mathrm{sn}(u,k); \qquad u = \int_0^{\mathrm{sn}} \frac{dy}{\sqrt{(1-y^2)(1-k^2y^2)}} \tag{8.39}$$

with $\mathrm{sn}^2 + \mathrm{cn}^2 = 1$, $k^2\mathrm{sn}^2 + \mathrm{dn}^2 = 1$. The approximate step function is

$$\epsilon_N(x) = x\sum_{\ell=1}^{N} \frac{b_\ell}{x^2 + c_{2\ell-1}}. \tag{8.40}$$

For $x \in (x_{\min}, x_{\max})$ define $\kappa = x_{\min}/x_{\max}$; $\kappa_p = \mathrm{cel}(\kappa, 1, 1, 1)$; and $u = \ell\kappa_p/(2N)$. Then the partial fraction coefficients are

$$c_\ell = \left[\frac{\mathrm{sn}(u,\kappa^2)}{\mathrm{cn}(u,\kappa^2)}\right]^2 x_{\min}^2$$
$$b_\ell = \frac{\prod_{i=1}^{N-1}(c_{2i} - c_{2\ell-1})}{\prod_{i=1;i\neq\ell}^{N}(c_{2i-1} - c_{2\ell-1})}. \tag{8.41}$$

The resulting $\epsilon_N(x)$ needs a small correction to remove an asymmetric deviation from $\epsilon(x) = 1$. The literature is silent on a preferred minimization criterion (absolute deviation, least squares, etc.). The extrema are located at $x_1 = x_{\min}$ and $x_2 = x_{\min}/\mathrm{dn}[\kappa_p/(2N)]$. Replacing b_ℓ by $2b_\ell/|\epsilon_N(x_1) + \epsilon(x_2)|$ to find the smallest absolute deviation is a good practical choice. The precision of the resulting approximation is

$$\frac{|\epsilon_N(x_1) - \epsilon(x_2)|}{|\epsilon_N(x_1) + \epsilon(x_2)|} \tag{8.42}$$

over the desired range, so it is straightforward to pick a desired accuracy for a known range simply by varying N. We illustrate an example of a Zolotarev approximation in Fig. (8.1).

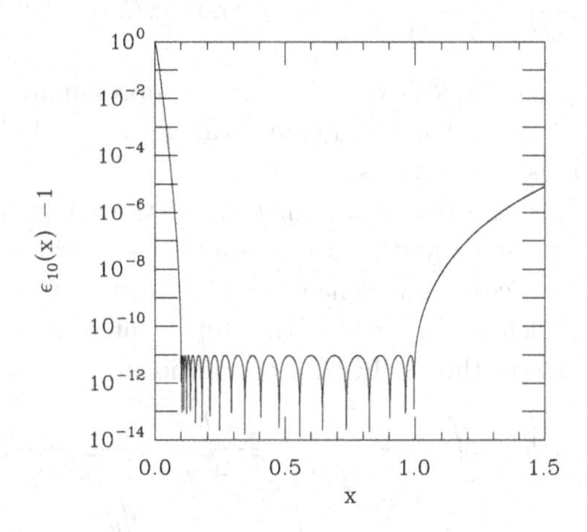

Fig. 8.1 An example of a tenth-order Zolotarev approximation to the step function, where the desired range is for $0.1 < x < 1$. We plot $\epsilon_{10}(x) - 1$ to display the error in the approximation.

With techniques for applying the overlap operator to vectors in hand, we turn to simulations with dynamical overlap fermions. At the time this section was written, only a few groups had published simulations with dynamical overlap fermions, so what we will say is tentative. The currently favored method uses hybrid molecular dynamics with the Φ algorithm.

As we have seen from Eq. (8.10), the Φ algorithm obtains the force on the gauge field from the derivative of the pseudofermion effective potential $S_f = \Phi^*[M(U)^\dagger M(U)]^{-1}\Phi$ with respect to the vector potential A or U. As long as $M(U)$ changes smoothly with U there is no problem. As we move along the molecular dynamics trajectory, however, one of the eigenmodes of the kernel operator might change sign. When that occurs, the spectrum of the overlap operator $M(U)$ changes discontinuously due to the appearance or disappearance of a zero mode. This discontinuity generates an infinite fermion force. Conventional MD stepping algorithms, such as the leapfrog, fail and we require special treatment. In particular we need a reversible MD treatment to carry us past the step.

A reversible molecular dynamics treatment of a discontinuous effective potential was proposed by Fodor *et al.* (2004). It is based on the analogy with classical particle motion in the presence of a step barrier. Imagine a unit-mass particle moving in two spatial dimensions (x, y), and approaching a potential step, with $V(x,y) = 0$, say, for $x < 0$ and $V(x,y) = V_0$ for $x > 0$. The component of momentum in the y direction is unchanged when the particle strikes the barrier, but there is an impulse in the x direction, which causes a discontinuous change in p_x. If the barrier is too high, the particle "reflects," $p_x \to -p_x$, but if the barrier is low enough, the particle crosses the barrier ("refracts") while p_x changes. The change is given by energy conservation, $\frac{1}{2}p_x^2 = \frac{1}{2}(p_x')^2 + V_0$.

In the QCD simulation, we must identify the MD time at which one of the eigenmodes of the kernel operator has a zero eigenvalue. We move along the trajectory until we arrive at that "crossing" time. Then we change the gauge field momentum H discontinuously, reflecting or refracting according to the size of the step in the potential ΔS_f. After reflecting or refracting, we return to our usual leapfrog integration until the end of the trajectory (or until we reach the next topological boundary).

We recall that the analog of the energy was $\frac{1}{2}H^2+S$. In terms of the unit normal vector to the surface of discontinuity N, there are two possibilities for changes in the gauge momenta, depending on the size of the momentum component along the normal. In compact notation we denote the inner product of two vector fields in the Lie Algebra of $SU(N)$ by

$$N^*H = \sum_{n\mu} \mathrm{Tr}(N^\dagger_{n\mu} H_{n\mu}) \; . \tag{8.43}$$

Then, if $(N^*H)^2 > 2\Delta S_f$, we have refraction:

$$\Delta H = -N\; N^*H + N\; \mathrm{sign}(N^*H)\; \sqrt{(N^*H)^2 - 2\Delta S_f}. \tag{8.44}$$

Alternatively, if $(N^*H)^2 \leq 2\Delta S_f$, we have reflection:

$$\Delta H = -2N(N^*H) \tag{8.45}$$

Here ΔS_f is the height of the discontinuity, which only comes from the fermionic part of the action. The discontinuity arises because one eigenmode of the kernel action h changes sign. (Recall Eq. (8.37).) Calling this eigenvector χ, the (unnormalized) normal vector N is the matrix element of the derivative of the kernel action with respect to A, computed in this special crossing state

$$N(x,\mu) = \chi^* \frac{\partial h}{\partial A(x,\mu)} \chi. \tag{8.46}$$

In practice, the exact step function of the overlap operator is replaced by some approximation, such as a rational functional approximation. To obtain an accurate approximation to the step function, it is also necessary to project out low modes of the kernel operator, as in Eq. (8.37). This introduces another complication for HMC: we have to be able to differentiate the projector. Evaluating the derivative requires determining the change in the projector under small perturbations in h. Labeling the projector on an eigenmode of $h(-r0)$ with eigenvalue λ as $P_\lambda = |\lambda\rangle\langle\lambda|$, first order perturbation theory gives (Narayanan and Neuberger, 2000)

$$\delta P_\lambda = \frac{1}{\lambda - h}(1 - P_\lambda)\delta h P_\lambda + P_\lambda \delta h \frac{1}{\lambda - h}(1 - P_\lambda). \tag{8.47}$$

In the force, this term is combined with an analytic differentiation of the rational functional approximation, using sources from which the low modes have been projected out, basically a mix of Eqs. (8.10) and (8.37).

8.7 Monte Carlo methods for fermions

Fermionic methods based on molecular dynamics evolution are not used because they work exceptionally well, but rather because they work at all. They suffer from rather long computer time correlations compared with Monte Carlo updating methods. (The reader can easily check this fact by simulating a pure gauge theory.) Comparisons for the pure gauge theory may not be relevant to fermionic systems like QCD, since the latter theories have propagating Goldstone bosons (the pions) which surely affect the performance of any algorithm. Even so, the desire to speed up fermionic simulations results in the creation of alternative update strategies, and it is worthwhile to survey possible directions these strategies can go.

We would like to briefly describe some alternative methods for dynamical fermion simulation. They do not really share very much in common (other than not being molecular dynamics), but here are their similarities: These methods are designed to make large changes in the gauge fields. Typically, they do not require computing a force vector, and so they do not require that the gauge connection used in the fermionic action to be differentiable. This makes them useful for fat link actions, since it is easier to design such actions without the constraint of differentiability.

8.7.1 *Multiboson method*

One such method is the multiboson method of Lüscher (1994). We will describe its implementation for two degenerate flavors of Wilson fermions. Our objective is to generate the determinant in Eq. (8.6). Using γ_5−hermiticity again, it can be rewritten as $\det(M^\dagger M) = \det(H^2)$ where the hermitian Dirac matrix is $H = \gamma_5 M$. We then approximate the inverse $1/H^2$ as a polynomial $P_n(H^2)$ of degree n in H^2. The polynomial factors as follows

$$\begin{aligned} P_n(H^2) &= \prod_{k=1}^{n} (H - \mu_k - i\nu_k)((H - \mu_k + i\nu_k) \\ &= \prod_{k=1}^{n} ((H - \mu_k)^2 + \nu_k^2). \end{aligned} \tag{8.48}$$

Then the desired determinant can be generated from a multiple integration over bosonic fields, one for each factor:

$$\det H^2 = \int \prod_{k=1}^{n} d\phi_k^\dagger d\phi_k \exp\left[-\sum_{k=1}^{n} \phi_k^\dagger ((H - \mu_k)^2 + \nu_k^2)\phi_k \right]. \tag{8.49}$$

Since the partition function involves only bosons, one can update with a combination of Monte Carlo and overrelaxation. The problem with this method is, of course, that as the fermion mass becomes small, the condition number of H^2 diverges, and the required order of the polynomial that approximates H^2 grows. (The reader should think about sparse matrix inversion *à la* conjugate gradient, where the inverse of H^2 is also a polynomial in H^2). Then the number of boson fields also grows, as does the cost of the simulation.

8.7.2 *Ratio of determinants*

Another class of updating algorithms begins with the partition function

$$Z = \int [dU] \exp[-S_G(U)] \det M(U), \tag{8.50}$$

proposes changes by looking only at $S_G(U)$, and then accepts or rejects them according to the ratio of the determinants. For example, one could perform a change of link variables that is microcanonical with respect to $S_G(U)$, that is, $S_G(U') = S_G(U)$, and then accept with an acceptance probability

$$P_{\rm acc}(U',U) = \min\left\{1, \frac{\det M(U')}{\det M(U)}\right\}, \tag{8.51}$$

or, alternatively, writing $S_G(U) = S_G^a(U) + S_G^b(U)$, microcanonically updating with respect to $S_G^a(U)$ and pushing $S_G^b(U)$ into the accept/reject step:

$$P_{\rm acc}(U',U) = \min\left\{1, \frac{\det M(U')}{\det M(U)} \exp\left[-S_G^b(U') + S_G^b(U)\right]\right\}. \tag{8.52}$$

These methods encounter two problems. The first problem is that the change in the fermionic determinant is usually large, except when the change in the gauge fields is small, so the acceptance rate is tiny. Most of the "strength" of the determinant comes from its ultraviolet part, and so this problem can be addressed by using fermions with fat links (this decouples them from the UV part of the gauge field), and by preconditioning the fermion action, essentially integrating out its UV degrees of freedom and replacing them by an effective gauge action, so that $S_G^a(U)$ in Eq. (8.52) contains a fermionic contribution.

The second problem is that the ratio of determinants usually has to be calculated using a noisy estimator, for example

$$P'_{\rm acc}(U',U) = \min\left\{1, e^{\xi^\dagger[M(U')-M(U)]\xi}\right\}, \tag{8.53}$$

where the vector ξ is generated according to the distribution

$$P(\xi) \propto e^{-\xi^\dagger M(U')\xi}. \tag{8.54}$$

If the change in the determinant is large, the variance of the noisy estimator diverges.

We can compute the variance in the stochastic estimator along the lines of Hasenfratz and Alexandru (2002): the determinant ratio is

$$r = \frac{\det M(U')}{\det M(U)} = \int d\xi^* d\xi \exp\left[-\xi^* \frac{M(U')}{M(U)}\xi\right], \tag{8.55}$$

which can be written as an expectation value in a sample of Gaussian random vectors

$$r = \int d\xi^* d\xi \exp\left[-\xi^*(\frac{M(U')}{M(U)} - 1)\xi\right] \exp(-\xi^*\xi) / \int d\xi^* d\xi \exp(-\xi^*\xi)$$
$$\equiv \left\langle \exp\left[-\xi^*\left(\frac{M(U')}{M(U)} - 1\right)\xi\right]\right\rangle_{\xi^*\xi}. \tag{8.56}$$

The estimator for the square of the determinant involves

$$\left\langle \exp\left[-\xi^*\left(2\frac{M(U')}{M(U)} - 1\right)\xi\right]\right\rangle_{\xi^*\xi} = \det\left[\left(2\frac{M(U')}{M(U)} - 1\right)^{-1}\right], \tag{8.57}$$

and the squared variance is

$$\sigma^2 = \det\left\{\left[2\frac{M(U')}{M(U)} - 1\right]^{-1}\right\} - \det\left\{\frac{M(U')}{M(U)}\right\}^2. \tag{8.58}$$

The problem is that if the matrix $M(U')/M(U)$ has an eigenmode less than or equal to $1/2$, the standard deviation diverges. A multiboson trick, writing

$$\det A = (\det A^{1/n})^n$$
$$= \left\langle \exp\left(-\sum_{j=1}^{n} \xi_j^*[A^{1/n} - 1]\xi_j\right)\right\rangle_{\xi_j^*\xi_j} \tag{8.59}$$

turns the divergent term in the standard deviation to $\det^{-n}(2A^{1/n} - 1)$. which is finite if none of the eigenvalues of A are less than 2^{-n}.

Finally, the cost of these methods scales roughly with the squared volume of the simulation, since each time one updates a "patch" of gauge fields (and the number of patches scales with the volume), the cost of computing the determinant goes as the volume. Nevertheless, simulations using variants of this algorithm are being used to produce interesting physics results. For examples, see (Hasenfratz and Knechtli, 2002; Hasenfratz and Alexandru, 2002; Hasenfratz *et al.*, 2005).

8.8 Conjugate gradient and its relatives

Much of the computational effort of a calculation involving fermions is spent solving sparse linear systems. The molecular dynamics updating requires solving $M^\dagger M X = \Phi$ for a random vector Φ and fermion matrix

M. Evaluating expectation values of hadronic correlators usually requires constructing quark propagators, that is, the inverse M^{-1} of M, obeying the equation

$$MM^{-1} = M^{-1}M = I. \tag{8.60}$$

Since M is a square matrix of dimension a few times the lattice volume, but has only order (volume) nonzero entries, this is a problem of sparse matrix inversion. It is usually not possible to construct or store $M^{-1}(x, y)$ for all lattice sites x and y, since the inverse is dense, requiring finding order (volume)^2 numbers. Instead, one typically constructs $g = M^{-1}h$ for some selected source vector h by solving

$$Mg = h, \tag{8.61}$$

leading us back again to a sparse matrix problem.

The conjugate gradient algorithm and its variants have proven to be the most successful sparse-solution methods for lattice QCD applications. The standard conjugate gradient algorithm (CG) requires a hermitian positive definite matrix. The biconjugate gradient method and variants are used for non-positive-definite matrices. The RHMC algorithm and overlap HMC algorithm require efficient multi-shift solvers, which are available for both conjugate and biconjugate gradient solvers. Here we give a brief introduction to these methods.

8.8.1 *Even-odd preconditioning*

A simple block LU preconditioning (Schur decomposition) takes advantage of the special structure of the staggered and Wilson Dirac matrices and cuts the problem size in half. Decomposed into subspaces of even and odd lattice sites, the Dirac matrix has the block structure

$$M = \begin{pmatrix} R_e & D_{eo} \\ D_{oe} & R_o \end{pmatrix}, \tag{8.62}$$

where the upper row and left column refer to even sites and the lower row and right column to odd. For staggered fermions $R_e = R_o = m$ and $D_{eo} = D = -D_{oe}^\dagger$. For Wilson fermions R_e and R_o are site-diagonal, mixing only colors and spins on each site. So they are easily inverted. They satisfy $\gamma_5 R \gamma_5 = R^\dagger$. The off-diagonal blocks satisfy $\gamma_5 D_{eo} \gamma_5 = D_{oe}^\dagger$. For either case we may write the Schur decomposition of M:

$$M = UAL = \begin{pmatrix} 1 & D_{eo}R_o^{-1} \\ 0 & 1 \end{pmatrix} \begin{pmatrix} R_e - D_{eo}R_o^{-1}D_{oe} & 0 \\ 0 & R_o \end{pmatrix} \begin{pmatrix} 1 & 0 \\ R_o^{-1}D_{oe} & 1 \end{pmatrix}. \tag{8.63}$$

The inverses of L and U are obtained trivially by reversing the sign of the off-diagonal block.

We then start from the linear system

$$Mg = UALg = h \tag{8.64}$$

and define $Lg = x$ and $U^{-1}h = b$. Then we must solve

$$\begin{pmatrix} A_{ee} & 0 \\ 0 & R_o \end{pmatrix} \begin{pmatrix} x_e \\ x_o \end{pmatrix} = \begin{pmatrix} b_e \\ b_o \end{pmatrix}, \tag{8.65}$$

where $A_{ee} = R_e - D_{eo}R_o^{-1}D_{oe}$ and we have split the vectors into even-site and odd-site components. For staggered fermions the matrix $A_{ee} = (m^2 + DD^\dagger)/m$ is hermitian positive definite. For Wilson fermions it is not. Solution on the odd sites is trivial, so the task reduces to solving a sparse system on the even sites. Once the solution x is known we finish with the trivial calculation $g = L^{-1}x$. If h has support only on odd sites, we can do a similar preconditioning leading to a linear system on only the odd sites.

For staggered fermions the matrix $M^\dagger M$ is block diagonal in the even-odd decomposition:

$$M^\dagger M = \begin{pmatrix} m^2 + DD^\dagger & 0 \\ 0 & m^2 + D^\dagger D \end{pmatrix}, \tag{8.66}$$

so the if b has support only on even sites, the linear system $M^\dagger Mx = b$ is same one we encountered in the preconditioned algorithm above.

8.8.2 *The conjugate gradient algorithm*

Since $M^\dagger M$ is hermitian positive definite, one could invert the Dirac matrix M by inverting $M^\dagger M$ and then multiplying by $M^\dagger$, $M = M^\dagger(M^\dagger M)^{-1}$. This is arguably the optimal choice for staggered fermions (de Forcrand, 1996).

For the linear system $Ax = b$, the conjugate gradient algorithm constructs a series of approximate solutions x_n in such a way that the residual vectors $r_n = b - Ax_n$ become progressively smaller. The algorithm starts from an arbitrary initial guess x_0, which gives an initial residual $r_0 = b - Ax_0$. The problem then reduces to one of solving

$$A(x - x_0) = r_0. \tag{8.67}$$

The algorithm does this by working with an n-dimensional Krylov space, K_n, consisting of the span of the vectors $\{r_0, Ar_0, A^2r_0, \ldots, A^{n-1}r_0\}$. The space is progressively enlarged as n is increased during the iteration. The approximate solution $x_n - x_0$ is the vector in the Krylov space that minimizes the Cartesian norm of the residual vector r_n.

Solving the hermitian positive definite linear system $Ap = r_0$ for $p = x - x_0$ is equivalent to minimizing the quadratic form

$$f(p) = \frac{1}{2}(p, Ap) - (r_0, p), \tag{8.68}$$

where (x, y) denotes the inner product of the vectors x and y (Golub and Van Loan, 1996; Press *et al.*, 2002). A possible linear search method starts from the origin, picks an initial search direction, say, $p_0 = r_0 \in K_0$, and minimizes $f(\alpha p)$ as a function of α, resulting in

$$\alpha_0 = \frac{(r_0, p_0)}{(p_0, Ap_0)}. \tag{8.69}$$

We arrive at the first approximate solution

$$x_1 = x_0 + \alpha_0 p_0 \tag{8.70}$$

with residual

$$r_1 = b - Ax_1 = r_0 - \alpha_0 Ap_0 \in K_1, \tag{8.71}$$

and we get an updated linear system $A(x - x_1) = r_1$. It is easy to show that $(r_1, r_0) = (r_1, p_0) = 0$. We select a new search direction p_1, this time in K_1, and repeat the process. To collect what we have so far, the recursion relation starts with x_0 and $p_0 = r_0 = b - Ax_0$ and proceeds for $n = 0, 1, \ldots$ with

$$p_n \in K_n \tag{8.72}$$

$$\alpha_n = \frac{(r_n, p_n)}{(p_n, Ap_n)} \tag{8.73}$$

$$r_{n+1} = r_n - \alpha_n Ap_n \in K_{n+1} \tag{8.74}$$

$$x_{n+1} = x_n + \alpha_n p_n, \tag{8.75}$$

so finally we have $x_n = x_0 + \sum_{k=0}^{n-1} \alpha_k p_k$. But how do we select an optimum new search direction $p_n \in K_n$ for $n > 0$? The conjugate gradient criterion specifies that the search vectors be "A-conjugate" *i.e.* mutually orthogonal with respect to the metric A:

$$(p_i, Ap_j) = 0 \quad \text{for } i \neq j. \tag{8.76}$$

It suffices to make the new search direction be the following linear combination of the previous search direction and the new residual:

$$p_{n+1} = r_{n+1} + \beta_n p_n \tag{8.77}$$

$$\beta_n = \frac{(r_{n+1}, r_{n+1})}{(r_n, r_n)}. \tag{8.78}$$

What is especially remarkable is that the residual vectors are then mutually orthogonal and the search vectors are orthogonal to all subsequent residuals:

$$(r_i, r_j) = 0 \quad (r_i, p_j) = 0 \quad \text{for } i > j. \tag{8.79}$$

Thus the residual vectors $(r_k, k = 0, \ldots, n-1)$ become an orthogonal basis for K_n. The search vectors p_k are likewise a basis, but orthogonal with respect to the metric A. These assertions can be proven easily by induction.

With this choice of search direction the expression for α_n can be rewritten

$$\alpha_n = \frac{(r_n, r_n)}{(p_n, Ap_n)}. \tag{8.80}$$

With exact arithmetic the process terminates eventually with $r_n = 0$, either when the Krylov space fills the complete vector space, or it fills an invariant subspace containing r_0. Then we have $\alpha_n = 0$.

With ordinary precision, when several hundred iterations are required, the algorithm suffers from an accumulation of round off errors. The late search directions begin to lose conjugacy with the early search directions and the algorithm begins to regenerate them. Another consequence is that the recursively developed residuals become an increasingly unreliable estimate of the true residual $b - Ax_n$. One may remedy both problems either by increasing the precision of the calculation or by restarting the algorithm, using the last best solution x_n as a starting guess. Typically, after a few iterations, the residuals drop well below the value they would have had without restarting.

8.8.3 *Biconjugate gradient*

Wilson-Dirac matrices are neither hermitian nor antihermitian. Even-odd preconditioning for the Wilson Dirac matrix does not produce a hermitian matrix. So we must turn to other solvers if we want to invert them directly. The biconjugate gradient algorithm is used for nonhermitian systems. It is quite similar to the conjugate gradient algorithm described above. It produces two sequences of residual vectors r_k and a dual vector $\hat{r}_k$ and two sequences of search vectors p_k and its dual $\hat{p}_k$. Starting from x_0, $p_0 = r_0 = b - Ax_0$ and an arbitrary $\hat{p}_0 = \hat{r}_0$ it generates the sequence from the algorithm

$$\alpha_n = \frac{\hat{r}_n, p_n}{\hat{p}_n, Ap_n} \tag{8.81}$$

$$r_{n+1} = r_n - \alpha_n A p_n \tag{8.82}$$

$$\hat{r}_{n+1} = \hat{r}_n - \alpha_n A^\dagger \hat{p}_n \tag{8.83}$$

$$x_{n+1} = x_n + \alpha_n p_n \tag{8.84}$$

$$\beta_n = \frac{(\hat{r}_{n+1}, r_{n+1})}{(\hat{r}_n, r_n)} \tag{8.85}$$

$$p_{n+1} = r_{n+1} + \beta_n p_n \tag{8.86}$$

$$\hat{p}_{n+1} = \hat{r}_{n+1} + \beta_n \hat{p}_n \tag{8.87}$$

$$x_{n+1} = x_n + \alpha_n p_n \tag{8.88}$$

for $n = 0, 1, \ldots$. The vectors satisfy the bi-orthogonality relations

$$(\hat{p}_i, Ap_j) = 0 \quad (p_i, A^\dagger \hat{p}_j) = 0 \tag{8.89}$$

$$(\hat{r}_i, r_j) = 0 \quad (r_i, \hat{r}_j) = 0 \tag{8.90}$$

$$(\hat{r}_i, p_j) = 0 \quad (r_i, \hat{p}_j) = 0 \tag{8.91}$$

for $i > j$.

Since A is nonhermitian there is no related minimization problem. The orthogonality relations assure that with exact arithmetic, the method terminates, but there is no assurance that the norm of the residual decreases monotonically. The coefficients α_n and β_n are not guaranteed to be ratios of positive values. Instabilities arise when the denominators are small, leading to fluctuating residuals. The stabilized method discussed below is preferable.

The method generates two Krylov spaces, one based on A and r_0 for the vectors r_n and p_n and another based on $A^\dagger$ and $\hat{r}_0$ for the dual vectors.

Thus we can write

$$r_n = P_n(A)r_0 \qquad \hat{r}_n = P_n(A^\dagger)\hat{r}_0, \tag{8.92}$$

where P_n is an nth degree polynomial, with coefficients determined by the recursion relation. The orthogonality relation between the residual vectors can be written as

$$(\hat{r}_i, r_j) = (P_i(A^\dagger)\hat{r}_0, P_j(A)r_0) = (\hat{r}_0, P_i(A)P_j(A)r_0). \tag{8.93}$$

This observation suggests an asymmetric approach in which only $r_k = P_k^2(A)$ is iterated and the dual vector $\hat{r}_0$ is fixed. This is the conjugate gradient squared algorithm, which suffers from similar instabilities, but which inspired the stabilized biconjugate gradient algorithm discussed next.

8.8.4 *Stabilized biconjugate gradient*

The stabilized biconjugate gradient algorithm (BiCGStab) (van der Vorst, 1992) resolves problems with the biconjugate gradient and conjugate gradient squared algorithms and remains a top choice (Frommer *et al.* (1994); de Forcrand (1996)). The BiCGγ_5 method exploits the special symmetry of the Dirac matrix and is also competitive (de Forcrand, 1996).

Like the other CG methods, the BiCGStab algorithm is a Krylov space solver. It is similar to the conjugate gradient squared algorithm mentioned above in that it does not evolve the dual vector $\hat{r}_0$. But it replaces the sequence of residual vectors $r_n = P_n^2(A)$ with the sequence $r_n = Q_n(A)P_n(A)$ where

$$Q_n(x) = \prod_{k=0}^{n}(1 - \omega_k x). \tag{8.94}$$

and at step n, ω_n is selected to minimize r_n. Thus we have the biorthogonality relation

$$(\hat{r}_i, r'_j) = 0 \tag{8.95}$$

where $\hat{r}_i = Q_i(A)\hat{r}_0$ and $r'_j = P_j(A)r_0$. As a result, with exact arithmetic the method terminates after a finite number of steps. But we don't actually produce these vectors in the iteration. Instead, we generate $r_n = Q_n(A)P_n(A)$.

Here is the BiCGStab algorithm. Start with $\hat{r}_0 = r_0 = p_0$. Then for $n = 0, 1 \ldots,$

$$\alpha_n = (\hat{r}_0, r_n)/(\hat{r}_0, Ap_n) \tag{8.96}$$

$$s = r_n - \alpha_n A p_n \tag{8.97}$$

$$t = As \tag{8.98}$$

$$\omega_n = \frac{(t, s)}{(t, t)} \tag{8.99}$$

$$r_{n+1} = (1 - \omega_n A)s \tag{8.100}$$

$$x_{n+1} = x_n + \omega_n s + \alpha_n p_n \tag{8.101}$$

$$\beta_n = \frac{(\hat{r}_0, r_{n+1})}{(\hat{r}_0, r_n)} \frac{\alpha_n}{\omega_n} \tag{8.102}$$

$$p_{n+1} = r_{n+1} + \beta_n (1 - \omega_n A) p_n. \tag{8.103}$$

8.8.5 *Shifted solvers*

The staggered fermion RHMC and overlap algorithm both require evaluating a series of the form

$$g = \left[\sum_{k=1}^{n} \frac{a_k}{M + b_k} \right] h = \sum_{k=1}^{n} a_k g_k, \tag{8.104}$$

where M is one of the standard Dirac matrices, h is a "source" vector, and a_k and b_k are constants. Thus we require the solution to a set of shifted linear systems:

$$(M + b_k) g_k = h. \tag{8.105}$$

The problem appears again in applications that require determining a series of quark propagators with different masses, but the same source vector.

Even-odd preconditioning works just as well with a shifted matrix, and the generic problem reduces to a shifted sparse system on even lattice sites:

$$(A + s)x = b. \tag{8.106}$$

If we start from a zero guess for the solution, the first residual is $r_0 = b$, independent of the shift s. Then the Krylov space for $A+s$ is the same as the one for A, permitting us to reuse the same space to solve simultaneously for multiple shift values. The algorithms CG-M, BiCGγ_5-M, and BiCGStab-M are short-recurrence methods that do just that (Frommer *et al.*, 1995; Jegerlehner, 1996, 1998). We refer the reader to the literature for details.

When the shift s is positive, the condition number of the matrix is decreased. Thus the smallest shift usually controls the number of iterations required for convergence. As we have remarked, when the number of iterations is high, round off errors accumulate and the method degrades. But requiring a zero starting guess rules out restarting from the previous best solution. Thus the only protection against round off error is to work at higher precision.

8.8.6 *Computing fermion eigenmodes*

Eigenmodes of the Dirac operator have many uses. Just as finding eigenvalues and eigenmodes of ordinary matrices is best left to canned routines, so also with eigenvalues and eigenfunctions of the lattice Dirac operator. The most widely used package we know of is the "Arpack" family (Maschhof and Sorensen, 1996).

If one wants eigenmodes of a hermitian operator whose spectrum is bounded from below ($D^\dagger D$, for example), there is a reasonably straightforward method of choice: the Ritz variational method. The idea is familiar: construct a variational bound on an eigenmode of the hermitian operator by diagonalizing the operator in a subspace of trial functions.

We denote the eigenvalues and eigenvectors of a hermitian matrix A by λ_k and v_k and order them so $\lambda_j \leq \lambda_k$ for $j < k$. The Ritz variational method for finding the kth eigenvalue and eigenvector minimizes

$$\mu(x) = \frac{(x, Ax)}{(x, x)} \tag{8.107}$$

subject to the constraint that $(x, v_j) = 0$ for $j < k$. The minimization can be done using a conjugate gradient method. The algorithm has strong similarities with the linear solver. Here we summarize the accelerated method of Kalkreuter and Simma (1996).

Let a gradient function be defined as

$$g(x) \equiv \frac{1}{2}\nabla_x \mu(x) = \frac{[A - \mu(x)]x}{(x, x)}. \tag{8.108}$$

It satisfies $(g(x), x) = 0$. It vanishes when x is an eigenvector and plays the role of the residual r in the linear solver. Starting from a nonzero initial guess x_0 and initial search direction along the gradient $p_0 = g_0 \equiv g(x_0)$, we minimize the rational function $\mu(x_0 + \alpha p_0)$ with respect to α, yielding α_0. This exercise is equivalent to finding the lowest eigenvalue and corresponding eigenvector of A in the two-dimensional space spanned by x_0 and $p_0 = g_0$, or, equivalently, the two-dimensional Krylov space K_2 spanned by $\{x_0, Ax_0\}$. The process is then repeated with a new search direction p_1 in the Krylov space K_3. After each such search, the improved eigenvector is updated using

$$x_{n+1} = x_n + \alpha_n p_n. \tag{8.109}$$

The gradient at each new position $g_{n+1} = g(x_{n+1})$ is, by construction, orthogonal to p_n as well as x_{n+1}. Rather than searching in the direction of the new gradient g_{n+1}, the new search direction is chosen somewhat arbitrarily to be a linear combination of g_{n+1}, p_n, and x_{n+1} that is orthogonal to x_{n+1}:

$$p_{n+1} = g_{n+1} + \beta_n \left[p_n - x_{n+1} \frac{(x_{n+1}, p_n)}{(x_{n+1}, x_{n+1})} \right]; \tag{8.110}$$

$$\beta_n = \frac{(g_{n+1}, g_{n+1})}{(g_n, g_n)}. \tag{8.111}$$

After the CG-like iterations have converged sufficiently, we have an approximation w_0 to the lowest eigenvector. We proceed to the next higher eigenvector, requiring that the initial vector and all search directions be orthogonal to w_0. The second CG-like iteration leads eventually to an approximation w_1 for this eigenvector. The process is repeated, always keeping the guess and search direction orthogonal to the previous w_k's. Eventually, we generate an orthogonal set of approximate eigenvectors $\{w_0, w_1, \ldots w_{N-1}\}$. At this point the Kalkreuter algorithm rediagonalizes A on this basis, yielding improved approximations w'_k and improved estimates of the eigenvalues. We then restart the entire method, using w'_k as the initial vector for each search.

For further details, we refer the reader to the literature.

Chapter 9

Data analysis for lattice simulations

Let us suppose that, after an enormous effort, the reader has collected a series of measurements of some observable O_i from a lattice simulation. Can she then proceed directly to the extraction of a physical observable from her data set? Unfortunately, the answer is usually no. Lattice data sets conceal correlations, which must be treated in a successful statistical analysis.

Correlations arise in lattice simulations from two sources. First, the ensemble of field configurations on which the observable is measured form a Markov chain with inherent correlations from one member to the next. The simulation algorithm produces a new configuration in the chain by evolving from the previous one. We will call this process "evolving in simulation time". One often refers to these correlations as "autocorrelations" in simulation time. When the chain comes into equilibrium, by definition the probability of finding a configuration with action $S(\phi)$ present in our sample set is $\exp[-S(\phi)]$. But one cannot assume successive measurements of O_i are statistically independent.

Second, many observables are not merely the volume average of a single quantity. Frequently we are interested in a correlation function $C(x, y) = \langle\phi(x)\phi(y)\rangle$. The function is measured at the same set of coordinate x, y pairs on each configuration. The resulting set of values is then averaged over the ensemble of configurations. The measured values fluctuate from configuration to configuration in a correlated way simply because each set comes from the same configuration. Thus one cannot assume that measurements at different coordinate pairs x, y are statistically independent.

9.1 Correlations in simulation time

Suppose we have measured A_i for a series of N equilibrium field configurations i, spaced at regular intervals $\Delta\tau$ in simulation time. We compute the usual mean $\langle A\rangle$, where $\langle\rangle$ means averaging over N values. The goal of the statistical analysis is to estimate the variance in that mean, correcting for autocorrelations. The naive sample variance is $\sigma_0^2 = \langle A^2\rangle - \langle A\rangle^2$. We define the scaled fluctuation $x_i = (A_i - \langle A\rangle)/\sigma_0$. The autocorrelation coefficient at time lag $\tau = \Delta\tau(i-j)$ is then defined as

$$C(\tau) = \langle x_i x_j\rangle\,. \tag{9.1}$$

Note that $C(0) = 1$. If the data points are uncorrelated, $C(\tau) = 0$ for $\tau \neq 0$. If not, one expects that $C(\tau)$ is a sum of decaying exponentials. One definition of autocorrelation time for the system is the time constant of the slowest exponential.

The naive variance of the mean of M successive measurements is given by $\sigma_{0,\mathrm{mean}}^2 = \sigma_0^2/M$. But it is biased by autocorrelations. How do we remove this bias? We must have a statistically independent sampling. If our N values span many autocorrelation times, we can construct an independent statistic by averaging over a subset of M successive values, $\frac{1}{M}\sum_i x_i$, provided M is much larger than the autocorrelation time. If we do this for many such subsets, each of them well separated in evolution time, we can compute the true variance in the mean over M successive sample values from

$$\sigma_{\mathrm{mean}}^2 = \left\langle \left(\frac{1}{M}\sum_i x_i\right)^2 \right\rangle = \frac{1}{M^2}\left(\sum_i \langle x_i^2\rangle + 2\sum_i \langle x_i x_{i+1} + \ldots\rangle\right). \tag{9.2}$$

Here the $\langle\rangle$ means averaging over many such subsets. From the definition of the autocorrelation coefficient, we then have

$$\sigma_{\mathrm{mean}}^2 = \sigma_{0,\mathrm{mean}}^2[1 + 2C(\Delta\tau) + 2C(2\Delta\tau) + \ldots]. \tag{9.3}$$

This result shows how to correct the naive variance for the effects of autocorrelations. It can be shown that the correction factor is greater than 1. In actual practice we take $M = N$, and, of course, to get the usual unbiased estimator of the naive sample variance σ_0^2 we must multiply by $N/(N-1)$.

If we can measure $C(\tau)$ up to a sufficiently large τ, we can use Eq. (9.3) to correct the naive variance. But $C(\tau)$ itself has statistical errors that may make it unreliable. Blocking is an alternative: group the data into blocks of n_b successive measurements and form the N/n_b block averages B_i. If n_b is long compared with the autocorrelation time, the blocked data will be less correlated. Of course, the variance of the mean of the N/n_b block averages B_i,

$$\sigma^2_{\text{mean}}(n_b) = \frac{n_b^2}{N^2}\sum_i (B_i - \langle A \rangle)^2 , \tag{9.4}$$

is an estimate of the variance σ^2_{mean} of the mean over all N measurements. But it is biased by residual autocorrelations between values close to either side of the boundaries between adjacent blocks. As we increase the block size, the bias decreases, and for sufficiently large n_b, the variance rises asymptotically to the unbiased variance of the mean as

$$\sigma^2_{\text{mean}}(n_b) = \sigma^2_{\text{mean}} - \alpha/n_b \tag{9.5}$$

for positive α. One can attempt a linear extrapolation in $1/n_b$ to "infinite block size". Since data is finite, the number of blocked averages N/n_b shrinks as n_b increases, and the statistical reliability of $\sigma^2(n_b)$ deteriorates with increasing block size, which introduces uncertainties in the extrapolation. [Recall that the fractional uncertainty on σ_{mean} is $\sqrt{2/(N/n_b)}$.] Ultimately, no matter what method we use, there is no statistical remedy if the autocorrelation time is comparable to or exceeds the total measurement time, since the information needed to compensate for it is simply not present in the data.

For correlation functions $C(x,y)$ the statistical complications of dealing with both autocorrelations in simulation time and correlations between different choices of x,y usually limit us to a more simplistic approach. We start by estimating the autocorrelation time by examining fluctuations in $C(x,y)$ at selected values of x,y (or possibly selected linear combinations of those values). From this information we choose a few reasonable blocking sizes n_b, and from blocked data, carry out the analysis described in the next section. The goal is, of course, to avoid contamination of our final results from autocorrelation. So the test is whether we see a statistically significant variation in the final results as n_b is varied.

Some calculations are so inexpensive that measurements can be made at generously spaced intervals with no significant autocorrelations. Quenched lattice simulations are an example. A good updating algorithm can make this possible. For spin models powerful "cluster" updating algorithms that change whole blocks of spins at once are very effective in reducing the autocorrelation time. Unfortunately, as of this writing, no such algorithm exists for QCD, with or without dynamical fermions.

9.2 Correlations among observables

9.2.1 *Correlated least chi square*

Let us imagine that we wish to measure a correlation function R_j as a function of a lattice parameter j, *e.g.*, lattice coordinate, constructed from a lattice operator O, and we expect that it is described by a model function f:

$$R_j = \langle O(j)O(0)\rangle = f(j, \alpha_a). \tag{9.6}$$

The model function is parameterized by a set of coefficients α_a, whose values and uncertainties we wish to determine from the data. We will do this by adjusting α_a in a conventional least chi square fit. But we must take into account possibly important correlations between measurements at different j. Let us suppose that we have N independent configurations (or N blocks of configurations that we assume to be independent). On the ith configuration we construct our correlator R_{ji}. The average value of the correlator over all configurations is denoted by

$$\overline{R_j} = \frac{1}{N}\sum_{i=1}^{N} R_{ji}. \tag{9.7}$$

To estimate the statistical uncertainty in each $\overline{R_j}$, and to find the correlations between them, define a correlation matrix

$$C_{ij} = \frac{1}{N(N-1)}\sum_{k=1}^{N}[R_{ik} - \overline{R_i}][R_{jk} - \overline{R_j}]. \tag{9.8}$$

The extra N in the denominator is introduced so that the diagonal elements of C_{ij} are just the variances in the means $\overline{R_j}$. The off-diagonal elements carry information about correlations. Apart from the presence of correlations, there is nothing very special about lattice data. One can transcribe most of the tools from any standard treatment of least chi square curve fitting. Here we summarize the results without detailed derivation.

Let us define $\hat{R}_i$ to be the result of averaging an infinitely large sample. If the theory were correct, then $\hat{R}_i$ would be equal to $f(i, \alpha_a)$ for a particular choice of the parameters α_a. Under a reasonably mild set of assumptions (essentially that the fluctuations of the averages $\overline{R_j}$ are Gaussian-distributed), the probability distribution for the $\overline{R}_j$'s that leads to the observed correlations must be

$$P(\overline{R}) \propto \exp\left[-\frac{1}{2}\sum_{ij}(\overline{R_i} - \hat{R}_i)[C^{-1}]_{ij}(\overline{R_j} - \hat{R}_j)\right]$$
$$= \exp\left\{-\frac{1}{2}\sum_{ij}\left[\overline{R_i} - f(i, \alpha_a)\right][C^{-1}]_{ij}\left[\overline{R_j} - f(j, \alpha_a)\right]\right\}. \quad (9.9)$$

To find the best fit, we maximize the probability distribution in Eq. (9.9) by varying the α_a's. This amounts to minimizing (twice the) exponent,

$$\chi^2 = \sum_{ij}\left[\overline{R_i} - f(i, \alpha_a)\right][C^{-1}]_{ij}\left[\overline{R_j} - f(j, \alpha_a)\right]. \quad (9.10)$$

Once the best fit α_a's are found, two questions remain: How good is the fit, and what are the uncertainties on the measured α_a's?

As is the case with uncorrelated measurements, the uncertainties in the α_a can be found in the usual way from the error matrix of the fit. And the probability distribution for the best fit values maps into a probability distribution for χ^2, which is just a distribution for a d dimensional vector, where d is the number of degrees of freedom–*i.e.*, the number of points to be fit minus the number of fit parameters. Each vector component is Gaussian distributed with a standard deviation of unity:

$$P(\chi^2)d\chi^2 = N(\chi^2)^{d/2-1}\exp(-\chi^2/2)d\chi^2. \quad (9.11)$$

With many degrees of freedom, we expect to get $\chi^2 = d \pm \sqrt{2d}$, so a "chi-squared per degree of freedom of about one" ($\chi^2/d \simeq 1$) would characterize a good fit. Goodness of fit is determined in the usual way from the "confidence level," the probability that a correct model with correct estimates of the data would give a lower chi-squared than was found.

9.2.2 *Truncating the correlation matrix*

Suppose R_j measures the correlation between operators separated by a coordinate distance proportional to j. In that case the correlation matrix is typically diagonally dominant with values that fall away from the diagonal [see Eq. (9.27) below]. An analysis of variance starts with a spectral decomposition of the correlation matrix. The largest eigenvalues account for most of the variance. The corresponding eigenvectors give the directions of the most important joint fluctuations in the R_j. They are called the "principal factors". If the components of the real eigenvectors are plotted against j, they typically resemble the normal modes of vibration of an open string. That is, the largest eigenvalue typically corresponds to all points fluctuating in the same direction. The second largest corresponds to a fluctuation with one node, etc. The smallest eigenvalue is associated with the most nodes. If we are fitting a correlation function to a single decaying exponential, we can see that, roughly speaking, the largest eigenvalue controls the uncertainty in the overall normalization and the next largest eigenvalue contributes primarily to the exponential decay rate. Typically the ratio of the largest to smallest eigenvalue is quite large. When we invert the correlation matrix to construct chi square, the smallest eigenvalues have the largest weight.

Not having enough data is a common problem in a lattice simulation, and it can cause "unusual" difficulties, since one might not have a good estimate of the correlation matrix. As a general rule, the large eigenvalue eigenmodes of the correlation matrix are well determined. The small eigenvalue modes, however, may not be. If so, chi square may suffer large fluctuations from their contribution. Because it is constructed from N outer products, the rank of the matrix C can be no greater than N. So in the extreme case of insufficient data, if the number of configurations N is less than the dimension of C, the smallest eigenvalues are zero and C simply has no inverse.

One possible remedy is to use singular value decomposition. Starting from the spectral decomposition

$$C_{ij} = \sum_k |\phi_i^k\rangle \lambda_k \langle \phi_i^k| \tag{9.12}$$

we approximate the inverse by a restricted sum over the modes, keeping only the ones with λ_k greater than a cutoff value Λ

$$[C^{-1}]_{ij}(\Lambda) = \sum_{\lambda_k > \Lambda} |\phi_k(i)\rangle \frac{1}{\lambda_k} \langle \phi_k(i)|. \tag{9.13}$$

(A common variant often used in the social sciences starts by similarly truncating the spectral decomposition of C_{ij}, but then, before inverting, restores only the diagonal values to their full measured variance (Harman,

1960). This procedure is justified by Occam's razor: we accept only the principal factors that the data require, and beyond that assume no further correlation.)

Having constructed C^{-1}, one can proceed to fit the data. Of course, Λ was arbitrarily chosen, and for the fit to make sense, the best fit values and their uncertainties should not depend on the choice of Λ.

9.2.3 *Jackknife and bootstrap methods*

The jackknife and bootstrap methods provide another way to take into account correlations. Starting from a sample of N values, the jackknife method removes J of them, leaving a sample size $N - J$. One carries out the full statistical analysis on the reduced set and records the best fit results for the parameters α_a. The process is then repeated, this time removing a difference set of J values. The repetition continues until the original sample is exhausted. If this is done systematically, we end up with $P = N/J$ sets of best fit values α_{aj} for $j = 1, \ldots, P$. The jackknife mean and variance for the parameters α_a are then constructed as follows:

$$\overline{\alpha_a} = \frac{1}{P} \sum_j \alpha_{aj}$$
$$\sigma_a^2 = (1 - 1/P) \sum_j (\alpha_{aj} - \overline{\alpha_a})^2. \tag{9.14}$$

The result for the average is obvious. The expression for the variance starts from the average of the squared deviation and multiplies it by $(N/J - 1)$ to compensate for the bias introduced by the reuse of data in constructing the samples. One can similarly construct a jackknife estimate of the full error matrix for the fit parameters.

The jackknife is a simple and effective method for taking into account the influence of correlations on the fitted parameters. It can also help account for autocorrelations. If one chooses an elimination set size J larger than the autocorrelation time, the best fit jackknife parameter means and variances should be stable within statistical uncertainties.

The bootstrap method has some similarities to the jackknife method. Starting from N sample values, the bootstrap method makes a random selection to build a new set with M values. In this case the selection is done without deletion, so it is possible that the new set has repetitions. In fact we could even have $M > N$. The full statistical analysis is carried out on the set M and the best fit results are recorded. The process is repeated

P times, where P is anything we like. We then have P sets of values for the parameters.

To simplify the following derivation, set $\overline{\alpha} = 0$. The jth bootstrap sample has entries α_{jk} with $k = 1$ to M. Its mean is

$$\alpha_j^b = \frac{1}{M} \sum_{k=1}^{M} \alpha_{jk} \tag{9.15}$$

and the bootstrap population standard deviation is

$$\sigma_b^2 = \frac{1}{P} \sum_{j=1}^{P} (\alpha_j^b)^2 = \frac{1}{PM^2} \sum_{jkk'} \alpha_{jk} \alpha_{jk'}. \tag{9.16}$$

Now in the bootstrap sample j the average number of repetitions of a given value α_{jk} is M/N, so the average of $\sum_{k'} \alpha_{jk}\alpha_{jk'}$ over random samples is $\sigma^2 M/N$ independent of j and k. That is to say, the original values are assumed independently distributed with mean adjusted to zero, so the cross correlations average to zero, but the correlation of a value with itself averages to σ^2, and the expected number of such self-correlations is M/N. So, inserting the average into the expression for σ_b^2 and summing over j,k, and k', we get

$$\sigma_b^2 = \frac{1}{M} \sigma^2 \times \frac{M}{N}. \tag{9.17}$$

This means that the variance in the mean is the same as the variance in the bootstrap means:

$$\sigma_{\text{mean}}^2 = \sigma_b^2. \tag{9.18}$$

We have ignored small correction factors for biases.

We know of no particular reason to prefer either method over the other. The jackknife and bootstrap methods are so similar in construction, as long as the sample size is adequate, that both methods extract the same information. Some practitioners use the bootstrap method to go beyond a calculation only of parameter variances and attempt to generate a probability distribution in parameter values, quoting asymmetric confidence ranges. We know of no solid statistical justification for this practice.

9.3 Fitting strategies

9.3.1 *Fitting range*

The fitting strategy depends on the physics objectives, of course. Let us consider the very common problem of extracting hadron masses from the correlation function of two operators. We average over transverse distances and express it as a function of only one variable, the Euclidean time. Let $O(t) = \sum_{\vec{x}} O(\vec{x}, t)$ and consider

$$R(t) = \langle O(t)O(0)\rangle \,. \tag{9.19}$$

If we write $O(\vec{x}, t) = \exp(Ht)O(\vec{x}, 0)\exp(-Ht)$, insert a complete set of states, and assume that the spectrum of H is given by a discrete set of energy levels, then

$$R(t) = \sum_n |\langle 0|O|n\rangle|^2 \exp(-m_n t). \tag{9.20}$$

The fitting function is (see Ch. 11)

$$f(t) = \sum_n A_n \exp(-m_n t) \tag{9.21}$$

plus possible boundary effects.

If we are interested only in the lightest mass in the channel, we could select the fitting range $t \in [t_{\min}, t_{\max}]$ with $t_{\min}$ high enough that the lowest mass dominates the correlation function. This process should be done with care, because setting $t_{\min}$ too high risks throwing away useful information, resulting in a larger variance in the mass value. So how high is enough? One way to set it is to fit with only one or a few masses, and vary the minimum $t = t_{min}$ until the chi square is good. As t_{min} rises, excited states become less important and one would expect that the fitting function would be a good representation of the data. Another way to set it is to compute the "effective mass" from $m_{\text{eff}}(t) = \log[R(t)/R(t+1)]$. If the fates are kind, a plot of the parameter $m_{\text{eff}}(t)$ *vs.* t will show a plateau where the lowest mass term dominates and the fitting range can be set. (Since the effective mass depends only on pairs of time slices, it does not incorporate as much information as a full-fledged fit. For this reason, the authors use it only as a rough guide.)

9.3.2 *Signal to noise ratio*

Clearly, excited state properties are more difficult to measure than ground state properties, because the lower mass states dominate the correlator exponentially at large t. States with vacuum quantum numbers are difficult because the vacuum contribution gives a constant function of t, under which the signal vanishes exponentially.

Physics issues often determine the intrinsic noisiness of an observable. The fluctuations in the correlator $R(t)$ are given by the correlations of the absolute squares of the operators:

$$\begin{aligned} N\sigma^2(t) &= \langle |O(t)|^2 |O(0)|^2 \rangle \\ &= \sum_n \langle 0| |O(t)|^2 |n\rangle \langle n| |O(0)|^2 |0\rangle \exp(-m_n t). \end{aligned} \tag{9.22}$$

It may happen that the relevant intermediate states in Eqs. (9.20) and (9.22) are different. For example, consider ordinary light quark spectroscopy. The operator O used in the correlator creates a single quark-antiquark pair. From Eq. (9.22) the noise in meson correlators comes from operators that create two quarks and two anti-quarks. The lightest state with this particle content is a two-pion state. Thus, the signal-to-noise ratio scale as

$$\frac{R(t)}{\sigma(t)} \simeq \sqrt{N} \exp[-(m_M - m_\pi)t]. \tag{9.23}$$

This is a constant for the pion itself, but falls exponentially for any other state. Similarly, baryon signals degrade as

$$\frac{R(t)}{\sigma(t)} \simeq \sqrt{N} \exp[-(m_B - (3/2)m_\pi)t]. \tag{9.24}$$

This happens because the operator creates three quarks, so its absolute square creates three quarks and three antiquarks. The lightest $(3q)(3\bar{q})$ state is a three-pion state. Nonzero momentum states (where m_n is replaced by $E_n(p) = \sqrt{p^2 + m_n^2}$) suffer a similar fate, because generally the lightest state coupling to the squared operator consists of two zero-momentum particles.

9.3.3 *Interpolating operator*

Although these general considerations tell us how the signal to noise ratio inevitably degrades with increasing t, it says nothing about the constant of proportionality in Eqs. (9.23-9.24). An insightful choice of the operator O can help. Any local operator with the correct channel quantum numbers can serve as an interpolating field, but an operator that "looks more like the hadron" may permit a smaller t_{min}, *e.g.*, an operator that distributes the quark creation operators over a volume of a cubic fermi, perhaps with some interesting spatial dependence.

There are other strategies. One can choose several interpolating operators O_i in the same channel and compute the full correlation matrix $R_{ij}(t)$ from the mixed correlation between operator O_i and O_j. If the hadron masses are distinct, the fitting functions have the factorized form

$$f_{ij}(t) = \sum_n b_{in} b_{jn} \exp(-m_n t). \tag{9.25}$$

This approach is closely analogous to the Rayleigh-Ritz variational treatment of the ground state wave functions and eigenenergies of a Hamiltonian. With a larger basis set the system can choose the linear combination of interpolating operators that most resembles the ground state. There may even be sufficient freedom to spare that the excited states can be modeled effectively.

9.3.4 *Asymmetric lattice*

Another strategy works with an asymmetric lattice in which the Euclidean time interval is much smaller than the spatial lattice constant. Of course, the physics is still the same, but, measured in lattice time units, all time variation is slowed down. The signal decays more slowly, and noise rises more slowly. But the correlation in measurements at adjacent times increases as well, reducing the statistical independence of the measurements. Thus this method is most effective in cases where those correlations are small to start with – for example for heavy quarks and glueball masses.

The method we used above in the operator analysis of variance can also tell us the origin of correlations in the data. Consider the correlation of fluctuations from different times,

$$\begin{aligned} NC_{t_1,t_2} &= \left\langle [R(t_1) - \overline{R(t_1)}][R(t_2) - \overline{R(t_2)}] \right\rangle \\ &= \langle O(t_1)O(0)O(t_2)O(0) \rangle - \langle O(t_1)O(0) \rangle \langle O(t_2)O(0) \rangle . \end{aligned} \tag{9.26}$$

When t_1 and t_2 become large, but $t_1 - t_2$ is fixed, this expression becomes

$$NR_{t_1,t_2} \simeq \langle O(t_1)O(t_2)\rangle \langle O(0)O(0)\rangle , \qquad (9.27)$$

and if we again assume the correlator of two O's involves exchange of a particle of mass m, we get the exponential decay $\exp[-m(t_1 - t_2)]$.

9.3.5 *Bayesian methods*

Limiting the fitting range to $t \geq t_{\min}$ in order to suppress the contribution from too many excited states risks throwing away useful information. Here Bayesian methods [introduced to the lattice community by Lepage *et al.* (2002)] can help. The method is based on Bayes' theorem about conditional probabilities.

Bayes theorem deals with a statistical ensemble whose members may have two attributes, say A or not A and B or not B. Consider a randomly selected member of the ensemble and let $P(A)$ be the unconditional probability that it has attribute A and $P(B)$ be the unconditional probability it has B. Let $P(A|B)$ be the conditional probability that any member has attribute A, given that it already has attribute B. Similarly define $P(B|A)$. Then a simple counting exercise shows that

$$P(A|B)P(B) = P(B|A)P(A). \qquad (9.28)$$

This theorem is then applied to curve fitting in the following manner. We let A be the probability that the parameters take on values in the neighborhood of $\{\alpha_a\}$ and B be the probability that the measured mean correlator values take on values in the neighborhood of $\{\overline{R_t}\}$. Our starting probability in Eq. (9.9), interpreted in this language, represents the probability we obtain the mean values $\{\overline{R_t}\}$, given that the parameters are $\{\alpha_a\}$. But when we do curve fitting, we want to turn this interpretation around. Our likelihood function gives the probability that the parameter values are $\{\alpha_a\}$, given that our measured values are $\{\overline{R_t}\}$. Bayes' theorem tells us how to do this reversal and get the likelihood function:

$$P(\{\alpha_a\}|\{\overline{R_t}\}) = P(\{\overline{R_t}\}|\{\alpha_a\})P(\{\alpha_a\})/P(\{\overline{R_t}\}). \qquad (9.29)$$

The left hand side is the likelihood function we use for optimizing the parameters. The first factor on the right is nothing but our original model probability function Eq. (9.9). The second factor is the probability that the parameter values are $\{\alpha_a\}$, independent of our measurement: that is to say, their *a priori* probability. This is where our prior knowledge must be encoded. The third factor $1/P(\{\overline{R_t}\})$ is interpreted as just a normalization constant, since everything is conditioned on having already obtained $\{\overline{R_t}\}$.

Thus from Bayes' theorem we are asked to fold in prior information into our likelihood function Eq. (9.9). It is rewritten

$$P_{\text{full}}(\{\alpha_a\}) \propto P(\{\alpha_a\}) P_{\text{prior}}(\{\alpha_a\}). \tag{9.30}$$

The operational question is how much freedom we allow ourselves in setting $P_{\text{prior}}(\{\alpha_a\})$. A purist might prefer to set the prior probability to 1 to avoid any prejudice. But even purists make hidden assumptions about the parameters. Negative hadron masses are nonsensical, so would be rejected. Even keeping only a few terms in the fitting function in Eq. (9.21) is equivalent to the prejudice that the coefficients of the neglected terms are zero.

The Bayesian modification offers the opportunity to build reasonable assumptions into parameter fitting. Such assumptions might be informed by a model calculation that predicts the mass differences between ground and excited states. A simple way to put these assumptions into practice is to formulate the prior probability as a Gaussian in selected parameters. Then the priors simply augment chi square:

$$\chi^2_{aug} = \chi^2 + \chi^2_{prior} \tag{9.31}$$

where

$$\chi^2_{prior} = \sum_n \frac{(A_n - A_n^0)^2}{\sigma^2_{A_n}} + \sum_n \frac{(m_n - m_n^0)^2}{\sigma^2_{m_n}} \tag{9.32}$$

The parameters A_n^0, m_n^0, σ_{A_n}, and σ_{m_n} are inputs ("prior parameters" or "priors") to the fit.

Now one does a fit as follows: choose a set of priors and find the set of parameters that minimize the augmented χ^2_{aug}. One could fit over a range $t \in [t_{\text{min}}, t_{\text{max}}]$ or fit to the full range. Instead of varying the range (t_{min} of the fit), one could just add more priors until the output fit parameters stabilize. One must also vary the prior variables and their spreads (the σ's). Some of the output parameters will depend sensitively on the choices for the priors. These parameters are not determined by the data. Some will be independent of the choice of prior. These are parameters that are fixed by the data.

We began this section assuming that the reader had a data set and wanted to extract physics results from it. We conclude by assuming that the reader has done a statistical analysis, has measured some quantities and now wants to write a paper. We hope that the reader has taken to heart all the cautionary remarks we have interjected, and has devised a set of checks on his results. Remember, if a result looks too good to be true, it probably is.

Chapter 10

Designing lattice actions

10.1 Motivation

Lattice calculations begin by picking a lattice discretization of one's desired theory. The resulting bare lattice action is a function of the cutoff a and a set of bare couplings $g(a)$. If the bare action is in the universality class of the desired theory, then as the lattice spacing is taken to zero, the calculational results are universal. At any nonzero a scaling violations spoil the universality of the results. Since the cost of a lattice simulation increases rapidly as the lattice spacing is reduced, it makes good sense to consider alternative discretizations with smaller scaling violations. Then one may do simulations at a larger lattice spacing. Techniques for doing this are called "improvement."

In principle actions that lie in the same universality class have an identical set of relevant and marginal operators. Only their irrelevant operators differ. Lattice operators are generally mixtures of operators of different dimensionality, and so different choices for lattice actions amount to taking actions with different mixtures of irrelevant operators. As a simple example, consider two definitions of the derivative operator

$$\Delta_1\phi = \frac{\phi(x+a)-\phi(x-a)}{2a} = \frac{\partial\phi(x)}{\partial a} + \frac{a^2}{6}\frac{\partial^3\phi(x)}{\partial a^3} + \ldots \tag{10.1}$$

and

$$\Delta_2\phi = \frac{\phi(x+2a)-\phi(x-2a)}{4a} = \frac{\partial\phi(x)}{\partial a} + \frac{2a^2}{3}\frac{\partial^3\phi(x)}{\partial a^3} + \ldots. \tag{10.2}$$

The leading terms are identical; the non-leading terms differ. One could cancel the a^2 terms with $\Delta_3\phi = 4/3\Delta_1\phi - 1/3\Delta_2\phi$. Order a^4 terms would remain. Because all discretization errors in the derivative up to $\mathcal{O}(a^3)$ have been eliminated, we call this an $\mathcal{O}(a^3)$ improved action, or say that we have an $\mathcal{O}(a^4)$ action. The generalization of this classical improvement to interacting theories, and the extension of improvement to the field variables themselves, forms the most-used improvement criterion, the Symanzik program (Symanzik, 1980, 1983a,b).

Of course, most of the improvement methods found in the literature do not come from some systematic analysis. If the reader will permit us to be unsystematic for a moment, the unwanted lattice artifacts that might require correcting include

- "Kinetic" violations: lattice artifacts can alter the quark dispersion relation (and hence the dispersion relation for bound states). They could also alter rotational invariance in, for example, the heavy quark potential. Typically, these effects are corrected with a more complicated lattice approximation to the derivative operator. As an example of the gain improved actions can bring, we show the heavy quark potential computed at tree level from the Wilson gauge action and from an improved action, the Lüscher-Weisz action, in Fig. 10.1.
- Violations of $SU(N_f) \times SU(N_f)$ chiral symmetry by either Wilson-type or staggered fermions: Typically, a more complicated gauge connection can help here. Fig. 10.2 illustrates such improvement in a perturbative mixing coefficient (which might be encountered in a lattice calculation of B_K) and for the additive mass renormalization of a clover fermion.

Both features contribute to scale violations in the hadronic spectrum or in matrix elements.

As we begin a more detailed discussion of improvement techniques, we must caution the reader: The lattice community generally agrees that some kind of improvement is a good thing. How it should be done and what constitutes a good test of improvement are subjects of considerable controversy. Presumably this is because, in the end, "improvement" is about dealing with irrelevant operators! Improvement may be motivated by purely theoretical considerations or by a practical need to repair a specific problem. Certainly, no improvement scheme fixes everything in a simulation. There is always a price to pay. At a minimum the cost of the simulation at fixed lattice spacing goes up. Moreover, as is often the case, an improvement scheme fixes one problem, but introduces another.

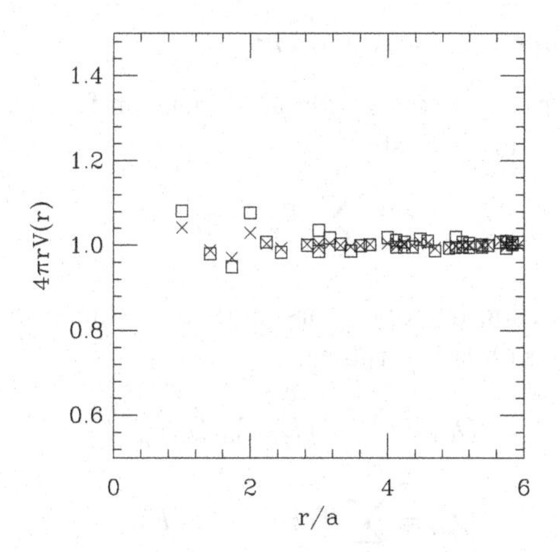

Fig. 10.1 A comparison of the heavy quark potential in free field theory from the Wilson gauge action (squares) and the Lüscher-Weisz action (crosses).

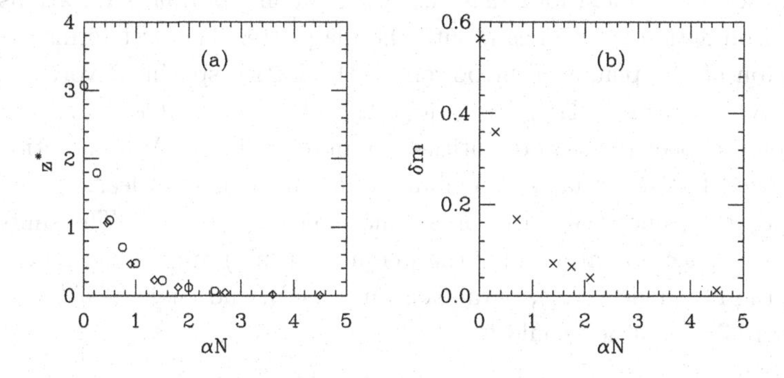

Fig. 10.2 Examples of chiral improvement from an APE-smeared fat link. The unimproved action has $\alpha N = 0$. The larger αN, the greater the smearing. (a) A perturbative calculation of the coefficient of an undesirable four-fermion operator that mixes the operator relevant to B_K with an opposite chirality operator. (b) The troublesome additive mass renormalization for quenched clover fermions using the Wilson gauge action at a coupling $\beta = 6.0$, as measured in a Monte Carlo simulation. The parameters α and N are defined in Eq. (12.28).

10.2 Symanzik improvement

Tree level Symanzik improvement is a good introductory example. Gauge fields require a slightly different treatment from fermion fields.

10.2.1 *Gauge action improvement*

Gauge invariance restricts the available dimension-four operators for the continuum gauge action to just one:

$$O_4 = \sum_{\mu\nu} \mathrm{Tr} F_{\mu\nu} F_{\mu\nu}. \tag{10.3}$$

There are no dimension-five operators, and three dimension-six operators, all of them breaking $O(4)$ invariance:

$$\begin{aligned} O_{6a} &= \sum_{\mu\nu} \mathrm{Tr} D_\mu F_{\mu\nu} D_\mu F_{\mu\nu} \\ O_{6b} &= \sum_{\mu\nu\rho} \mathrm{Tr} D_\mu F_{\nu\rho} D_\mu F_{\nu\rho} \\ O_{6c} &= \sum_{\mu\nu\rho} \mathrm{Tr} D_\mu F_{\mu\rho} D_\nu F_{\nu\rho}. \end{aligned} \tag{10.4}$$

Symanzik improvement for gauge actions commonly produces variations on the Wilson action, so we begin with the plaquette. The first terms in the expansion of the plaquette in powers of the lattice spacing involve all of the above operators: $\mathrm{Tr} U_p = N + (1/2)a^4 O_4 + a^6 \sum r_i O_{6i} + \ldots$. Thus one would expect physical quantities computed with the Wilson action to have $\mathcal{O}(a^2)$ lattice artifacts. To eliminate them requires at least three lattice operators, since there are three dimension-six operators. The simplest choice of lattice operators after the plaquette ("pl") are the perimeter-six loops, the rectangle ("rt"), parallelogram ("pg"), and chair ("ch"), shown in Fig. 10.3. An action could be

$$S(u) = \beta \sum_i c_i O_i \tag{10.5}$$

where $O_i = (1/3)\mathrm{ReTr}(1 - U_i)$. It is convenient (for now) to normalize $c_{\mathrm{pl}} + 8c_{\mathrm{rt}} + 8c_{\mathrm{pg}} + 16c_{ch} = 1$ so the coefficient of $F_{\mu\nu}F_{\mu\nu}$ is unity.

To fix the coefficients, we must choose an improvement criterion. The first such calculation involved correlation functions of Polyakov loops, Wilson loops that cross the box and are joined periodically. The improvement criterion was to enforce the absence of $O(a^2)$ corrections to the scattering amplitude. As a second choice, one could require that the potential between static charges reproduces the continuum lowest order perturbative result $[V(r) = 1/r]$ up to terms of order a^4. In either case, at tree level, one finds $c_{\mathrm{pg}} = 0$, $c_{\mathrm{rt}} + c_{ch} = -1/12$.

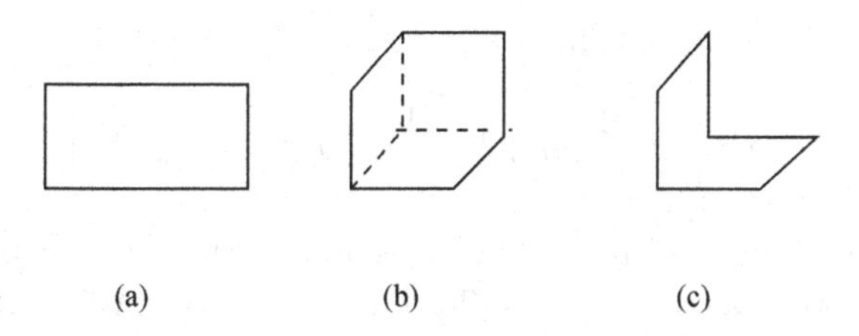

Fig. 10.3 The three perimeter-six loops: (a) "rectangle," (b) "parallelogram," (c) "chair."

It is convenient to choose $c_{ch} = 0$ since the multiplicity of "chair" loops per site is higher than that of the other operators, and this is what is usually done in practice. Thus, at tree level, the improved gauge action is characterized by the plaquette, with weight $c_{\rm pl} = 5/3$, and the rectangle, with $c_{\rm rt} = -1/12$. This is called the "tree-level Lüscher-Weisz" action (Lüscher and Weisz, 1985b) in the literature.

Improvement can be extended to higher order, where now $c_j = c_j^{(0)} + g^2 c_j^{(1)}$. Lüscher and Weisz (1985a) carried the Polyakov loop correlator calculation to next order and computed the coefficients. We will defer writing it explicitly until the end of the next section.

Beyond one loop, lattice perturbation theory becomes arduous, and the authors are not aware of higher order improvement calculations. Instead, perturbative improvement is often replaced by various kinds of nonperturbative improvement, which we will describe below.

10.2.2 *Fermion action improvement*

We already described much of the Symanzik program for Wilson-type fermions in Ch. 6. As we recall, doubling complicates the program for fermions. The Wilson term prevents doubling, but by itself it adds an $\mathcal{O}(a)$ term to the action. Instead, introducing the Wilson term through a field redefinition gives $\mathcal{O}(a)$ improvement. The redefinition is typically done in the continuum. Write the continuum action in terms of continuum fields ψ_c and $\bar{\psi}_c$ and a continuum Dirac operator $M_c = \not{D} + m_c$:

$$S = \int d^4x \, \bar{\psi}_c M_c \psi_c. \tag{10.6}$$

Now perform a field redefinition $\psi_c = \Omega_c \psi$, $\bar{\psi}_c = \bar{\psi}\bar{\Omega}_c$, so the action density becomes $\bar{\psi} M_\Omega \psi$ where $M_\Omega = \bar{\Omega}_c M_c \Omega_c$. The usual choice is $\Omega_c = \bar{\Omega}_c =$

$1-(ra/4)(\not{D}-m_c)$, for which

$$M_\Omega = \not{D} + m_c\left(1+\frac{1}{2}ram_c\right) - \frac{1}{2}ra\left[D^2 + \frac{1}{2}\sigma_{\mu\nu}F_{\mu\nu}\right]. \tag{10.7}$$

with terms of $\mathcal{O}(a^2)$ omitted. The Jacobian of the transformation contributes, effectively, a correction of $\mathcal{O}(a^2)$ to the action. A field redefinition does not change the quark spectrum, but it has introduced a second derivative, along with an accompanying anomalous magnetic moment term with a fixed coefficient. Now we discretize $\not{D}$, D^2 and $F_{\mu\nu}$ to any desired order in a^n, and add corrections from the Jacobian, if needed, to produce a lattice action that is correct to that order. Using the naive derivative, the Wilson term, and the clover term, we arrive again at the clover action with $O(a^2)$ errors. If all we want are spectral quantities, we are done: we have an undoubled lattice action.

Continuum off-shell observables, such as the propagator $\langle\psi_c\bar{\psi}_c\rangle$ or current $\bar{\psi}_c\Gamma\psi_c$, are also transformed under the field redefinition. The lattice transformation is written in terms of Ω, a discretized version of Ω_c. Thus the current

$$J_\Gamma = \bar{\psi}_c\Gamma\psi_c, \tag{10.8}$$

written in terms of the transformed fields ψ and $\bar{\psi}$, becomes the "improved current"

$$J_\Gamma = \bar{\psi}\left[1-\frac{ra}{4}(\not{D}-m_c)\right]\Gamma\left[1-\frac{ra}{4}(\not{D}-m_c)\right]\psi, \tag{10.9}$$

or, using the equation of motion, for on-shell hadronic matrix elements,

$$J_\Gamma = \bar{\psi}\left(1-\frac{ra}{2}\not{D}\right)\Gamma\left(1-\frac{ra}{2}\not{D}\right)\psi. \tag{10.10}$$

For a nice introduction to operator improvement, see Heatlie *et al.* (1991).

Making more accurate discretizations in Eq. (10.7) gives an action with $O(a^4)$ classical errors, called the "D234 action" (Alford *et al.*, 1997).

10.2.3 *Nonperturbative improvement*

Nonperturbative improvement involves parameterizing both the action and the improved operators, and tuning or fitting the parameters to eliminate lattice artifacts. As long as one restricts oneself to $O(a)$ improved actions, there is only one interesting action parameter to tune, the clover term. The operators have additional parameters. Rather than using the full improved current, Eq. (10.9), it is customary to expand to lowest order in a and employ a simpler expression. For example, the flavor non-singlet axial vector current becomes

$$A_\mu^{a,imp} = Z_A[(1 + b_A a m_q) A_\mu^a + c_A a \partial_\mu P^a], \tag{10.11}$$

where the derivative is, of course, approximated by some lattice difference. In this expression, the axial and pseudoscalar currents are just

$$A_\mu^a(x) = \bar{\psi}(x)\gamma_\mu\gamma_5 \frac{1}{2}\tau^a \psi(x) \tag{10.12}$$

and

$$P^a(x) = \bar{\psi}(x)\gamma_5 \frac{1}{2}\tau^a \psi(x). \tag{10.13}$$

(τ^a is an isospin generator.) The technical problem now is to separate the effects of action parameters and operator ones.

The most extensive studies use simulations in a box with fixed boundary conditions for the gauge fields, the Schrödinger functional formalism (Lüscher *et al.*, 1997b). The improvement criterion uses the PCAC quark mass, defined by a ratio of correlators,

$$m(x_1, x_2) \equiv \frac{1}{2} \frac{\langle \partial_\mu A_\mu^a(x_1) O^a(x_2) \rangle}{\langle P^a(x_1) O^a(x_2) \rangle} + O(a), \tag{10.14}$$

The quark mass should be a constant, and so parameters are tuned to minimize the variation of $m(x_1, x_2)$ within the volume of the box. One can choose to improve only at $m_q = 0$; this simplification eliminates the b_A coefficient. The numerator of Eq. (10.14) involves two terms. The c_A term can be eliminated from the action-matching part of the calculation by requiring that a linear combination of m's and pseudoscalar correlators, rather than m itself, not vary. Then after c_{SW} is fixed, the procedure can be repeated for c_A.

Symanzik improvement for fermions is generally not pursued beyond $O(a^2)$. The problem is that at dimension 6, four fermion operators appear in the effective action. It is difficult to simulate an action with these operators, because when a fermion action is no longer bilinear in the field variables, it is not possible to integrate out the fermions to produce a functional determinant without introducing auxiliary fields.

10.3 Tadpole improvement

Tadpole improvement is a variation on Symanzik improvement in which the matching coefficients are computed using a heuristic implementation of perturbation theory called "tadpole improved perturbation theory" (Lepage and Mackenzie, 1993). In Ch. 12, we learn that lattice perturbation theory is performed by expanding the link variable as

$$U_\mu(x) = \exp[iagA_\mu(x)] \to 1 + iagA_\mu(x) - a^2g^2A_\mu(x)^2/2 + \ldots. \quad (10.15)$$

The coupling g and lattice spacing a seem to be correlated, but in lattice perturbation theory a class of diagrams have ultraviolet divergences (regulated on the lattice) whose divergence is canceled by the explicit a dependence in the vertex. Thus these terms are suppressed only by powers of g, not by powers of a. They tend to be large. These are the QCD tadpoles. One is shown in Fig. 10.4.

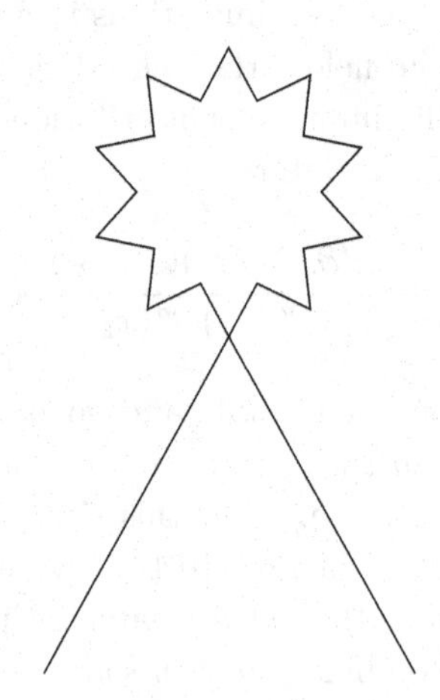

Fig. 10.4 A "tadpole diagram" for the fermionic self energy.

Tadpoles come from the high-momentum part of loop integrals, hence from the high momentum part of the gauge field. We should be able to remove them from the theory. Speaking poetically, one can imagine performing an RG calculation in a smooth gauge and integrating out the UV modes. The link would be replaced by its IR part, a "tadpole factor" u_0

would parameterize the UV part, and Eq. (10.15) would be replaced by

$$U_\mu(x) \to u_0 \exp[iagA_\mu(x)] \to u_0[1 + iagA_\mu(x) - a^2g^2A_\mu(x)^2/2 + \ldots] \tag{10.16}$$

where now A_μ has support only from IR modes.

The tadpole factor u_0 depends on the parameters of the theory and can be measured in a simulation. There are many possible choices (which give slightly different versions of tadpole improvement). One definition of u_0 sets it to the average value of $\mathrm{Tr}U_\mu(x)/N$ in Landau gauge. Another common definition uses the plaquette expectation value:

$$u_0 = \left(\frac{1}{N}\langle \mathrm{Tr}U_p\rangle\right)^{1/4}. \tag{10.17}$$

Tadpole improvement replaces all links U_μ in lattice expressions with U_μ/u_0, where u_0 is measured in the simulation. Removing the tadpoles reduces lattice renormalizations. Because u_0 is measured in the simulation, it also might remove some of the nonperturbative renormalization of lattice operators.

As an example, consider the tadpole-improved Wilson gauge action

$$S = \sum \frac{1}{g^2}(\mathrm{Tr}U_p + h.c.) \to \sum \frac{1}{\hat{g}^2 u_0^4}(\mathrm{Tr}U_p + h.c.). \tag{10.18}$$

Perturbation theory in $\hat{g}^2 = g^2/u_0^4$ has no tadpoles, so the perturbation series shows improved convergence. (We will see an example below).

Tadpole improvement is often used in lattice perturbation theory calculations, but it is most commonly encountered as a way of approximately implementing the Symanzik improvement program. One takes an action, whose coefficients have been computed in perturbation theory, and recomputes the coefficients using one of the tadpole improvement schemes.

Let us see how this works for clover fermions. (The use of the Wilson gauge action, and $a = 1$, is implied.) We recall that the fermion Lagrange density, setting $r = 1$ but keeping c_{SW} as a free parameter, is

$$\begin{aligned}\mathcal{L}_F = {} & m\bar{\psi}(x)\psi(x) \\ & -\frac{1}{2}\sum_\mu[\bar{\psi}(x)(1+\gamma_\mu)U_\mu\psi(x+\hat{\mu}) + \bar{\psi}(x)(1-\gamma_\mu)U_\mu^\dagger\psi(x-\hat{\mu}) \\ & -2\bar{\psi}(x)\psi(x)] - \frac{1}{4}c_{SW}\bar{\psi}(x)\sigma_{\mu\nu}F_{\mu\nu}\psi(x).\end{aligned} \tag{10.19}$$

We rescale the action by multiplying by $2\kappa = 1/(m+4)$, or, equivalently, rescale the field to $\psi' = \sqrt{m+4}\psi$, to get

$$\mathcal{L}_F = \bar{\psi}'(x)\psi'(x) \\ -\kappa \sum_\mu [\bar{\psi}'(x)(1+\gamma_\mu)U_\mu\psi'(x+\hat{\mu}) + \bar{\psi}'(x)(1-\gamma_\mu)U_\mu^\dagger\psi'(x+\hat{\mu})] \\ -\frac{1}{2}\kappa c_{SW}\bar{\psi}'(x)\sigma_{\mu\nu}F_{\mu\nu}\psi'(x). \tag{10.20}$$

Tadpole improvement amounts to making the replacement $U_\mu \to U_\mu/u_0$ in the action. With the usual "clover" definition for $F_{\mu\nu}$, the clover term picks up a factor of $1/u_0^4$, and the tadpole-improved lattice action is (dropping the primes)

$$\mathcal{L}_F = \bar{\psi}(x)\psi(x) - \frac{\kappa}{u_0}\sum_\mu [\bar{\psi}(x)(1+\gamma_\mu)U_\mu\psi(x) + \bar{\psi}(x)(1-\gamma_\mu)U_\mu^\dagger\psi(x)] \\ -\frac{1}{2}\frac{\kappa}{u_0}\frac{c_{SW}}{u_0^3}\bar{\psi}(x)\sigma_{\mu\nu}F_{\mu\nu}\psi(x). \tag{10.21}$$

Now imagine using this action in a simulation. In the derivative term, we merely redefine κ/u_0 to be a new hopping parameter κ' . The simulation uses a clover term $c'_{SW} = c_{SW}/u_0^3$. The tree level value of c_{SW} is unity, hence, the tree-level tadpole-improved clover coefficient is $1/u_0^3$.

Now we go beyond tree level. Let us first do a tadpole-improved prediction of the additive mass renormalization, $\delta m_c = 1/(2\kappa_c) - 4$, for Wilson fermions. At tree level, we expect that $\kappa'_c = \kappa_c u_0 = 1/8$, or $\kappa_c = 1/(8u_0)$. Thus at this level,

$$\delta m_c a = \frac{1}{2\kappa_c} - 4 = -4(1-u_0). \tag{10.22}$$

In first order perturbation theory, $1/(2\kappa_c) - 4 = -5.457\alpha$, and the perturbative expansion for the plaquette gives $u_0 = 1 - g^2/12 = 1 - \pi\alpha/3$. So to get an expression that combines nonperturbative and first order contributions, we subtract the first order part of u_0 and add the first order part of δm_c:

$$\delta m_c a = \left[-4(1-u_0) + \frac{4\pi}{3}\alpha\right] - 5.457\alpha = -4(1-u_0) - 1.269\alpha, \tag{10.23}$$

where the term in brackets is zero, perturbatively. The perturbative expansion is better behaved, in the sense that 1.269 is a smaller number than 5.457. The scale of α, q^*, in the language of Ch. 12, must be recomputed, of course. If that is done, we find a lower value of q^* than for the non-tadpole-improved calculation. This is consistent with the idea that what is left after tadpoles are removed is more infrared. Equation (10.23) actually does give a fairly good reproduction of lattice data [see the discussion in Lepage and Mackenzie (1993)].

Now we use tadpole-improved perturbation theory to predict the value of the clover term that improves the action. It was computed in perturbation theory (Lüscher and Weisz, 1996; Wohlert, 1987) as

$$c_{SW} = 1 + c_{SW}^{(1)} g_0^2, \tag{10.24}$$

with g_0 the bare lattice coupling constant and $c_{SW}^{(1)} = 0.2659$. Since $u_0 = 1 - g^2/12$, the tadpole-improved prediction is

$$c_{SW} = \frac{1}{u_0^3}\left[1 + \frac{g_0^2}{u_0^4}\left(c_{SW}^{(1)} - \frac{1}{4}\right)\right], \tag{10.25}$$

so the coefficient of g_0^2 is reduced to 0.0159.

To illustrate the use of tadpole-improved perturbation theory, we show in Fig. 10.5 a comparison of the clover coefficient c_{SW} for the ordinary clover action in quenched approximation with the Wilson gauge action. We compare ordinary perturbation theory, the one-loop tadpole formula, and an analytic interpolating formula for the nonperturbative determination from Lüscher *et al.* (1997a)

$$c_{SW} = \frac{1 - 0.656g_0^2 - 0.152g_0^4 - 0.054g_0^6}{1 - 0.922g_0^2}. \tag{10.26}$$

Panel (a) shows that the tadpole-improved formula captures most of the nonperturbative contribution, up to $g_0^2 \sim 1$. However, panel (b) is potentially more realistic. It shows the same result, but in terms of $\beta = 6/g_0^2$. The value $\beta = 6$ corresponds to a lattice spacing of about 0.1 fm, whereas $\beta = 6.5$ corresponds to a lattice spacing of about 0.045 fm. These are lattice spacings at which realistic simulations are performed. Is the fact that tadpole improvement misses about half the excess over the pure perturbative formula important? We suspect that for most simulations, the answer is no.

For Wilson-type fermions, tadpole improvement of the clover term, *i.e.* replacing $c_{SW} = 1$ by $c_{SW} = 1/u_0^3$, slightly increases the lower limit of bare quark masses below which exceptional configurations cause trouble.

Tadpole improvement is a major part of modern simulations. What it lacks in rigor (compared with a full nonperturbative determination of

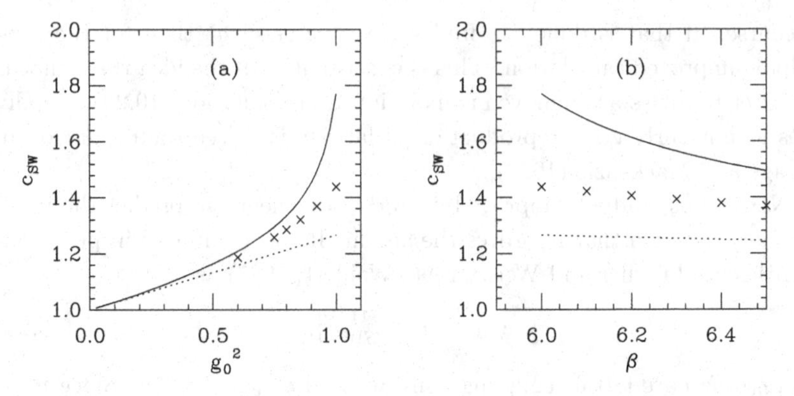

Fig. 10.5 A comparison of a nonperturbative determination of the clover coefficient c_{SW} (the solid line) with the tadpole-improved expectation (crosses) and the naive perturbative prediction (the dotted line) as a function of the bare lattice coupling, in panel (a), and versus $\beta = 6/g^2$, in panel (b).

action coefficients) it more than repays in ease of use. It would be daunting, indeed, to determine all the coefficients of some high-order NRQCD action nonperturbatively. And, by and large, tadpole improved actions show fairly good scaling behavior.

A "standard action," which we recommend for use in simulations, is the "tadpole-improved Lüscher-Weisz action," (Lüscher and Weisz, 1985a; Alford *et al.*, 1995) composed of the plaquette, the rectangle, and the parallelogram loops. It uses the one-loop perturbative coefficients of Lüscher and Weisz, augmented by tadpole improvement. In common usage the normalization convention for the gauge coupling differs from that of the Wilson action. (Namely, g_0^2 is $10/\beta_{\mathrm{pl}}$, not $6/\beta_{\mathrm{pl}}$.) The action is

$$\begin{aligned} S[U] = \beta_{\mathrm{pl}} \sum \frac{1}{3} \mathrm{ReTr}(1 - U_{\mathrm{pl}}) \\ + \beta_{\mathrm{rt}} \sum \frac{1}{3} \mathrm{ReTr}(1 - U_{\mathrm{rt}}) \\ + \beta_{\mathrm{pg}} \sum \frac{1}{3} \mathrm{ReTr}(1 - U_{\mathrm{pg}}) \end{aligned} \tag{10.27}$$

with couplings

$$\begin{aligned} \beta_{\mathrm{rt}} &= -\frac{\beta_{\mathrm{pl}}}{20\, u_0^2} \left(1 + 0.4805\, \alpha_s\right), \\ \beta_{\mathrm{pg}} &= -\frac{\beta_{\mathrm{pl}}}{u_0^2}\, 0.03325\, \alpha_s. \end{aligned} \tag{10.28}$$

The parameters $u_0 \equiv \langle \mathrm{Tr} U_{\mathrm{pl}}/3 \rangle^{1/4}$ and $3.068\alpha_s = -\ln\langle \mathrm{Tr} U_{\mathrm{pl}}/3 \rangle$ are determined self-consistently in the simulation. The coefficients in Eq. (10.28) are a combination of the tree-level and one-loop Symanzik couplings and the perturbative expansion of the tadpole; we have introduced $1 = (1 - (3.068/2)\alpha_s)/u_0^2$ to rescale the coefficients of the rectangle and parallelogram. Because u_0 is supposed to be determined self-consistently, one may need to do tuning runs to determine it before beginning a big simulation. In our experience, these runs do not have to be very accurate.

10.4 Renormalization-group inspired improvement

Symanzik improvement attempts to correct through some order a^n. But what is the best that one can hope to do? We recall the answer to that question was given in Ch. 4: Think of improvement in the context of the renormalization group. We want an action without cutoff effects. Such actions live, in principle, along the renormalized trajectory of some renormalization group transformation.

One could obtain such an action by the following procedure. Imagine having a set of field variables $\{\phi\}$ defined with a cutoff a. Introduce some coarse-grained variables $\{\Phi\}$ defined with respect to a new cutoff a', and integrate out the fine-grained variables to produce a new action

$$e^{-\beta S'(\Phi)} = \int d\phi \, e^{-\beta[T(\Phi,\phi)+S(\phi)]}, \tag{10.29}$$

where $\beta T(\Phi, \phi)$ is the blocking kernel that functionally relates the coarse and fine variables. Repeat this procedure many times. The action for the blocked variables will flow toward the fixed point of the RG transformation and then out along the renormalized trajectory, eventually reducing to an action consisting only of relevant and marginally relevant operators. Simulations done with the blocked action should show only the scaling violations associated with the strength of these operators, which in QCD are the quark mass terms and the running coupling.

Unfortunately, integrating Eq. (10.29) for QCD generates ever more complicated actions, which in practice have to be severely truncated. The resulting actions do not live on any renormalized trajectory. In the literature, such actions are typically called "renormalization-group inspired." The authors are aware of only two such gauge actions that have received much attention. They both consist of the plaquette and the rectangle, *i.e.*

Eq. (10.27) with $\beta_{\rm pg} = 0$, $\beta_{\rm pl} = \beta c_0$, and $\beta_{\rm rt} = \beta c_1$, with the normalization condition $c_0 + 8c_1 = 1$. The Iwasaki action (Iwasaki, 1983) has $c_1 = -0.331$; the DBW2 action (Takaishi, 1996) has $c_1 = -1.4088$.

The critical surface for asymptotically free theories is at at $\beta = \infty$. In that limit, as first pointed out by Hasenfratz and Niedermayer (1994), the blocking transformation of Eq. (10.29) becomes a steepest-descent equation

$$S'(\Phi) = \min_{\phi}[T(\Phi, \phi) + S(\phi)] \tag{10.30}$$

that can be used to find the "fixed point action"

$$S_{FP}(\Phi) = \min_{\phi}[T(\Phi, \phi) + S_{FP}(\phi)]. \tag{10.31}$$

These actions have several noteworthy properties. First, not only are they classically perfect actions (*i.e.*, no $\mathcal{O}(a^n)$ scaling violations for any n), but they are also one-loop quantum perfect (Hasenfratz and Niedermayer, 1997). That is, as one moves out the renormalized trajectory,

$$\frac{1}{g^2} S_{RT}(g^2) = \frac{1}{g^2}[S_{FP} + O(g^4)]. \tag{10.32}$$

Physically this happens because the original action has no irrelevant operators, and they are only generated through loop graphs. Thus these actions are an extreme realization of the Symanzik program. Second, because these actions are at the fixed point, they have scale invariant classical solutions. This fact can be used to define a topological charge operator on the lattice in a way that is consistent with the lattice action (Blatter *et al.*, 1996). And third, fixed-point fermion actions obey the Ginsparg-Wilson relation (Hasenfratz *et al.*, 1998) and are thus chiral.

Fixed point actions have been used in simulations of $d = 2$ sigma models (Hasenfratz and Niedermayer, 1994). A completely constructed fixed-point action for QCD is too complicated to use in simulations, but approximate fixed-point actions have seen a fair amount of use. These actions are "engineered" in the following way. One picks a favorite blocking kernel, which has some free parameters, and solves Eq. (10.31). Then one tunes the parameters in the kernel to optimize the action for locality. This action is truncated for use in simulations. Typical fixed-point actions are hypercubic (the lattice Dirac operator joins sites offset by $n_\mu = (\pm1, 0, 0, , 0), (\pm1, \pm1, 0, 0) \ldots (\pm1, \pm1, \pm1, \pm1))$. They typically have $O(a^2)$ lattice artifacts.

10.5 "Fat link" actions

Most of the time, lattice actions are made gauge invariant using gauge connections that are simple products of link variables along the shortest path; for example, in the nearest-neighbor derivative,

$$\bar{\psi}(x)\psi(x+\hat{\mu}) \to \bar{\psi}(x)U_\mu(x)\psi(x+\hat{\mu}). \tag{10.33}$$

However, in the late 1990's, people began to explore alternative connections, in which the $U_\mu(x)$ was replaced by a sum of terms, each one of which was a product of link variables over a different paths. The sum of terms connecting, for example, points x and y —call it $\Sigma(x,y)$)—could be replaced by a single unitary variable $V(x,y)$, which could be chosen by maximizing $\mathrm{Tr}V(x,y)\Sigma(x,y)^\dagger$. These connections are called "fat links." People discovered that non-chiral actions with fat links had much smaller chiral symmetry violations than equivalent non-chiral actions using ordinary "thin" links as connections. The additive mass renormalization was commonly reduced by an order of magnitude. Staggered actions with fat links showed much smaller taste symmetry breaking.

It is easy to understand these results in the context of perturbation theory. The thin-to-fat link transformation acts as a form factor on the fermion-gauge field coupling. This suppresses tadpoles dynamically. For staggered fermions, it also suppresses taste-changing interactions. On the other hand fat links do not improve "kinetic" properties of actions (such as dispersion relations or the restoration of rotational invariance).

Fat links decouple ultraviolet fluctuations of the gauge field from the fermions. In addition to the good features that we have already described, their use also seems to reduce the number of conjugate gradient iterations needed to invert the Dirac operator. However, nothing is free. Fat links distort the heavy quark potential at short distances.

For example, in the nonrelativistic quark model for a meson consisting of a heavy quark and a light antiquark, the decay constant is proportional to the bound state wave function at zero separation. The fat link softens the potential, the wave function spreads out, and its value at the origin falls. This is, of course, only a scale violation, but it could be a big scale violation and might spoil the calculation. APE-smeared links with heavy smearing produce a large distortion of this type. There are other choices for fat links. Della Morte *et al.* (2005b) studied static-light systems with the HYP link of Hasenfratz and Knechtli (2001) for the gauge connections and found they improved the signal. These fat links smooth the gauge field only within one hypercube, while APE-smearing extends over longer distances.

Dynamical fermion algorithms were difficult to marry to fat links. The problem is that in the standard molecular dynamics algorithm, one needs to differentiate the action – and hence, the fat link – with respect to the link variable, $\partial V(x,y)/\partial U_\mu(z)$. One way to do this is to use the sum of paths $\Sigma(x,y)$ as the fat link, and differentiate by "following the paths." The problem with this solution is that it strongly constrains the amount of fattening one can achieve. Recently, differentiable fat links have been proposed and used. As this subject is still evolving, we will merely mention some of the popular alternatives: the "stout link" of Morningstar and Peardon (2004), and the "FLIC" link of the Adelaide group (Kamleh *et al.*, 2002).

Fat links can be embedded in the Symanzik program as follows. In momentum space the fat link's gauge field $B_\mu(q)$ is related to the thin link's gauge field $A_\mu(q)$ through a form factor $\tilde{h}_{\mu\nu}(q)$

$$B_\mu(q) = \sum_\nu \tilde{h}_{\mu\nu}(q) A_\nu(q) \ , \tag{10.34}$$

and $\tilde{h}_{\mu\nu}(q) \sim 1 + O(a^2q^2) + O(a^4q^4) + \dots$. To achieve full a^2 improvement, one would tune the fat link parameterization to eliminate the $O(a^2q^2)$ term.

To date, the most heavily used (measured by CPU hours consumed) fat link action is the "AsqTad" action developed by the MILC collaboration (Orginos *et al.*, 1999) and by Lepage (1999). It is a staggered fermion action with a nearest and third-nearest neighbor interaction

$$\Delta_\mu - \frac{a^2}{6}(\Delta_\mu)^3. \tag{10.35}$$

The third-nearest neighbor coupling is a thin link, but the nearest-neighbor interaction uses the fat link

$$V_\mu(x) = c_1 U_\mu(x) + \sum_\nu \Big[w_3 S^{(3)}_{\mu\nu}(x) + \\ + \sum_\rho \Big(w_5 S^{(5)}_{\mu\nu\rho}(x) + \sum_\sigma w_7 S^{(7)}_{\mu\nu\rho\sigma}(x) \Big) + w_L S^{(L)}_{\mu\nu}(x) \Big] \tag{10.36}$$

where

$$\begin{aligned} S^{(3)}_{\mu\nu}(x) &= U_\nu(x)U_\mu(x+\hat\nu)U^\dagger_\nu(x+\hat\mu) \\ S^{(5)}_{\mu\nu\rho}(x) &= U_\nu(x)S^{(3)}_{\mu\rho}(x+\hat\nu)U^\dagger_\nu(x+\hat\mu) \\ S^{(7)}_{\mu\nu\rho\sigma}(x) &= U_\nu(x)S^{(5)}_{\mu\rho\sigma}(x+\hat\nu)U^\dagger_\nu(x+\hat\mu) \\ S^{(L)}_{\mu\nu}(x) &= U_\nu(x)S^{(3)}_{\mu\nu}(x+\hat\nu)U^\dagger_\nu(x+\hat\mu). \end{aligned} \tag{10.37}$$

The coefficients are $c_1 = 8w_5 = 48w_7 = 1/8\ w_L = -1/16$, and $w_3 = -5/16$. In practice, they are all tadpole-improved.

Chapter 11

Spectroscopy

11.1 Computing propagators and correlation functions

A great many lattice calculations begin with spectroscopy. Masses are determined from the asymptotic behavior of Euclidean-time correlation functions. The general correlator is written

$$C_{ij}(t) = \mathrm{Tr}[O_i(t)O_j(0)e^{-\beta H}]/Z \tag{11.1}$$

where $Z = \mathrm{Tr}e^{-\beta H}$ and $O_i(t)$ are interpolating operators, *i.e.*, operators that create the state of interest from the vacuum. Making the replacement

$$O_i(t) = e^{Ht}O_i e^{-Ht} \tag{11.2}$$

and inserting a complete set of energy eigenstates into Eq. (11.1) yields

$$C_{ij}(t) = \sum_{m,n} \langle m| O_i |n\rangle \langle n| O_j |m\rangle \, e^{-E_n t} e^{-E_m(\beta-t)}/Z. \tag{11.3}$$

This expression is valid for any temperature $T = 1/\beta$. We have assumed that the spectrum of energy eigenstates is discrete. If t and $\beta - t$ are large (zero temperature limit), the correlation function is dominated by the lowest states. Keeping only the vacuum state "0" and first excited state "1" and taking the special case of a diagonal correlator ($i = j$) gives

$$\begin{aligned} C_{ii}(t) \doteq \; & |\langle 0| O_i |0\rangle|^2 + |\langle 0| O_i |1\rangle|^2 \left[e^{-E_1 t} + e^{-E_1(\beta-t)}\right] \\ & + \mathcal{O}[e^{-(E_2-E_1)t}] + \mathcal{O}[e^{-(E_2-E_1)(\beta-t)}]. \end{aligned} \tag{11.4}$$

The first term is the vacuum disconnected part. Correlation functions are usually defined with it removed. If the operator has no vacuum expectation value, it is zero anyway. If it has a vacuum expectation value, removing it typically degrades the signal to noise ratio. Figure 11.1 gives an example of a correlator with no vacuum disconnected part. If the operator has a zero momentum component, E_1 is the mass of the lightest state it excites. If the operators O_i have poor overlap with the lightest state, the correction terms have a relatively large coefficient and a reliable value for the mass can be extracted only at a large time t.

If also $t \ll \beta$ the diagonal vacuum-connected correlator has the asymptotic form

$$C_{ii}(t) \simeq |\langle 0| O_i |1\rangle|^2 e^{-E_1 t}, \tag{11.5}$$

thus allowing the energy to be extracted from an "effective mass" formula

$$m_{\text{eff}}(t) = \log[C(t)/C(t+1)]. \tag{11.6}$$

The asymptotic value of this function is E_1.

Just as with the Rayleigh-Ritz variational method, including more operators in the correlator helps in extracting higher-lying states. If there are N interpolating operators, we may consider N excited states:

$$C_{ij}(t) \simeq \sum_{m=1}^{N} g_{im} e^{-E_m t} g_{jm}^*, \tag{11.7}$$

where $g_{jm} = \langle 0| O_j |m\rangle$. For convenience we take O_i to be hermitian, so C_{ij} is also hermitian. The generalized effective mass formula in matrix form is then

$$g\Lambda g^{-1} = C(t+1)C^{-1}(t) \tag{11.8}$$

where $\Lambda = \text{diag}\{e^{-E_m}\}$. So the effective energies are obtained from eigenvalues of the channel transfer matrix $C(t+1)C^{-1}(t)$ in the limit of large t. We can also formulate the analysis as a generalized eigenvalue problem

$$C(t+1)x_m = e^{-E_m} C(t) x_m \tag{11.9}$$

where the eigenvector x_m is the mth column of $g^{\dagger -1}$. As the space of interpolating operators is increased, we expect, in analogy with Rayleigh-Ritz, that the lowest-lying eigenvalues converge first.

As we remarked in Ch. 9, the effective mass analysis works only with pairs of time slices, so it does not incorporate as much information as a full-fledged fit. For this reason the authors do not recommend its exclusive use.

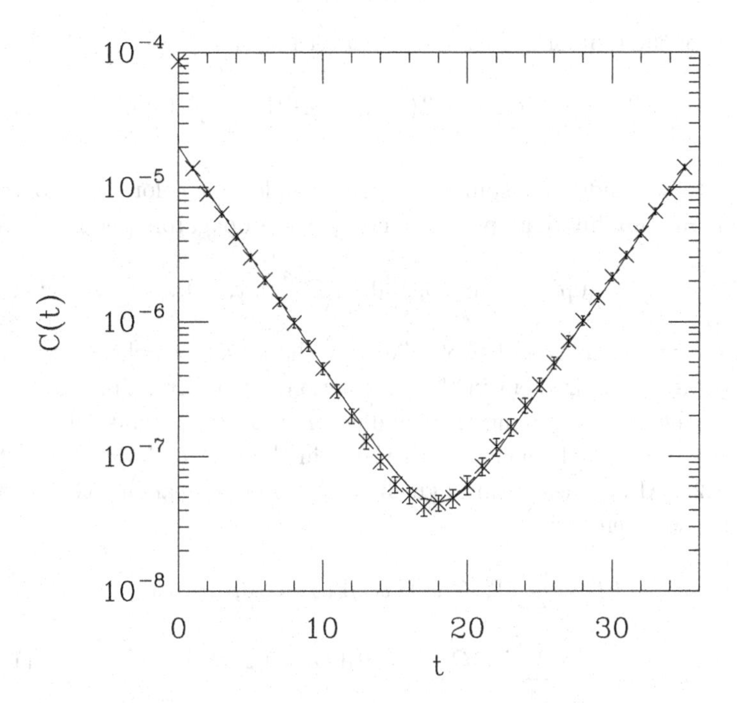

Fig. 11.1 An obviously very nice looking lattice correlator and its fit. In this case $\beta = N_t a = 32a$.

Most of the interesting observables involve valence fermions. The interpolating operators create or annihilate fermion fields from the vacuum. Let's suppose we wanted to measure the mass of a meson in the Wilson formalism. Then we might consider measuring a correlation function

$$C(t) = \sum_x \langle J(x,t) J(0,0) \rangle \tag{11.10}$$

where

$$J(x,t) = \bar{\psi}(x,t) \Gamma \psi(x,t) \tag{11.11}$$

and Γ is a Dirac matrix. We say that the source of the meson is at time 0 and the sink is at time t. The intermediate states $|n\rangle$ that saturate $C(x,t)$ are the hadrons that the current J can create from the vacuum: the pion, for a pseudoscalar current, the rho, for a vector current, and so on. Written in terms of the fermion fields, the correlator is

$$C(t) = \sum_x \langle 0|\bar{\psi}_i^\alpha(x,t)\Gamma_{ij}\psi_j^\alpha(x,t)\bar{\psi}_k^\beta(0,0)\Gamma_{kl}\psi_l^\beta(0,0)|0\rangle \tag{11.12}$$

with a Roman index for spin and a Greek index for color. We contract creation and annihilation operators into quark propagators [see Eq. (3.83)]

$$\langle 0|\psi_j^\alpha(x,t)\bar{\psi}_k^\beta(0,0)|0\rangle = G_{jk}^{\alpha\beta}(x,t;0,0). \tag{11.13}$$

There are two ψ fields and two $\bar{\psi}$ fields in the meson correlator. So there are two ways to pair them in the contraction. One way pairs the ψ with the $\bar{\psi}$ in the source, forcing the same contraction in the sink. The other way pairs the ψ in the source with the $\bar{\psi}$ in the sink and vice versa. See Fig. (11.2). Also remembering sign changes from interchanging Grassmann variables, we then have

$$\begin{aligned} C(t) &= \sum_x \mathrm{Tr}\left[G(x,t;x,t)\Gamma\right] \mathrm{Tr}\left[G(0,0;0,0)\Gamma\right] \\ &- \sum_x \mathrm{Tr}\left[G(x,t;0,0)\Gamma G(0,0;x,t)\Gamma\right] \end{aligned} \tag{11.14}$$

where the trace runs over spin and color indices. If the meson is a flavor nonsinglet, the former contraction gives zero. Baryon correlators are constructed similarly. We see that the space-time "Feynman rule" for Fig. (11.2) associates a valence quark line from (x',t') to (x,t) with $G_{jk}^{\alpha\beta}(x,t;x',t')$, but does not display gluon and sea quark lines.

The computation of the second (quark-line connected) term in Eq. (11.14) typically proceeds as follows. The Dirac equation (sparse matrix problem) is solved with a delta function source at $(0,0)$ to get $G(x,t;0,0)$ for all sink locations (x,t). This calculation is done for each source color and spin. The path-reversed propagator is found immediately from the identity $G(0,0;x,t) = \gamma_5 G^*(x,t;0,0)\gamma_5$. The computation of the first term in Eq. (11.14) is considerably more difficult, since it requires $G(x,t;x,t)$ for every point (x,t), *i.e.*, a sparse matrix solution starting from every lattice point. One strategy uses a stochastic estimator with random sources R_t defined for each spatial site x on the time slice t. We then have

$$\sum_x \mathrm{Tr}\left[G(x,t;x,t)\Gamma\right] = \langle R_t G\Gamma R_t\rangle\,, \tag{11.15}$$

where the expectation value represents an average over a large number of random source vectors. One sparse matrix solution is required for each random vector R_t.

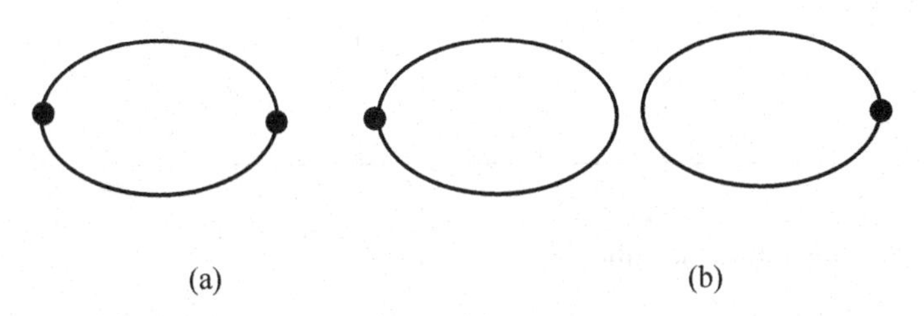

Fig. 11.2 Connected (a) and disconnected (b) quark diagrams corresponding to the two terms in Eq. (11.20).

11.2 Sewing propagators together

In this section we examine the construction of mesonic correlators with specific spin and parity quantum numbers in the Wilson and staggered fermion formalism.

11.2.1 *Mesonic correlators – Wilson fermion formalism*

A typical mesonic interpolating operator is written as $\chi_\Gamma = c_\Gamma \bar{q}\Gamma q$, where q and $\bar{q}$ are Grassmann fermion fields. The Dirac matrix Γ is chosen to have space-time symmetries appropriate to the desired channel (pseudoscalar for $\Gamma = \gamma_5$, vector, for $\Gamma = \gamma_i$, and so on. We use the conventional hermitian basis for the Euclidean Dirac matrices. We postpone a discussion of flavor and a possible spatial wave function. With the suppressed summation over color (to make a color singlet) and Dirac components displayed, the operator is

$$\chi_q = c_\Gamma \bar{q}\Gamma q = \sum_c \sum_{s,s'} c_\Gamma \bar{q}^c_s(x)\Gamma_{ss'}q^c_{s'}(x). \tag{11.16}$$

It is convenient to work with hermitian operators so that $\chi_\Gamma^\dagger = \chi_\Gamma$. Since $\chi_\Gamma^\dagger = c_\Gamma^* \bar{q}\gamma_0\Gamma^\dagger\gamma_0 q$, we require $c_\Gamma\Gamma = c_\Gamma^*\gamma_0\Gamma^\dagger\gamma_0$. See Table 11.1. We are not including any minus signs associated with interchanging the order of the Grassman q and $\bar{q}$.

Thus, recalling the minus sign for the closed fermion loop, our flavor

Table 11.1 Coefficients for hermitian mesonic correlators

Γ	$\Gamma' = \gamma_0 \Gamma^\dagger \gamma_0$	c_Γ to compensate
1	1	1
γ_5	$-\gamma_5$	i
γ_0	γ_0	1
γ_i	$-\gamma_i$	i
$\gamma_0\gamma_5$	$\gamma_0\gamma_5$	1
$\gamma_i\gamma_5$	$-\gamma_i\gamma_5$	i

nonsinglet correlators become

$$\begin{aligned}
\langle i\bar{\psi}(x)\gamma_5\psi(x) i\bar{\psi}(0)\gamma_5\psi(0)\rangle &= \langle \mathrm{Tr}\,[\gamma_5 G(x,0)\gamma_5 G(0,x)]\rangle \\
\langle \bar{\psi}(x)\gamma_0\psi(x)\bar{\psi}(0)\gamma_0\psi(0)\rangle &= -\langle \mathrm{Tr}\,[\gamma_0 G(x,0)\gamma_0 G(0,x)]\rangle \\
\langle i\bar{\psi}(x)\gamma_i\psi(x) i\bar{\psi}(0)\gamma_i\psi(0)\rangle &= \langle \mathrm{Tr}\,[\gamma_i G(x,0)\gamma_i G(0,x)]\rangle \\
\langle \bar{\psi}(x)\gamma_0\gamma_5\psi(x)\bar{\psi}(0)\gamma_0\gamma_5\psi(0)\rangle &= -\langle \mathrm{Tr}\,[\gamma_0\gamma_5 G(x,0)\gamma_0\gamma_5 G(0,x)]\rangle \\
&= \langle \mathrm{Tr}\,[\gamma_0\gamma_5 G(x,0)\gamma_5\gamma_0 G(0,x)]\rangle \\
\langle i\bar{\psi}(x)\gamma_i\gamma_5\psi(x) i\bar{\psi}(0)\gamma_i\gamma_5\psi(0)\rangle &= \langle \mathrm{Tr}\gamma_i\gamma_5 G(x,0)\gamma_i\gamma_5 G(0,x)\rangle, \quad (11.17)
\end{aligned}$$

with an obvious generalization for non-diagonal operators (correlators of χ_Γ and $\chi_{\Gamma'}$).

It is sometimes useful to construct interpolating operators as multiplets of the symmetries that mix them, so the states they generate belong to the same multiplet. For example, suppose one wanted to study pseudoscalar and scalar mesons in a theory with an $SU(2)_L \otimes SU(2)_R$ symmetry. The operators creating the real $O(4)$ vector multiplet $\vec{\phi} = (\pi^a, f_0)$ and its opposite parity partner multiplet $\vec{\phi}' = (a_0^a, \eta)$ can be defined as follows:

$$\pi^a = i\sum_{ij} \bar{q}\gamma_5 \tau^a_{ij} q_j \qquad f_0 = \sum_i \bar{q}_i q_i \tag{11.18}$$

$$a_0^a = -\sum_{ij} \bar{q}_i \tau^a_{ij} q_j \qquad \eta = i\sum_i \bar{q}_i \gamma_5 q_i. \tag{11.19}$$

Here the τ^a's ($a = 1,2,3$) are the $SU(2)$ generators. Under a conventional vector $SU(2)$ flavor rotation the π^a's rotate among themselves, as do the a_0^a's, and the f_0 and η are left invariant. Under an axial $SU(2)$ flavor rotation the π's mix with the f_0 and the a_0's with the η. Under a global $U(1)$ axial rotation, the $\vec{\phi}$ and $\vec{\phi}'$ vectors mix. As we noted above, the flavor singlet correlator has two contributions, one, a valence quark-line disconnected piece in which the quark and antiquark in the source annihilate and the sink is obtained by pair production, and the other, a valence quark-line connected piece in which the quark and antiquark lines both pass from source to sink. With the above definitions we can now see the flavor factors:

$$\langle \eta(x)\eta(y)\rangle = -4\langle \mathrm{Tr}\gamma_5 G(x,x)\mathrm{Tr}\gamma_5 G(y,y)\rangle + 2\langle \mathrm{Tr}\gamma_5 G(x,y)\gamma_5 G(y,x)\rangle \tag{11.20}$$

Finally, we remark that the quark and antiquark fields in the interpolating operator need not appear at the same space-time point and they may have nonzero momentum. In principle χ_Γ can create all hadrons with the selected quantum numbers, ground states and excited states of all momenta. Suppose we are interested only the ground state. Euclidean propagation eventually always suppresses excited states, no matter how we construct the operator, but in the process the signal to noise ratio drops. If we choose an interpolating field that "looks" more like the hadron, we can get a better signal: a mesonic operator could be

$$\chi_\Gamma(x,t) = \sum_{x_1,x_2} \Phi(x_1,x_2;x,t)\bar{q}(x_1,t)\Gamma q(x_2,t) \tag{11.21}$$

where x_1, x_2, and x are spatial vectors. The function $\Phi(x_1,x_2;x,t)$ should be chosen to look like one's favorite model of a hadronic wave function. Without special care, however, the correlation function (we take a diagonal one for example),

$$\begin{aligned} C(x,t;y,0) &= \langle c_\Gamma \Phi(x_1,x_2;x,t)\bar{q}(x_1,t)\Gamma q(x_2,t) \\ &\quad c_\Gamma^* \Phi^\dagger(x_3,x_4;y,0) c_\Gamma \bar{q}(x_3,0)\Gamma q(x_4,0)\rangle \\ &= |c_\Gamma|^2 \langle \mathrm{Tr}\Phi(x_1,x_2;x,t)\Phi^\dagger(x_3,x_4;y,0) \\ &\quad \Gamma G(x_2,t;x_3,0)\Gamma G(x_4,0;x_1,t)\rangle . \end{aligned} \tag{11.22}$$

is not gauge invariant, in which case the gauge field integration gives zero. It is usually worthwhile to persevere, and add extra ingredients to our trial interpolating operators. One possible solution is to fix to some smooth gauge and just take $\Phi(x_1,x_2;x,t)$ to be some smooth function. (So as not to interfere with time propagation, the gauge choice for this purpose should not constrain the time-like gauge links.) The authors have had much success using Coulomb gauge and a $\Phi(x_1,x_2;x,t)$ which is a product of Gaussians of width about one fermi,

$$\Phi(x_1,x_2;x,t) = \phi(x_1 - x)\phi(x_2 - x). \tag{11.23}$$

Another possibility is to let $\Phi(x_1,x_2;x,t)$ be a function of the link fields so that Eq. (11.22) is gauge invariant. The simplest method includes a gauge connection (product of gauge link matrices) between the points x_1 and x_2. Another method defines

$$\Phi(x_1, x_2; x, t) = \phi(x_2; x, t)^\dagger \phi(x_1; x, t) \tag{11.24}$$

and constructs the ϕ's "diffusively." For example, if

$$(\vec{D}^2 + \lambda^2)\phi(\vec{x}; \vec{x}_0, t) = \delta(\vec{x} - \vec{x}_0)\delta(t - t_0) \tag{11.25}$$

where $\vec{D}$ is a covariant spatial derivative, Eq. (11.22) is gauge invariant and $\Phi(x_1, x_2; x, t)$ has an extension roughly $1/\lambda$ in space. This is called "Jacobi smearing" in the literature.

To calculate the correlator in Eq. (11.22) it is necessary to construct the fermion propagator (solve a sparse matrix problem) from the source point $(x_3, 0)$ to the sink point (x_2, t) for as many values of x_3 as required by the wave function. This can be very expensive. In practice, factorized wave functions like Eqs. (11.23-11.24) are almost always used. Then the correlator can be written as

$$C(x, t; y, 0) = |c_\Gamma|^2 \langle \mathrm{Tr}\Phi(x_1, x_2; x, t)\Gamma\mathcal{G}(x_2, t; y, 0)\Gamma\mathcal{G}(y, 0; x_1, t)\rangle. \tag{11.26}$$

where $\mathcal{G}(x_2, t; y, 0) = \sum_x G(x_2, t; x, 0)\phi(x - y)$ obeys

$$(D + m)x_2, t; x', t'\mathcal{G}(x', t'; y, t_0) = \phi(x_2 - y)\delta(t - t_0). \tag{11.27}$$

The calculation of the correlator then proceeds as if point sources were used.

To study states of nonzero momentum it is usually more convenient to project onto a state of definite momentum at the sink operator. This is done by inserting the appropriate phase factor $\exp(i\vec{p} \cdot \vec{x})$.

11.2.2 *Baryonic correlators – Wilson fermion formalism*

The interpolating operators in common use for baryons are quark trilinears. There are a number of choices, which can be derived from the standard quark model wave functions described in Eqs. (2.10) and (2.9).

Consider, first, the proton. Start by combining the valence up and down quarks to form a color antitriplet diquark that is a scalar in spin and parity and a singlet in flavor. For a Grassmann Pauli spinor χ with component $\chi_\uparrow$ and $\chi_\downarrow$ and transpose χ^T the spin singlet combination is

$\chi^T c\chi = \chi_\uparrow\chi_\downarrow - \chi_\downarrow\chi_\uparrow$ for $c = i\sigma_2$. The nonrelativistic diquark wave function

$$\epsilon_{\alpha\beta\gamma}\chi_d^{\alpha T}(x,t)c\chi_u^\beta(x,t). \tag{11.28}$$

antisymmetrized over color indices α and β has the quantum numbers we need. Combining it with a third, up quark, we get the proton wave function in nonrelativistic language, corresponding to Eq. (2.10),

$$\epsilon_{\alpha\beta\gamma}\chi_d^{\alpha T}(x,t)i\sigma_2\chi_u^\beta(x,t)\chi_u^\gamma \tag{11.29}$$

The relativistic generalization is

$$p(x,t) = \epsilon_{\alpha\beta\gamma}[d^{\alpha T}(x,t)C\gamma_5 u^\beta(x,t)]u^\gamma(x,t). \tag{11.30}$$

where now $C = \gamma_0\gamma_2$ in the Euclidean Bjorken-Drell convention. With this convention, in the nonrelativistic limit the upper components of the spinors survive and this expression reduces to the previous expression.

There is an alternative relativistic expression. We can drop the γ_5 and start with a pseudoscalar diquark, finishing with a γ_5 multiplying the third quark to restore the correct parity:

$$p'(x,t) = \epsilon_{\alpha\beta\gamma}[d^{\alpha T}(x,t)Cu^\beta(x,t)]\gamma_5 u^\gamma(x,t). \tag{11.31}$$

In this case in the nonrelativistic limit the diquark inner product is small, because it combines upper and lower components of the spinor.

The proton correlator

$$C(x,t) = \langle\bar{p}(x,t)p(0,0)\rangle \tag{11.32}$$

is computed in the same manner as the meson correlators. After integration over the Grassmann variables, it becomes the product of three quark propagators with spins and colors combined at the source and sink according to the spin-color construction of the interpolating operators.

The other members of the spin 1/2 baryon octet are similarly constructed. For the Σ^0 hyperon the operator corresponding to Eq. (11.30) is

$$\begin{aligned}\Sigma^0(x,t) &= \epsilon_{\alpha\beta\gamma}[u^{\alpha T}(x,t)C\gamma_5 s^\beta(x,t)]d^\gamma(x,t)\\ &+ \epsilon_{\alpha\beta\gamma}[d^{\alpha T}(x,t)C\gamma_5 s^\beta(x,t)]u^\gamma(x,t).\end{aligned} \tag{11.33}$$

The analog of the alternate operator Eq. (11.31) is obvious. The octet Λ

hyperon is similarly

$$\begin{aligned}\Lambda(x,t) &= 2\epsilon_{\alpha\beta\gamma}[u^{\alpha T}(x,t)C\gamma_5 d^\beta(x,t)]s^\gamma(x,t) \\ &+ \epsilon_{\alpha\beta\gamma}[u^{\alpha T}(x,t)C\gamma_5 s^\beta(x,t)]d^\gamma(x,t) \\ &- \epsilon_{\alpha\beta\gamma}[d^{\alpha T}(x,t)C\gamma_5 s^\beta(x,t)]u^\gamma(x,t).\end{aligned} \tag{11.34}$$

To get an interpolating operator for the Δ^+ with spin projection $+\frac{3}{2}$, we start with a spin-1 diquark as a flavor triplet. In nonrelativistic form it is

$$\epsilon_{\alpha\beta\gamma}\chi_d^{\alpha T}(x,t)c\sigma_-\chi_u^\beta(x,t) + (u \leftrightarrow d) \tag{11.35}$$

where $\sigma_- = \sigma_1 - i\sigma_2$ selects the spin projection +1. The second term with u and d interchanged is identical to the first because the Grassmann spinors anticommute. The relativistic generalization with the third quark included and flavor symmetrized is

$$\begin{aligned}\Delta_{+3/2}(x,t) &= 2\epsilon_{\alpha\beta\gamma}[d^{\alpha T}(x,t)Ci\gamma_- u^\beta(x,t)]u^\gamma(x,t) \\ &+ \epsilon_{\alpha\beta\gamma}[u^{\alpha T}(x,t)Ci\gamma_- u^\beta(x,t)]d^\gamma(x,t),\end{aligned} \tag{11.36}$$

where $\gamma_- = \gamma_1 - i\gamma_2$. The alternate relativistic form is

$$\begin{aligned}\Delta'_{+3/2}(x,t) &= 2\epsilon_{\alpha\beta\gamma}[d^{\alpha T}(x,t)C\gamma_5 i\gamma_- u^\beta(x,t)]\gamma_5 u^\gamma(x,t) \\ &+ \epsilon_{\alpha\beta\gamma}[u^{\alpha T}(x,t)C\gamma_5 i\gamma_- u^\beta(x,t)]\gamma_5 d^\gamma(x,t)].\end{aligned} \tag{11.37}$$

For a more complete discussion including a treatment of excited baryons and a wider variety of interpolating operators see Basak *et al.* (2006) and Leinweber *et al.* (2005).

11.2.3 *Mesonic correlators – staggered fermion formalism*

In the staggered fermion formalism each quark flavor comes in a multiplet of four tastes. The extra taste quantum number complicates the construction of mesonic interpolating operators. Each quark-antiquark meson comes with sixteen tastes. Their masses become degenerate in the continuum limit. The rigorous treatment of zero momentum meson interpolating operators was developed by Golterman and Smit (1984); see also Kilcup and Sharpe (1987). Since the Dirac classification of bilinears is more familiar, we prefer to use it, instead, as a heuristic for discovering the combinations that produce the desired states (Kluberg-Stern *et al.*, 1983).

11.2.3.1 *Flavor nonsinglet mesons*

We start by writing the desired bilinear for the interpolating operator in the spin-taste basis

$$\chi_{ff'ST} = c_{ST}\bar{\psi}_y^{f\beta b}\Gamma_S^{\beta\alpha}\Gamma_T^{*ba}\psi_y^{f'\alpha a} \tag{11.38}$$

with flavor indices f, f' and implicit color indices. Here Γ_S operates on the Dirac spin components, represented by Greek letters, and Γ_T, on the taste components, represented by the low Roman alphabet. The coordinate y labels the 2^4 staggered hypercube. There are sixteen choices for Γ_T, corresponding in the continuum to a degenerate multiplet. Recognizing that hermiticity involves both spin and taste, we may use the same procedure as for Wilson fermions to construct hermitian interpolating operators. Flavor indices can be added as well.

Let us first consider free field meson propagators. To convert the spin-taste bilinear to the staggered-fermion single-component basis, we use Eq. (6.52), which gives

$$\chi_{ff'ST} = \sum_{\eta\eta'} \bar{\chi}^f_{2y+\eta'} g_{ST}(\eta', \eta)\chi^{f'}_{2y+\eta}/16, \tag{11.39}$$

where

$$g_{ST}(\eta', \eta) = \frac{1}{4}\mathrm{Tr}\left(\Omega^\dagger_{\eta'}\Gamma_S\Omega_\eta\Gamma^\dagger_T\right). \tag{11.40}$$

The simplest operators are "diagonal" with $S = T$. For this case $g_{SS}(\eta', \eta) = g_S(\eta)\delta_{\eta'\eta}$. We call these the "local" mesonic operators. When $S \neq T$ the operators connect $\bar{\chi}$ and χ at different points in the hypercube, so they are called "non-local". The full taste multiplet for a given meson requires nonlocal operators as well as local. When S and T have different numbers of γ_0 factors, the operator straddles two time slices. If we want to avoid such "time splitting" we simply replace Γ_S with $\gamma_0\Gamma_S$. The replacement does not change the quantum numbers of the channel.

For an important example of a diagonal operator, if $\Gamma_S = \Gamma_T = \gamma_5$ then $g_5(\eta) = (-)^\eta$, where $(-)^\eta = (-)^{\eta_0+\eta_1+\eta_2+\eta_3}$. For $f = d$ and $f' = u$, this is the Goldstone pion operator,

$$\chi_{du55}(y) = \sum_\eta (-)^\eta \bar{\chi}^d_{2y+\eta}\chi^u_{2y+\eta}. \tag{11.41}$$

The zero momentum meson correlator induced by this operator is

$$C_{\pi^+,5}(2t) = \sum_{\vec{y}} \langle \bar{\chi}_{ud55}(y)\chi_{du55}(0)\rangle$$
$$= -\sum_{\vec{y}} \sum_{\eta\eta'} \mathrm{Tr}\left[G(2y+\eta',\eta)G(\eta,2y+\eta')(-)^{\eta-\eta'}\right] \quad (11.42)$$

where $y = (\vec{y}, t)$, the trace is over color, and $G_{ij}(x', x)$ is the staggered quark propagator from x to x'. The path reversal symmetry gives $G(x, x') = G^\dagger(x', x)(-)^{x-x'}$, so the Goldstone boson propagator is then

$$C_{\pi^+,5}(2t) = -\sum_{\vec{y},\eta\eta'} \mathrm{Tr}\left[G(2y+\eta',\eta)G^\dagger(2y+\eta',\eta)\right]. \quad (11.43)$$

This result is not entirely satisfactory, however, since the interpolating operator occupies two adjacent time slices and the time interval between source and sink ranges imprecisely between $2t-1$ and $2t+1$. It is much preferable to use single-time-slice operators in a hadronic correlator. For a general operator ST we first remove time splitting, if necessary, by changing Γ_S to $\Gamma_S\gamma_0$. Then we remove the sum over multiple time slices in Eq. (11.43) by replacing the interpolating operator with the combination

$$\Gamma_S \otimes \Gamma_T^* \pm \Gamma_S\gamma_0\gamma_5 \otimes (\Gamma_T\gamma_0\gamma_5)^* \quad (11.44)$$

in the bilinear. If we use the notation $S05T05$ to denote the second Dirac matrix kernel, we have $g_{ST}(\eta) = g_{S05,T05}(\eta)(-)^{\eta_0}$ where η_0 is the time coordinate of the hypercube. Adding or subtracting cancels one of the time slice contributions inside the hypercube. These are then single-time-slice operators. Since the spin and taste content of the meson is determined by the Dirac matrix kernel, changing the kernel in this way changes the spin, parity, and taste content. Thus the single-time-slice operators generate two mesons of opposite parity, one of them with a correlation function decaying exponentially in the usual way as $\exp(-Et)$ and the other, oscillating as $(-)^t \exp(-E't)$.

To illustrate a single-time-slice operator, consider the correlator for a flavor-nonsinglet, taste-singlet scalar meson, *e.g.*, the a_0. The kernel of the local interpolating operator is then $\Gamma_S \otimes \Gamma_T^* = 1 \otimes 1$. The single-time-slice operators are, then, $1 \otimes 1 \pm \gamma_0\gamma_5 \otimes (\gamma_0\gamma_5)^*$. The second term generates a pseudoscalar meson with the same flavor assignment, in this case a pion. The other pion above, the Goldstone state, has a different taste assignment, so has a different, lower mass at nonzero lattice spacing.

The single-time-slice operators for the Goldstone pion itself are $\gamma_5 \otimes \gamma_5 \pm \gamma_0 \otimes \gamma_0$. They can be combined into a single correlator at even or odd times t as follows

$$C_{\pi+,5}(t) = -\sum_{\vec{x}} \mathrm{Tr}\left[G(x,0)G^\dagger(x,0)\right]. \qquad (11.45)$$

where now $x = (\vec{x}, t)$.

When the quark and antiquark have the same mass, the parity partner operator $\gamma_0 \otimes \gamma_0$ is the density for a conserved charge. The zero momentum component is the charge itself, which does not excite states from the vacuum. Since it is not a taste singlet, it cannot excite a purely gluonic state and so has a zero vacuum expectation value. Therefore the single-time-slice operators for the Goldstone pion couple to only the pion itself.

Carrying the single-time-slice construction to an extreme, we may make a local operator that lives only on the origin of the hypercube. Its kernel is $\sum_S \Gamma_S \otimes \Gamma_S^*$. Such an operator is a favorite broad-spectrum source for the local staggered meson states, since it excites all of them at once. Because the staggered quantum numbers are conserved, it suffices to let the sink operator select the desired spin and parity.

A further effective strategy splits the source points for quark and antiquark in the same manner as described for Wilson fermions. For staggered fermions, the spin and taste assignments are preserved under even relative displacements of the quark and antiquark. The popular "wall source" for zero momentum mesons puts the quark source on all hypercube origins and an uncorrelated antiquark on all hypercube origins as well. (The authors are unaware of constructions like Eqs. (11.23-11.24) for staggered fermions.)

The diagonal $S = T$ correlators based on the local single-time-slice interpolating operators and the single broad-spectrum wall source have the form

$$C_{duSS}(t) = \sum_{\vec{x},\vec{x}',\vec{x}''} \mathrm{Tr}\left[G(x;2\vec{x}',0)G^\dagger(x;2\vec{x}'',0)\right] g_S(x), \qquad (11.46)$$

where the phases and mesonic content are given in Table 11.2.

The local mesonic operators are gauge invariant, so they may be used unchanged in the interacting case. Nonlocal operators involve products of $\bar{\chi}(2y+\eta')$ and $\chi(2y+\eta)$ at different lattice sites, so must be accompanied by an appropriate gauge connection or defined in a smooth gauge, just as we did for the point-split Wilson fermion operators above.

To illustrate nonlocal operators, let us examine the full taste multiplet

Table 11.2 Phases for the local staggered meson operators and the parity partner contributions for the case of equal quark and antiquark masses and nonsinglet flavor. The Dirac-basis taste assignment is indicated in parentheses. Here $i \in \{1,2,3\}$ and j,k are the other two indices in the set.

S	$g_S(x)$	mesons
5	1	$\pi(5)$
i	$(-)^{x_i}$	$\rho(i) \pm b_1(jk)$
i5	$(-)^{x_j+x_k}$	$a_1(i) \pm \rho(i0)$
1	$(-)^{x_1+x_2+x_3}$	$a_0(1) \pm \pi(05)$

for the pion. To assure a non-vanishing result, we assume that the operators are defined in Coulomb gauge. We use the Dirac bilinear heuristic to construct the interpolating operator. Preferably, our point-split operators don't straddle two time slices. So if the bilinear $\gamma_5 \otimes \Gamma_T^*$ is split in the time direction, we take $\gamma_0\gamma_5 \otimes \Gamma_T^*$ instead, which also generates the pion. Accordingly, the single-time-slice operators we want are either $\gamma_5 \otimes \Gamma_T^* \pm \gamma_0 \otimes (\Gamma_T\gamma_0\gamma_5)^*$ or $\gamma_0\gamma_5 \otimes \Gamma_T^* \pm I \otimes (\Gamma_T\gamma_0\gamma_5)^*$, whichever occupies a single time slice. The corresponding Dirac kernels $g_{5,T}(\eta',\eta) \pm g_{0,T05}(\eta',\eta)$ or, as the case may be, $g_{05,T}(\eta',\eta) \pm g_{I,T05}(\eta',\eta)$ in the staggered single-component basis connect pairs of sites (η',η) in a 2^3 spatial cube. The path length of the displacement between the points is constant for a given taste. Thus $T = 5$ and 05 are the local pions we have seen already. The three displacement 1 states are $T = i5$ and jk. The three displacement 2 states are $T = i$ and $i0$, and finally the single displacement 3 taste-singlet state is $T = I$ and 0. The resulting taste multiplet is illustrated in Fig. 6.3.

The point-split combinations produced by the spin-taste heuristic are confined to a hypercube, so have "one-sided" displacements. Golterman and Smit (1984) define their operators in terms of symmetric shifts, so, for example, for a unit displacement in the i direction, they make the replacement

$$\bar{\chi}(x)\chi(x+\hat{\imath}) \to \bar{\chi}(x)D_i\chi(x) = \bar{\chi}(x)[U_i(x)\chi(x+\hat{\imath}) + U_i^\dagger(x-\hat{\imath})\chi(x-\hat{\imath})]. \tag{11.47}$$

For multiple displacements the symmetric shift is repeated. This replacement does not alter the spin content of the excited meson.

11.2.3.2 *Flavor singlet mesons*

The correlators for flavor singlet mesons involve two sets of Wick contractions as illustrated in Fig. 11.2. To get the correct relative weighting of the two contributions requires special care because of the taste degree of freedom. We use the replica trick (Aubin and Bernard, 2003). We introduce n_r species each of u, d, and s quarks and calculate for integer n_r. In the answer we then set $n_r = 1/4$ to compensate for four tastes. For the taste-singlet σ meson channel the interpolating operator is then

$$\chi_{\sigma I} = \frac{1}{\sqrt{2n_r}} \sum_{r=1}^{n_r} [\chi_{urur II} + \chi_{drdr II}] \tag{11.48}$$

where fr denotes the rth repetition of flavor f. The single time-slice correlator is then

$$C_{\sigma I}(t) = -\sum_{\vec{x}} (-)^x \mathrm{Tr}\left[G(x,0)G^\dagger(x,0)\right] + 2n_r \sum_{\vec{x}} \left[\mathrm{Tr}G(x,x)\mathrm{Tr}G^\dagger(0,0)\right]. \tag{11.49}$$

Finally, we put $n_r = 1/4$ to get the correct weighting.[1] This correlator has oscillating and non-oscillating contributions, corresponding to the taste-singlet σ channel and the taste 05 η meson channel. The latter is an unphysical state, unmixed by the axial vector anomaly. Thus, it is nearly degenerate with the taste 05 pion.

11.2.4 *Baryonic correlators – staggered fermion formalism*

Taste multiplicity complicates the analysis of the staggered baryon spectrum considerably. Ideally, we would construct an interpolating operator that excites only one taste state for a given baryon. Then we avoid the complications of sorting out additional spectral components in the correlators produced by taste splitting. As of this writing, a simple heuristic, such as the spin-taste construction that served us so well for mesons, has not been developed for baryons. This is a subject that deserves greater attention.

The classic reference for staggered baryons is Golterman and Smit (1985). They analyzed the staggered lattice symmetries of zero momen-

[1] Another way to get the correct weighting is to introduce a source term for the desired meson in the fermionic action, integrate out the fermions, take the fourth root of the resulting determinant, and then define the correlator as the functional derivative of the sources.

tum baryons in terms of the lattice geometric time-slice group (GTS) (the group of lattice rotations and shifts that governs lattice selection rules) and wrote the interpolating operators for the nucleon and Δ that have been widely used in the literature.

To appreciate the complexity one need only consider that flavor $SU(3)$ becomes flavor-taste $SU(12)$, leading to 572 spin 1/2 states and 364 spin 3/2 states (Bailey and Bernard, 2005). The decomposition of the mixed permutation symmetry **572** of $SU(12)$ into $SU(3)$ flavor times $SU(4)$ taste is given by

$$\mathbf{572_M} \to (\mathbf{10_S}, \mathbf{20_M}) \oplus (\mathbf{8_M}, \mathbf{20_S}) \oplus (\mathbf{8_M}, \mathbf{20}_M) \oplus (\mathbf{8_M}, \bar{\mathbf{4}}_\mathbf{A}) \oplus (\mathbf{1_A}, \mathbf{20_M}), \tag{11.50}$$

where the notation for irreducible representations (irreps) is (flavor, taste). In the continuum limit with degenerate u and d quarks the first two multiplets $(\mathbf{10_S}, \mathbf{20_M}) \oplus (\mathbf{8_M}, \mathbf{20_S})$ contain 120 baryons that are degenerate with the nucleon. At nonzero lattice spacing this degeneracy is lifted, leaving seven distinct multiplets according to the decomposition of the corresponding spin-taste combinations into the **8** and **16** of the GTS group (Bailey and Bernard, 2005).

$$\begin{aligned} (\frac{1}{2}, \mathbf{20_M}) &\to 3 \cdot \mathbf{8} \oplus \mathbf{16} \\ (\frac{1}{2}, \mathbf{20_S}) &\to \mathbf{8} \oplus 2 \cdot \mathbf{16} \end{aligned} \tag{11.51}$$

(In the first set each nucleon is replicated four times and in the second set, twice for a total of $240 = 2 \cdot 120$ states, including isospin and spin.)

The problem is that states of the same GTS symmetry may very well mix. So just constructing a baryon operator on the basis of its GTS symmetry does not guarantee it won't excite several likenesses of the baryon we want, each of a slightly different mass. Furthermore, the decomposition for the spin 3/2 state also contains **8**'s and **16**'s, so we need to make a special effort to disentangle Δ's and nucleons.

With these caveats in mind we turn to the traditional construction. It builds the nucleon interpolating operator in the simplest way as an **8** of the GTS. Take $x = (2\vec{y}, t)$ and define

$$\chi_N(x) = \sum_{abc} \epsilon_{abc} \chi_x^a \chi_x^b \chi_x^c, \tag{11.52}$$

where the color indices are shown explicitly and summed to form a color singlet. The Grassmann fields χ_x^a anticommute. Notice that each quark is created at the origin of a 2^3 cube. The other members of the **8** are created at the other sites of the cube. They are degenerate, so any one will do. This operator happens to belong to the $(\frac{1}{2}, \mathbf{20_M})$.

The corresponding correlator at zero momentum is

$$C_N(t) = \sum_{\vec{y}} \langle \bar{\chi}_N(2\vec{y}, t)\chi_N(0,0)\rangle \,. \tag{11.53}$$

The nucleon wave function is better described by separated quarks, so a potentially better correlator starts from a wall source and lets the sink operator project the desired state. There are various ways to do this. There is the "corner wall" source analogous to our broad-spectrum meson source:

$$\chi_{\text{corner}}(t) = \sum_{\vec{x}_1, \vec{x}_2, \vec{x}_3} \epsilon_{abc}\chi^a_{x_1}\chi^b_{x_2}\chi^c_{x_3} \tag{11.54}$$

where $x_i = (2\vec{x}_i, t)$. Of course, one must fix the spatial gauge first. This operator creates quarks spread uniformly across the time slice.

Then there are other wall sources (Gupta *et al.*, 1991). An even wall source $\chi_{\text{even}}(t)$ is constructed in the same way as the corner source, but sums over all $x_i = (2\vec{x}_i + \eta_i, t)$ for $\eta_1 + \eta_2 + \eta_3$ even. A third choice for a baryon source $\chi_{\text{even-odd}}(t)$ sums all eight η_i. In both cases the sink operator is the corner operator $\chi_N(t)$, so we are still projecting onto an **8** of the GTS.

As we cautioned above, the state we construct in this way contains both nucleons and Δ's. At large t the lightest state, the nucleon, dominates the correlator, but we then must contend with a rapidly degrading signal-to-noise ratio.

As we mentioned above, special measures are required to create the Δ without also creating the nucleon. Golterman and Smit (1985) observed that, unlike the nucleon, the Δ contains an $\mathbf{8'}$ of the GTS. So an interpolating operator belonging to this irrep does the job. It is

$$\chi_\Delta(x) = \epsilon_{abc}D_1\chi^a_x D_2\chi^b_x D_3\chi^c_x, \tag{11.55}$$

where D_i is the symmetric shift of Eq. (11.47). This operator is traditionally used as the sink together with source $\chi_{\text{even-odd}}(0)$.

The most promising staggered baryon method at present builds them from multiple sources and extracts states from a multichannel fit.

11.3 Glueballs

Lattice glueball calculations face two challenges, namely, constructing a good interpolating operator and overcoming a noisy signal in an interesting channel (the scalar glueball).

Glueball interpolating operators are (usually) linear combinations of selected loops, *i.e.*, traces of link variables around closed paths. The loops are combined to produce operators that transform irreducibly under the cubic group in representations that are "like" angular momentum eigenstates. For example, one might use the plaquette as the basis for an interpolating operator. Calling the plaquette oriented in the ij plane U_{ij}, an interpolating operator for a scalar glueball might be

$$O(x,t) = \sum_{i,j=x,y,z;i<j} \mathrm{ReTr} U_{ij}(x). \tag{11.56}$$

A tensor glueball operator might be

$$O_2(x) = \mathrm{ReTr}\left[(U_{xy}(x) + U_{yz}(x) - 2U_{xz}(x)\right]. \tag{11.57}$$

A possible pseudoscalar operator could be a lattice analog of $F_{\mu\nu}\tilde{F}_{\mu\nu}$:

$$O_{PS}(x) = \mathrm{Tr} U_\mu(x) U_\nu(x+\hat{\mu}) U_\mu^\dagger(x+\hat{\nu}) U_\nu^\dagger(x) U_\rho(x) U_\sigma(x+\hat{\rho}) U_\rho^\dagger(x+\hat{\sigma}) U_\sigma^\dagger(x) \tag{11.58}$$

with $\mu \neq \nu \neq \rho \neq \sigma$.

Glueball signals are noisy. The mass comes (as usual) from the asymptotic behavior of

$$C(t) = \sum_{x,x',t} \langle O(\vec{x},t) O^\dagger(\vec{x}',0)\rangle = \sum_n |\langle 0|O|j\rangle|^2 \exp(-m_j t). \tag{11.59}$$

The scalar channel has a vacuum disconnected piece, so one must determine the mass from a fit to $C(t) \simeq C_0 + C_1 \exp(-m_1 t)$. In the other channels, the correlator corresponding to the noise is the correlator of the square of the interpolating field, which has nonzero overlap with the vacuum. Thus even though the signal is a sum of falling exponentials, the noise is a constant in time. (See Ch. 9 for more details.)

Two classes of techniques improve the signal. The first is to use an operator whose size is "physical." It turns out just using loops with greater perimeters is insufficient by itself. It helps also to replace the link variables with "fat links" (See Section 10.5). In contrast to their use in actions, where the fattening should extend over a distance on the order of the cutoff, here the fattening should extend out to some physical distance, on the order of a fermi. Then the operators resemble smoke rings. For a classic early use of these ideas, see Michael and Teper (1989).

The second signal-improvement method involves using an asymmetric space-time lattice. The scalar glueball signal is noisy because it dies exponentially compared to the signal. Making the lattice spacing in the temporal direction smaller flattens the exponential (since the lattice mass is the product of the physical mass and the temporal lattice spacing a_t), and improves the signal to noise ratio – exponentially. Going against this improvement, the signal on successive time slices becomes strongly correlated. The best glueball masses available as of today were obtained using this trick. See Morningstar and Peardon (1999).

11.4 The string tension

Another gluonic observable is the Wilson loop, $W(L,T)$, whose expectation value at large T gives the static potential. Like the nonscalar glueballs, the noise associated with $W(L,T)$ remains a constant in T, while the signal decays exponentially. Like the glueball case, the key to a good signal involves replacing as many of the link variables as possible with smeared links. If the chosen smearing extends over one lattice spacing, then it cannot be applied to the time-oriented links, since the correspondence of path integral to transfer matrix is lost, but if the smearing does not extend beyond a hypercube, the time links can also be smeared. Also as in the case of glueball calculations, an asymmetric space-time lattice produces a better signal than a symmetric lattice.

Smearing is apt to distort the potential at short distances. That can be seen from the perturbative analysis of smearing through form factors from Sec. 10.5: the "smeared potential" is given by a convolution of the form factors with the static gluon propagator,

$$V_s(\vec{r}) = \int d^3\vec{r}_1 d^3\vec{r}_2\, h_{00}(\vec{r}_1)h_{00}(\vec{r}_2)G_{00}(\vec{r}+\vec{r}_1-\vec{r}_2). \tag{11.60}$$

If the form factor can be computed, Eq. (11.60) can be used to fit the short distance part of the potential.

One generally sees information about the potential presented after a fit to some functional form. A typical parameterization is

$$V(r) = -\frac{c}{r} + \sigma r \tag{11.61}$$

or

$$V(r) = -cV_s(\vec{r}) + \sigma r \tag{11.62}$$

where σ is the string tension.

Because of the unfavorable signal to noise ratio at large distances, the string tension is difficult to determine. A shorter distance observable that replaces it is the "Sommer parameter" r_0 (Sommer, 1994), which is defined through the force,

$$-r^2 \frac{\partial V(r)}{\partial r}|_{r=r_0} = C \tag{11.63}$$

where C is some constant. The original Sommer radius assumes $C = 1.65$ and the physical value of r_0 (from fits to the heavy quark potential) is $r_0 \sim 0.46$ fm. Other choices for C are sometimes seen in the literature.

Chapter 12

Lattice perturbation theory

12.1 Motivation

Since the lattice was introduced to do nonperturbative calculations, why would one want to do perturbation theory with a lattice regulator? There are two main uses.

The first use is to test the results of a simulation. If the coupling constant is small, and one believes that nonperturbative effects are not important in some process, then the computer simulation is just reproducing results that could be obtained using perturbation theory. One checks a numerical result against an analytic one.

The second use is much more important in practice. A great many interesting quantities are not spectral. They may be matrix elements or coupling constants, whose values depend on the renormalization scheme or regularization scale. Since a continuum scheme, such as modified minimally subtracted ($\overline{MS}$) dimensional regularization, is a more standard reference point, one converts the result in the lattice scheme to the continuum one. The conversion typically requires lattice perturbation theory.

12.2 Technology

To perform a perturbation theory calculation, one first needs to make a perturbative expansion of the action into free field pieces (quadratic terms) and interaction vertices (everything else), from which the Feynman rules can be constructed. The dynamical variables for QCD, of course, are the quark fields $\psi(x)$ and $\bar{\psi}(x)$ and the vector potentials $A_\mu(x)$.

12.2.1 *Free fermion and gluon propagators*

To illustrate the straightforward process, take the lattice action

$$S = a^4 \sum_x \mathcal{L}(\bar{\psi}, \psi, U) \tag{12.1}$$

and replace the link field by an expansion in terms of gauge fields

$$U_\mu(x) = \exp[igaA_\mu(x)] = 1 + igaA_\mu(x) - \frac{1}{2}g^2a^2A_\mu(x)^2 + \ldots \tag{12.2}$$

where $A_\mu(x) = \sum_a (\lambda^a/2)A_\mu^a$ gives the decomposition into color components. The action has an expansion in powers of A, beginning with a quadratic term, and $\bar{\psi}$ and ψ, again beginning with a bilinear term. Next we transform to momentum space. In terms of the integral over the four-dimensional Brillouin zone,

$$\int_q \equiv \int_{-\pi/a}^{\pi/a} \frac{d^4q}{(2\pi)^4}, \tag{12.3}$$

the quark and antiquark fields are written as

$$\psi(x) = \int_q \psi(q)e^{iqx}; \qquad \bar{\psi}(x) = \int_q \bar{\psi}(q)e^{-iqx} \tag{12.4}$$

and the vector potential is

$$A_\mu(x) = \int_q A_\mu(q)e^{iq(x+a\hat{\mu}/2)}. \tag{12.5}$$

We have shifted the origin for the Fourier transform of the gauge field to beautify intermediate stages of calculations. The action can be written $S = S_0^F + S_0^G + gS_1 + g^2S_2 + \ldots$. The zeroth order terms give us free field theory. For example, the free fermion action is

$$S_0^F = \sum_{x,y} \bar{\psi}(x)\Delta(x-y)\psi(y), \tag{12.6}$$

which becomes in momentum space

$$S_0^F = \int_{pp'} (2\pi)^4\delta^4(p-p')[\bar{\psi}(p')d(p,m)\psi(p)], \tag{12.7}$$

with

$$\Delta(r) = \int_k e^{ikr}d(k,m). \tag{12.8}$$

Likewise, the free gauge boson action is

$$S_0^G = -\frac{1}{2}\int_{pp'}(2\pi)^4\delta^4(p+p')[A_\mu^a(p')D_{\mu\nu}^{ab}(p)A_\nu^b(p)]. \tag{12.9}$$

The fermion propagator is the inverse of $d(p,m)$. We have seen examples of these propagators in Ch. 6. Both the fermion and gauge boson propagators are always diagonal in color. (For the gauge boson, $D_{\mu\nu}^{ab} = \delta^{ab}D_{\mu\nu}$.) Just as in a continuum theory, the gauge boson action cannot be inverted to give the propagator without fixing the gauge. A conventional choice for a gauge fixing term is [introducing $\hat{k}^\mu = 2/a \sin(ak_\mu/2)$]

$$S_{\text{gf}} = -\frac{1}{2}\sum_{\mu\nu}\int_k \text{Tr}\frac{1}{\xi}\hat{k}_\mu\hat{k}_\nu A_\mu^a(-k)A_\nu^a(k). \tag{12.10}$$

Then the gauge boson propagator is found by solving the field equation

$$\sum_\nu\left[\frac{1}{\xi}\hat{k}_\mu\hat{k}_\nu + D_{\mu\nu}(k)\right]G_{\nu\tau}(k) = \delta_{\mu\tau}. \tag{12.11}$$

As an example, consider the Wilson gauge action

$$S_0^G = \frac{1}{2g^2}\sum_x\sum_{\mu\neq\nu}\text{Tr}[1 - P_{\mu\nu}(x) + h.c.] \tag{12.12}$$

where the plaquette $P_{\mu\nu}(x)$ has an expansion whose quadratic term is

$$\text{Tr}[1 - P_{\mu\nu}(x)] = \frac{1}{2}\text{Tr}\left[\frac{\lambda^a}{2}F_{\mu\nu}^a(x)\right]^2 \tag{12.13}$$

and to lowest order in g,

$$F_{\mu\nu}^a(x) = A_\mu^a(x) + A_\nu^a(x+\hat{\mu}) - A_\mu^a(x+\hat{\nu}) - A_\nu^a(x). \tag{12.14}$$

The momentum space version of this expression is

$$F_{\mu\nu}^a(k) = \hat{k}_\mu A_\nu^a(k) - \hat{k}_\nu A_\mu^a(k). \tag{12.15}$$

With $\text{Tr}(\lambda^a\lambda^b) = 2\delta_{ab}$, the action is

$$S_0^G = \frac{1}{4}\int_k F_{\mu\nu}^a(-k)F_{\mu\nu}^a(k), \tag{12.16}$$

so the vector potential's inverse propagator before gauge fixing is

$$D_{\mu\nu}^{ab}(k) = \delta^{ab}[\delta_{\mu\nu}\hat{k}^2 - \hat{k}_\mu\hat{k}_\nu]. \tag{12.17}$$

Then the gauge boson propagator is

$$G^W_{\mu\nu}(k) = \frac{1}{\hat{k}^2}\left[\delta_{\mu\nu} - \xi\frac{\hat{k}_\mu \hat{k}_\nu}{\hat{k}^2}\right]. \tag{12.18}$$

Although general formulas applicable for most commonly used actions can be found in an Appendix of Weisz (1983), the authors have to confess that when they do a perturbative calculation with a complicated gauge action, rather than using them, they just invert Eq. (12.11) numerically "on the fly."

12.2.2 *Quark-gluon vertices*

Next we turn to interactions. When we rewrite the link field in terms of the vector potential, a typical term in the lattice action with Dirac structure Γ, such as

$$S_i = \sum_x \bar{\psi}(x)\Gamma \dots U_\mu(x+y)\dots\psi(x+z) \tag{12.19}$$

will expand into

$$S_i = \sum_x \bar{\psi}(x)\Gamma \dots A_\mu(x+y)\dots\psi(x+z), \tag{12.20}$$

which will have a momentum space expression

$$S_i = \int_{p_1 p_2 k} \bar{\psi}(p_1)\Gamma\psi(p_2)A_\mu(k)(2\pi)^4\delta^4(k+p_2-p_1)e^{iky}e^{iqz}e^{ik_\mu a/2}. \tag{12.21}$$

Collecting all such terms and weighting them as they appear in the action (with a factor ± 1 for U_μ versus $U^\dagger_\mu$) gives us the first order gauge boson - fermion coupling term

$$S_1 = g\int_{p_1 p_2 k} \bar{\psi}(p_2)V^{(1)}_\mu(p_1,p_2,k)\psi(p_1)A_\mu(k)(2\pi)^4\delta^4(k+p_1-p_2). \tag{12.22}$$

We can repeat this analysis in ever higher order. The next term in the action is

$$S_2 = g^2\int_{p_1 p_2 k_1 k_2} (2\pi)^4\delta^4(k_1+k_2+p_1-p_2)\ \bar{\psi}(p_2)V^{(2)}_{\mu\nu}(p_1,p_2,k_1,k_2)\psi(p_1) \times A_\mu(k_1)A_\nu(k_2) \tag{12.23}$$

and so on.

As an example of lattice Feynman rules, let us suppose that we are working with the usual clover action. It is left as an exercise to verify the following formulas.

$$V_\mu^{(1,a)}(p_1,p_2,k) = \frac{\lambda^a}{2}\left[i\gamma_\mu \cos\left(\frac{(p_1+p_2)_\mu a}{2}\right) + \sin\left(\frac{(p_1+p_2)_\mu a}{2}\right)\right]$$

$$V_{\mu\nu}^{(2,ab)}(p_1,p_2,k_1,k_2) = \delta_{\mu\nu}\frac{a}{2}\left\{\frac{\lambda^a}{2},\frac{\lambda^b}{2}\right\}\left[-\cos\left(\frac{(p_1+p_2)_\mu a}{2}\right)\right.$$
$$\left. + i\gamma_\mu \sin\left(\frac{(p_1+p_2)_\mu a}{2}\right)\right]. \qquad (12.24)$$

There is also the clover term,

$$V_\mu^{(1,\mathrm{cl},a)} = \frac{1}{2}\frac{\lambda^a}{2} c_{SW}\gamma_\mu \sum_{\nu\neq\mu}\gamma_\nu \sin k_\nu a \cos\frac{k_\mu a}{2}. \qquad (12.25)$$

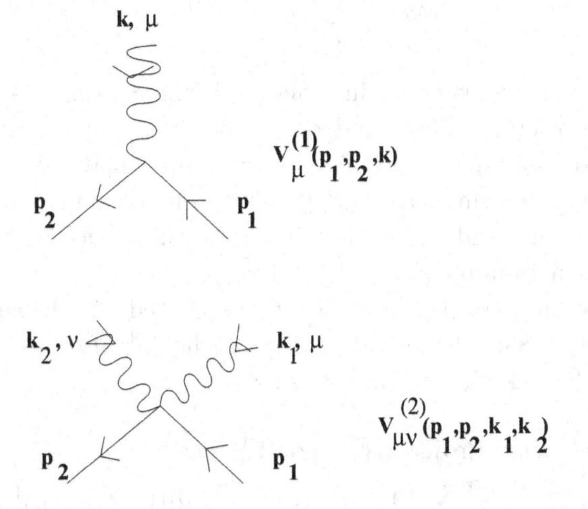

Fig. 12.1 Three-point and four-point vertices, showing our convention for momentum flows.

12.2.3 *Fat links*

The "fat link" refinement is becoming increasingly popular (see Sec. 10.5), wherein a single gauge link is replaced by a link derived from a linear combination of more wide-ranging gauge connections between the same two points. It is easiest to discuss the case in which the linear combination is projected onto the gauge group, making it unitary. For smooth fields unitary fat links have an expansion $W_\mu(x) = 1 + iaB_\mu(x) + \ldots$, whereas the original thin links have an expansion $U_\mu(x) = 1 + iaA_\mu(x) + \ldots$. For computations of 2- and 4-quark operator renormalization/matching constants at one loop, which we will soon discuss, only the linear part of the relation between fat and thin links is needed, and it can be parameterized as

$$B_\mu(x) = \sum_{y,\nu} h_{\mu\nu}(y) A_\nu(x+y) \; . \tag{12.26}$$

Quadratic terms in (12.26), which could only be relevant for tadpole graphs, appear as commutators and therefore do not contribute, since tadpoles are symmetric in the two gluons (Patel and Sharpe, 1993; Bernard and DeGrand, 2000; Lee and Sharpe, 2002). In momentum space, the convolution of Eq. (12.26) becomes a multiplicative form factor

$$B_\mu(q) = \sum_{\nu} \tilde{h}_{\mu\nu}(q) A_\nu(q) \; . \tag{12.27}$$

The reader could think of fat-link action Feynman rules as being constructed in two levels: First, find the vertices for actions with ordinary thin links, and then replace the thin link by a unitary fat link. Each quark-gluon vertex gets a form factor $\tilde{h}_{\mu\nu}(q)$, where q is the gluon momentum. If all gluon lines start and end on fermion lines, then, effectively, the gluon propagator changes into $G_{\mu\nu} \longrightarrow \tilde{h}_{\mu\lambda} G_{\lambda\sigma} \tilde{h}_{\sigma\nu}$.

Again, as an example, consider the so-called "APE-blocked link" (Falcioni *et al.*, 1985; Albanese *et al.*, 1987). The link after $n+1$ smearings is related to the link after n smearings by

$$\begin{aligned} W_\mu^{(n+1)}(x) = \mathrm{Proj}_{SU(3)}\{&(1-\alpha) W_\mu^{(n)}(x) \\ &+ \frac{\alpha}{6} \sum_{\nu \neq \mu} [W_\nu^{(n)}(x) W_\mu^{(n)}(x+\hat{\nu}) W_\nu^{(n)}(x+\hat{\mu})^\dagger \\ &+ W_\nu^{(n)}(x-\hat{\nu})^\dagger W_\mu^{(n)}(x-\hat{\nu}) W_\nu^{(n)}(x-\hat{\nu}+\hat{\mu})]\}. \end{aligned} \tag{12.28}$$

The fat link $W_\mu^{(n+1)}(x)$ is projected back onto $SU(3)$ after each step, and $W_\mu^{(0)}(n) = U_\mu(n)$ is the original link variable. The momentum-space smearing factor for one level of smearing is

$$\tilde{h}_{\mu\nu}(q) = f(q) \left[\delta_{\mu\nu} - \frac{\hat{q}_\mu \hat{q}_\nu}{\hat{q}^2} \right] + \frac{\hat{q}_\mu \hat{q}_\nu}{\hat{q}^2} \; , \tag{12.29}$$

with $f(q) = 1 - \frac{\alpha}{6}\hat{q}^2$. Because of the simple transverse/longitudinal decomposition of Eq. (12.29), it is easy to iterate this blocking. After N smearings, $\tilde{h}_{\mu\nu}(q)$ becomes $\tilde{h}^N_{\mu\nu}(q)$, which is just $\tilde{h}_{\mu\nu}$ with f replaced by f^N.

12.2.4 *Quark self energy and tadpole*

Armed with these preliminaries, we turn to a calculation of amplitudes. In particular, we consider the one-loop quark self energy and the one-loop radiative correction to the quark-gluon vertex. The relevant graphs are shown in Fig. 12.2.

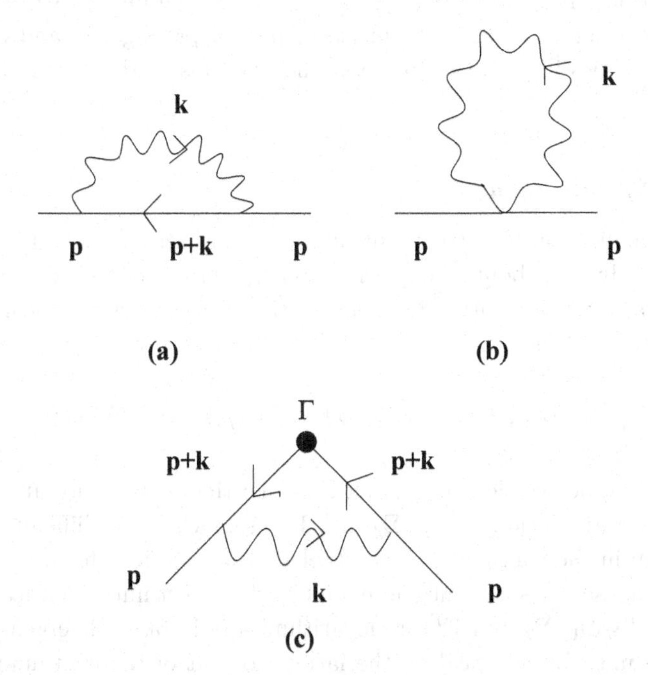

Fig. 12.2 The three one-loop diagrams (a) "sunset" and (b) "tadpole" fermion self energy and (c) vertex renormalization, showing our convention for momentum flows.

The "sunset" graph, Fig. 12.2a, uses first order vertices,

$$\Sigma_a = g^2 \int_k V^{(1)}_\mu(p, p+k, k) S(p+k) V^{(1)}_\nu(p+k, p, -k) G_{\mu\nu}(k). \qquad (12.30)$$

The "tadpole" graph, Fig. 12.2b, is not found in most continuum actions.

It is

$$\Sigma_b = -\frac{g^2}{2}\int_k V^{(2)}_{\mu\nu}(p,p,k,-k)G_{\mu\nu}(k). \quad (12.31)$$

Momenta are labeled as shown in the figure.

For small fermion momentum p and small mass m, the quark self energy is parameterized as

$$\Sigma(p,m) = \Sigma_0 + i\,\not{p}\Sigma_1 + m\Sigma_2. \quad (12.32)$$

The term Σ_0 is (minus) the additive mass renormalization and Σ_1 is the wave-function renormalization, needed for all external lines in vertex functions. To extract Σ_1 and Σ_2, one must expand the propagators and vertices in a power series in p and m, perform the integrals, and keep the leading term.

12.2.5 *Vertex graph*

The renormalization of currents involves both the vertex graph Fig. 12.2c and (if we only care about zero quark mass) Σ_1. This process is customarily evaluated at zero momentum insertion at the vertex. Just to be complete, we write its amplitude,

$$V^\Gamma = \int_k V^{(1)}_\mu(p,p{+}k,k)S(p{+}k)\Gamma S(p{+}k)V^{(1)}_\nu(p{+}k,p,-k)G_{\mu\nu}(k). \quad (12.33)$$

Because lattice vertices are so much messier than their continuum counterparts, the evaluation of Σ_1, Σ_2 and V^Γ is much more difficult on the lattice than in the continuum. Nevertheless, the generic behavior of a lattice process resembles its continuum analog. The most important feature is that typically Σ_1, Σ_2 and V^Γ are logarithmically infrared divergent. This divergence must be identical on the lattice to that of the continuum, because when pa for some generic momentum p flowing in the graph is much less than unity, the lattice Feynman rules reduce to continuum ones. The presence of the infrared divergence means that the calculation has to be regularized somehow. Possible choices include giving the gluon propagator a small mass, or keeping the quark mass, or working at nonzero external momenta.

As an example of this procedure, consider a the correction to the scalar current for massless quarks and work in the continuum (we will keep the Γ

for the vertex for the first line). We give the gluon a mass λ and find

$$V^\Gamma \sim \int_k \frac{1}{k^2+\lambda^2} \frac{\mathrm{Tr}\gamma_\nu(\not p + \not k)\Gamma(\not p + \not k)\gamma_\nu}{[(p+k)^2]^2} \tag{12.34}$$

Letting $p=0$, $\Gamma=1$, $d^4k \sim k^2dk^2$ and doing the trace, we find

$$V^1 \sim \int_0^{(\pi/a)^2} \frac{dk^2}{k^2+\lambda^2} \sim \log\frac{\pi^2}{a^2\lambda^2}, \tag{12.35}$$

which is what we wanted to show.

12.2.6 *Ultraviolet divergences*

The Σ_0 part of Σ_b has a rather peculiar behavior. The vertex scales as a, meaning that from the point of view of the action, this $\bar\psi A^2\psi$ interaction is an irrelevant operator. However, in Σ_b the gluon propagator scales as $1/k^2$, and so Σ_b scales as

$$\Sigma_b \sim g^2 a \int^{\pi/a} kdk \sim \frac{1}{a}. \tag{12.36}$$

Notice how the quadratic divergence combines with the explicit a−dependence of the irrelevant operator to give an overall linear divergence. In a lattice calculation with lattice units, we would work with the dimensionless combination am_q rather than with the quark mass itself. The quantity $a\Sigma_b$, which is a pure number, would be regarded as a finite additive renormalization of the lattice quark mass. These "tadpole contributions" play an important role in lattice perturbation theory, and will reappear when we describe improved actions in Ch. 10.

12.2.7 *Automation*

The standard method for carrying out a lattice perturbation theory calculation is to construct the integrand analytically as a combination of terms multiplying Dirac matrices, and then to project out the desired Dirac structure, producing a single scalar expression. This has to be done cautiously when infrared divergences are present. We will describe one way of doing this in the next section, which is specific to the situation we will discuss there (computing matching coefficients between lattice and continuum-regulated quantities).

Often the analytic form of the solution is uninteresting. Then the finite integral is simply evaluated numerically. A good way to do this is via a Monte Carlo routine such as VEGAS (Lepage, 1980). Modern lattice actions have become increasingly more complicated, and state-of-the-art lattice perturbation theory calculations are often done with considerable help from symbolic manipulation codes or scripting languages. For an example of such a construction, see Nobes and Trottier (2004) or Hart *et al.* (2005).

12.3 The scale of the coupling constant

Suppose we have computed a perturbative prediction for some lattice quantity,

$$\langle O_i \rangle = O_0 + \frac{g^2}{4\pi} O_1 + \ldots . \tag{12.37}$$

How do we interpret our result? What is g^2? This question has two parts. First, how do we define a coupling constant? And second, since the coupling constant of QCD is a scale-dependent quantity, presumably we want to replace the expansion coefficient by something scale dependent, as in $\alpha_s(q^*)$. What is the scale? The solution to this problem (unfortunately) involves at least a two-loop calculation because $\alpha(q) = \alpha(q_0)/[1 - \beta_0 \alpha(q_0) \log(q^2/q_0^2)]$, so $\alpha(q) - \alpha(q_0)$ is order α^2 Absent such a calculation, one could let q^* vary over some "reasonable" range ($1 < q^* a < \pi$ at lattice spacing a, for example), and see how the answer varies. Alternatively, one could construct some physically motivated ansatz for the scale. The best known example of such a construction is that of Lepage and Mackenzie (1993), which builds on the continuum approach of Brodsky *et al.* (1983).

The method begins with our first question. What g^2 should we use? The obvious answer is to use the bare coupling $g(a)^2 = 6/\beta$, at which the simulation is performed. However, direct comparisons of lattice data (say, for the plaquette) reveal that for standard lattice actions it is necessary to go to tiny values of g^2 before perturbation theory can account for lattice data.

Of course, the bare coupling is just one choice for an expansion parameter. Recall the textbook calculation of the spectrum of the hydrogen atom. One needs a value for the electric charge e. Where did it come from? It wasn't a bare parameter. It was determined by comparing an experimental measurement (presumably some very low energy scattering experiment) with a theoretical calculation done to some order in perturbation theory, just as in Eq. (12.37). We could repeat such a calculation using simulation data in place of a real experiment, and define a coupling through the heavy quark potential,

$$V(q) = -C_F \frac{4\pi\alpha_V(q)}{q^2} \tag{12.38}$$

with no higher order corrections. The "V" subscript specifies the "potential" scheme in which the coupling constant is defined and C_F is the fundamental Casimir constant [4/3 for $SU(3)$]. We could relate it, via perturbative calculations, to the $\overline{MS}$ or lattice couplings: the connection would be

$$\alpha_i = \alpha_V(q)[1 - \alpha_V(q)\frac{\beta_0}{4\pi}\ln(\frac{\pi^2}{a^2q^2}) + C_i] + \dots \tag{12.39}$$

where β_0 is the lowest order coefficient of the beta function $(11 - 2N_f/3)$ and C_i is some constant term.

This leaves the choice of scale. To estimate it, imagine that one has some process parameterized by a one loop integral, and assume that its higher order behavior is dominated by gluonic vacuum polarization. We can sum that behavior by pulling the coupling constant inside the integral and letting it run. This amounts to the replacement of integrals

$$\alpha \int I(q)d^4q \to \int \alpha(q)I(q)d^4q. \tag{12.40}$$

However, we want to represent the process by a running coupling constant times the original integral, so

$$\alpha(q^*) \int I(q)d^4q \approx \int \alpha(q)I(q)d^4q. \tag{12.41}$$

When we relate the running coupling constants at the two scales as $\alpha(q) = \alpha(q^*) - \beta_0/(4\pi)\alpha(q^*)^2 \log(q^2/q^{*2}) + \dots$ and compare the various expressions for what we assume to be the same quantity, we see that self consistency requires that the coefficient of $\alpha(q^*)^2$ vanish, or

$$\log(q^*) = \frac{\int d^4q \log(q)I(q)}{\int d^4q\, I(q)}. \tag{12.42}$$

The ratio is like the average of $\log q$ weighted by a distribution $I(q)$. The ratio is often written as $\langle\langle\log(q)\rangle\rangle$. Thus we interpret Eq. (12.37) as

$$\langle O_i \rangle = O_0 + \frac{g^2(q^*)}{4\pi} O_1 + \ldots . \tag{12.43}$$

While this procedure is plausible, it is still only a prescription. However, it can be tested. Determine $\alpha_V(q^*)$ in a simulation from one observable, and use it to predict the values of other observables. In practice, the determining observable is often taken to be the shortest distance observable, with the idea that its nonperturbative part is much smaller than its perturbative component. On the lattice one may choose the expectation value of a single plaquette. For the Wilson gauge action in a quenched simulation, this procedure gives the coupling at its q^* scale from the relation

$$-\ln\left\langle \frac{1}{3}\mathrm{Tr}U_{\mathrm{plaq}} \right\rangle = \frac{4\pi}{3}\alpha_V(3.41/a)[1 - 1.19\alpha_V] + \ldots . \tag{12.44}$$

(The leading term is easy to find: In perturbation theory, the plaquette expectation value is $\mathrm{Tr}U_{\mathrm{plaq}} = 1 - g^2/3$. The scale $q^* = 3.41/a$ comes from evaluating Eq. (12.42).) The observables that are tested could be the values of other Wilson loops, or the additive mass renormalization for some non-chiral fermion, or a renormalization factor that can be determined nonperturbatively.

It can happen that the denominator of Eq. (12.42) is close to zero, and the calculation of q^* using Eq. (12.42) produces absurd results. In that case, one can substitute the higher order expression of Hornbostel *et al.* (2003),

$$\log(q^{*2}) = \langle\langle\log(q^2)\rangle\rangle \pm [-\sigma^2]^{1/2} \tag{12.45}$$

with

$$\sigma^2 = \langle\langle\log^2(q^2)\rangle\rangle - \langle\langle\log(q^2)\rangle\rangle^2 \tag{12.46}$$

and

$$\langle\langle\log^2(q^2)\rangle\rangle = \frac{\int d^4q \, \log^2(q^2) I(q)}{\int d^4q \, I(q)} \tag{12.47}$$

is the weighted average analogous to Eq. (12.42).

This prescription is well defined for finite, pure lattice expressions, but when the operator has an anomalous dimension, there appears to be no general agreement about the prescription at the moment. [For one choice, see DeGrand (2003).]

Chapter 13

Operators with anomalous dimension

We are now ready to begin describing a rather technical subject, namely, how to relate a matrix element computed on the lattice to its value expressed in some continuum regularization scheme. The UV divergences are common to all schemes, but in subtracting them one also subtracts finite parts, which introduces an arbitrariness in the prescription. Two different schemes are then related by a finite renormalization.

Suppose we have a renormalized operator defined in one scheme (perhaps the continuum $\overline{MS}$ scheme) with a regularization point μ. It is proportional to the bare operator O_0,

$$O_1(\mu) = Z_1(\mu) O_0. \tag{13.1}$$

In a second scheme (perhaps the lattice), with a regularization point $1/a$, we have a lattice regulated operator, which is also proportional to the bare operator

$$O_a(1/a) = Z_a(1/a) O_0 \tag{13.2}$$

Consider now matrix elements of the operator in either scheme. Evaluating them in terms of the matrix elements of the bare operator, we find

$$\langle \alpha | O(\mu) | \beta \rangle_1 = \mathcal{Z}(\mu, a) \langle \alpha | O(1/a) | \beta \rangle_a. \tag{13.3}$$

The (finite) quantity $\mathcal{Z}(\mu, a) = Z_1(\mu)/Z_a(1/a)$ is called a "matching factor." It is needed to convert lattice measurements of matrix elements of operators into their values in a continuum regularization.

13.1 Perturbative techniques for operator matching

Usually, part of the calculation of $\mathcal{Z}$ involves perturbation theory, so let us continue the discussion in that context. Begin a calculation working in some continuum regularization scheme, and compute $Z_1(\mu)$. This is done by considering an amputated Green's function. Let us suppose we do that calculation using dimensional regularization in the $\overline{MS}$ scheme. If we specialize to the case where O involves n_F fermion legs, then we compute (in momentum space, in $4-2\epsilon$ dimensions)

$$\Gamma \equiv \frac{G_O^{(n_F)}(p_1,\ldots,p_{n_F},\mu,\epsilon)}{\prod^{n_F} G^{(2)}(p_i,\mu,\epsilon)} \equiv \frac{Z_O G_{O0}^{(n_F)}}{Z_Q^{n_F/2}}. \tag{13.4}$$

Next, compute Z_O in perturbation theory from all the one particle irreducible graphs involving the vertex, plus the counterterm for the operator that removes its UV divergence.

For a quark bilinear Z_O is given by the graph in Fig. 12.2c, and Z_Q from Fig. 12.2b. The result of the calculation for Z_O is

$$Z_O - 1 = \frac{g^2 C_F}{16\pi^2}\left[-A\left(\frac{1}{\epsilon}+\log 4\pi+\gamma_E\right) - \gamma_O \log\frac{\lambda^2}{\mu^2} + C_{\overline{MS}}\right], \tag{13.5}$$

where C_F is the fundamental Casimir, 4/3 in $SU(3)$, and the other constants depend on the operator O. The term proportional to A is the UV divergence. As we recall, the graph has an infrared divergence, which had to be regularized. This gives the logarithm. The regulator λ^2 is a typical momentum scale in the problem, *e.g.*, the squared momentum of the external legs or a regulating gluon or quark mass. To renormalize, we define the counter term to cancel the A term. In the $\overline{MS}$ scheme the extra factors of $\log 4\pi + \gamma_E$ are absorbed in the regularization.

The calculation of Z_Q is similar. The quark propagator is defined as $Z_Q/(i\not p + m)$, which we can rewrite in terms of the self energy as

$$G^{(2)} = \frac{1}{i\not p + m_0} + \frac{1}{i\not p + m_0}\Sigma(p)\frac{1}{i\not p + m_0} + \ldots. \tag{13.6}$$

With $\Sigma(p) = i\not p b\Sigma_1 + m_0 b\Sigma_2$, this series sums to

$$G^{(2)} = \frac{1}{i\not p(1-b\Sigma_1) + m_0(1-b\Sigma_2)}, \tag{13.7}$$

where $b = g^2 C_F/16\pi^2$. Continuing, but with massless quarks for simplicity, we have to leading order

$$Z_Q = \frac{1}{1 - b\Sigma_1} \simeq 1 + b\Sigma_1 \tag{13.8}$$

and after we calculate Σ_1, we find

$$Z_Q - 1 = \frac{g^2 C_F}{16\pi^2}\left[-B\left(\frac{1}{\epsilon} + \log(4\pi) + \gamma_E\right) - \gamma_Q \log\frac{\lambda^2}{\mu^2} + S_{\overline{MS}}\right]. \tag{13.9}$$

Again, choose the counterterm to eliminate the divergence along with the $\log(4\pi) + \gamma_E$, leaving the logarithm and constant term. Then we combine this expression with the one for Z_O to give the $\overline{MS}$ renormalization factor

$$Z_1(\mu) = Z_O/Z_Q = 1 + \frac{g^2 C_F}{16\pi^2}\left[-\gamma \log\frac{\lambda^2}{\mu^2} + C_{\overline{MS}} - S_{\overline{MS}}\right]. \tag{13.10}$$

Next, we repeat the calculation with a lattice regulator. With the same choice for the momentum regulator for the infrared divergences, we compute the same graphs. In the lattice calculation UV divergences are automatically absent, because momentum integrals run over a finite range, $-\pi/a < q_i < \pi/a$, but, as we saw in the last chapter, we still encounter the same IR divergences as in the continuum. They can be controlled by the following trick: Rescale the lattice momenta to be dimensionless (defining $k_i = q_i a$) and write the lattice integral as

$$I_{\text{latt}} = 16\pi^2 \int_{-\pi}^{\pi} \frac{d^4k}{(2\pi)^4} I(k, ap, am, a\lambda). \tag{13.11}$$

If $I_{\overline{MS}}$ has an $A\log(\mu^2/\lambda^2)$ term, I_{latt} has an $A\log[1/(\lambda^2 a^2)]$ IR divergence, too. It can be separated out by writing the integrand as

$$\begin{aligned} I_{\text{latt}} &= 16\pi^2 \int_{-\pi}^{\pi} \frac{d^4k}{(2\pi)^4}\left[I(k, ap, am, a\lambda) - A\frac{\theta(\pi^2 - k^2)}{k^2(k^2 + a^2\lambda^2)}\right] \\ &\quad + 16\pi^2 \int \frac{d^4k}{(2\pi)^4} A \frac{\theta(\pi^2 - k^2)}{k^2(k^2 + a^2\lambda^2)} \\ &\equiv J + A\log\frac{\pi^2}{a^2\lambda^2} \end{aligned} \tag{13.12}$$

The first term of Eq. (13.12) is IR finite, and one can set $\lambda = 0$ in it before evaluating it (perhaps by a numerical integration). Repeating this procedure for all necessary graphs, we get the complete lattice renormalization constant

$$Z_a(1/a) = 1 + \frac{g^2 C_F}{16\pi^2}\left[-\gamma \log(a^2\lambda^2) + C_{\rm latt} - S_{\rm latt}\right]. \qquad (13.13)$$

The constant terms in lattice regularization are different from those of the continuum. Thus the complete lattice to continuum matching factor is

$$\mathcal{Z}(\mu, a) = 1 + \frac{g^2 C_F}{16\pi^2}[\gamma \log(\mu^2 a^2) + C_{\overline{MS}} - C_{\rm latt} - S_{\overline{MS}} + S_{\rm latt}]. \qquad (13.14)$$

Notice that the scale λ^2 has disappeared. The $\gamma \log \mu^2 a^2$ term represents the running of the operator from its value at scale $1/a$ to scale μ. (In contrast to the usual perturbative calculations, which are usually concerned only with how operators run, here the difference between the constant terms from the two schemes is also important.)

Notice that the coupling constant in these expressions requires interpretation. We discussed this point in Sec. 12.3.

We have assumed that a single bare operator is mapped under renormalization into a single renormalized operator. This is not always the case. If not, there is a mixing matrix in each scheme

$$\begin{aligned} O_i(1) &= Z_{ij}(1) O_{0,j} \\ O_i(2) &= Z_{ij}(2) O_{0,j} \end{aligned} \qquad (13.15)$$

and the matching formula also involves a matrix

$$\langle \alpha | O_i(1) | \beta \rangle = Z_{ik}(1) Z_{kj}(2)^{-1} \langle \alpha | O_j(2) | \beta \rangle. \qquad (13.16)$$

The quark mass is a coupling constant, not a matrix element, so the matching technology is a little different. The propagator Eq. (13.6) has a pole at $-i\not{p} = m_R = Z_m m_0$, where we saw $Z_m \simeq 1 - b(\Sigma_2 - \Sigma_1)$. In the continuum, Z_m is UV divergent and, as we see from Eq. (13.9), it must be renormalized by subtracting the $1/\epsilon + \gamma + \ln(4\pi)$ term. We do that to give $Z_m(\mu)m(\mu) = m_R$ where m_R is physical (it is the location of the pole in the propagator), and $Z_m(\mu)$ and $m(\mu)$ are now finite (though μ dependent). The identical calculation on the lattice would again give $m_R = Z_m^{\rm latt}(a)m(a)$ where $m(a)$ is the bare mass. One may then determine a quark mass (in some regulator scheme) from a lattice simulation in two steps. First, one does a simulation at some lattice spacing, determining the bare lattice mass $m(a)$ from a fit to spectroscopy for some observable particle. Then the ratio of lattice and continuum Z-factors gives the running mass at scale μ

through

$$m(\mu) = \frac{Z_m^{\text{latt}}}{Z_m^{\text{cont}}} m(a) = Z(\mu, a) m(a). \tag{13.17}$$

13.2 Nonperturbative techniques for operator matching

Notice that the matching procedure had three parts. We had to fix the renormalization scheme (describe how to define a coupling constant), specify the scale at which the coupling is defined, and determine a numerical value for the coupling at that scale. The choice of coupling is arbitrary, and, at least at first order, the choice of scale is, too. This means that any perturbative calculation is intrinsically ambiguous; how well a calculation works usually comes down to the practical question, "How big is the effect you have calculated?" or the question whose answer is probably less well known, "How big will the next order be?" The matching technique described in the last section would be expected to work well when the Z factor was close to unity. To achieve this result requires that both the lattice and continuum Z factors ought to be close to unity, which is best achieved when the matching point's momentum scale is large and the lattice spacing is small.

One may not want to deal with such ambiguity. With lattice methods one can replace perturbative calculations of matching factors with results from simulations. The uncertainties in the answers may then be dominated only by numerical considerations. Here are some strategies.

13.2.1 *Methods for approximations to conserved currents*

Often we are interested in a conserved or partially conserved continuum current. The Z factor is then unity. The same is true for the corresponding conserved lattice current. But conserved lattice currents tend to be computationally unwieldy, and a simpler, non-conserved current V_μ^i may have $Z \neq 1$. One can easily determine its Z factor nonperturbatively from a ratio of matrix elements involving it and its conserved lattice counterpart V_μ^C,

$$Z_{V^i} = \frac{\langle h|V_\mu^C|h\rangle}{\langle h|V_\mu^i|h\rangle}, \tag{13.18}$$

and then use the simpler current in the rest of the calculation.

A variation on this method can be used with overlap fermions (or other fermions with a chiral symmetry) to compute Z_A. As in the continuum, the product $m_q\bar{\psi}\psi$ has no anomalous dimension, so $Z_m Z_S = 1$. Because overlap fermions are chiral, a Ward identity sets Z_P, the pseudoscalar renormalization factor, equal to Z_S the scalar current renormalization factor. Then one can do two calculations of the pseudoscalar decay constant, one from the pseudoscalar density

$$m_{PS}^2 f_{PS} = 2Z_m m_q Z_P \langle 0|\bar{\psi}\gamma_5\psi|PS\rangle \tag{13.19}$$

and one from the axial vector current

$$m_{PS} f_{PS} = Z_A \langle 0|\bar{\psi}\gamma_0\gamma_5\psi|PS\rangle . \tag{13.20}$$

(We have inserted the Z factors so that the left hand sides of these expressions are the continuum values, which are scheme-independent because of PCAC.) Since $Z_P Z_m = 1$, Eq. (13.19) gives f_{PS} directly, which can be used in Eq. (13.20) to give Z_A in terms of the measured axial vector current matrix element.

13.2.2 *Regularization-independent scheme*

The most commonly used nonperturbative method for computing matching factors is called the "regularization independent" or RI scheme (Martinelli *et al.*, 1995). One computes quark and gluon Green's functions in a smooth gauge, regulated by giving all external lines a large Euclidean squared momentum p^2 and uses combinations of them to determine the Z's.

Let us specialize to bilinear fermionic operators and call the operator

$$O_\Gamma = \bar{\psi}\Gamma\psi, \tag{13.21}$$

where Γ could be a Dirac matrix or something more complicated. The matching factor is defined by computing the ratio of the matrix elements of the operator between off-shell single particle momentum eigenstates in the full and free theories

$$Z_\Gamma^{RI}(\mu)\langle p|O_\Gamma|p\rangle_{p^2=\mu^2} = \langle p|O_\Gamma|p\rangle_0 . \tag{13.22}$$

or, equivalently but more conveniently,

$$Z_\Gamma^{RI}(\mu)\frac{1}{12}\mathrm{Tr}\langle p|O_\Gamma|p\rangle_{p^2=\mu^2}\langle p|O_\Gamma|p\rangle_0^{-1} = 1, \tag{13.23}$$

where the factor of 1/12 counts Dirac spins and colors. The free matrix element $\langle p|O_\Gamma|p\rangle_0$ is just Γ after amputation, so the second equation is a useful projector. The matrix element of the operator is proportional to an amputated Green's function $\Lambda_\Gamma(p)$ containing the operator

$$\langle p|O_\Gamma|p\rangle = Z_Q^{RI}\Lambda_\Gamma(p) \tag{13.24}$$

where as always Z_Q^{RI} is the quark field renormalization factor in the regularization-independent scheme. Thus we determine $Z_\Gamma^{RI}(\mu)$ from

$$\frac{1}{12}Z_\Gamma^{RI}(\mu)Z_Q^{RI}(\mu)\mathrm{Tr}\Lambda_\Gamma(p)\Gamma|_{p^2=\mu^2} = 1. \tag{13.25}$$

There is no good way to compute the amputated Green's function directly on the lattice, so it is obtained from the unamputated Green's function $G_\Gamma(p)$, which contains incoming and outgoing quark propagators $S(p)$, via

$$\Lambda_\Gamma(p) = S^{-1}(p)G_\Gamma(p)S^{-1}(p). \tag{13.26}$$

Finally, there are two ways to define Z_Q. In the original RI scheme,

$$\frac{1}{Z_Q^{RI}} = \frac{-i}{12}\mathrm{Tr}\frac{\partial S^{-1}(p)}{\partial \not p}. \tag{13.27}$$

It is hard to take derivatives on the lattice, so in practice the RI' scheme is generally used instead: it defines Z_Q from a projection against the free propagator $S_0(p)$:

$$Z_Q^{RI'} = \frac{1}{12}\mathrm{Tr}S(p)S_0^{-1}(p). \tag{13.28}$$

In a confining theory, interpreting any gauge-dependent correlator in conventional field-theoretic language is risky. Confinement together with the necessary gauge fixing spoils the canonical Lehmann-Symanzik-Zimmermann reduction of the asymptotic in and out states. One hopes that in some class of smooth gauges the results are unambiguous. As a further precaution, one avoids small p where confinement is certainly important, and, of course, one avoids large pa where discretization artifacts appear. Thus to measure the Z factor with this method, one requires a plateau in the observable versus p^2, where the coupling is small enough that one may at least hope that the lattice result gives a good approximation to a perturbative expansion to some useful, nontrivial order.

How do we implement this procedure in a simulation? We begin by gauge fixing the links to Landau gauge (by maximizing $\sum_\mu \mathrm{Tr}\,\mathrm{Re} U_\mu$). Then there are at least two ways to get G_O. The first way is to compute the correlator

$$\begin{aligned} G_\Gamma(x,y) &= \langle \bar{\psi}(x) O_\Gamma(0) \psi(y) \rangle \\ &= \frac{1}{N} \sum_{i=1}^{N} S_i(x,0) \Gamma S_i(0,y) \end{aligned} \tag{13.29}$$

averaged over N configurations. Here $S(x,0)$ is the quark propagator from a point source at the origin. Using $S_i(x,y) = \gamma_5 S(y,x)^\dagger \gamma_5$, we have

$$\begin{aligned} G_\Gamma(p) &= \int d^4x\, d^4y\, e^{-ip(x-y)} G_\Gamma(x,y) \\ &= \frac{1}{N} \sum_{i=1}^{N} \int d^4x\, S_i(x,0) e^{-ipx} \Gamma \int d^4y\, \gamma_5 S_i(y,0)^\dagger \gamma_5 e^{ipy} \\ &= \frac{1}{N} \sum_{i=1}^{N} S_i(p) \Gamma \gamma_5 S_i(p)^\dagger \gamma_5. \end{aligned} \tag{13.30}$$

One also needs the average propagator, which we compute by Fourier transforming $S(x,0)$ configuration by configuration,

$$S(p) = \frac{1}{N} \sum_{i=1}^{N} S_i(p). \tag{13.31}$$

This is a matrix at every value of p, which we invert to construct the amputated Green's function

$$\Lambda_\Gamma(p) \equiv S(p)^{-1} G_\Gamma(p) S^{-1}(p), \tag{13.32}$$

which is then traced in Eq. (13.25).

The method just described produces G_Γ at all values of p, but the signal may not be as clean as one wants. A second method (Gockeler *et al.*, 1999) produces G_Γ one momentum at a time, but it is claimed to have a much better signal-to-noise ratio. One begins with the operator at site z defined as $O_\Gamma(z) = \bar{\psi}(z) \Gamma \psi(z)$ and computes the volume average

$$G_\Gamma(p) = \frac{1}{V} \sum_{xyz} e^{-ip(x-y)} S(x,z) \Gamma S(z,y), \tag{13.33}$$

which after a straightforward passage of steps is

$$G_\Gamma(p) = \frac{1}{V}\sum_z \gamma_5 \mathcal{G}(z;p)^\dagger \gamma_5 \Gamma \mathcal{G}(y;p), \tag{13.34}$$

where

$$\mathcal{G}(z;p) = \sum_x S(z,x)e^{ipx}. \tag{13.35}$$

To find the Fourier-transformed propagators, we recall that S is the inverse of the Dirac operator D, $\sum_z D(y,z)S(z,x) = \delta_{y,x}$, so we solve with a plane wave source

$$\sum_z D(y,z)\mathcal{G}(z;p) = \exp(ipy). \tag{13.36}$$

Either way, an entirely numerical simulation produces $Z_O[\mu a, g(a)]$ from a simulation at lattice spacing a with a bare coupling $g(a)$. In a typical analysis, one would produce a plot of $Z_O[\mu a, g(a)]$ as a function of $p^2 = \mu^2$. The high momentum part of the plot would presumably be contaminated by lattice artifacts. The low momentum part would presumably be sensitive to nonperturbative physics. For example, the low-momentum pseudoscalar current ($\Gamma = \gamma_5$) couples to the pion, which is not part of the spectrum that we expected to see when we matched free field behavior to perturbation theory. The pion propagator would produce a $1/q^2$ contribution, where $q = 2p$ from the kinematics of the Green's function. It would have to be subtracted. If the calculation were successful, one might hope to see either a plateau versus μ^2 for the matching factors of operators that do not have anomalous dimensions (the vector or axial vector currents), or some smooth logarithmic dependence on μ for operators with an anomalous dimension.

In continuum phenomenology the $\overline{MS}$ regularization scheme is popular, not *RI* or *RI'*. What we have done gives us a matrix element in one of these schemes from a lattice calculation of the matrix element,

$$\langle\alpha|O_\Gamma|\beta\rangle_{RI'} = Z^{RI'}\langle\alpha|O_\Gamma|\beta\rangle_{\text{lattice}}. \tag{13.37}$$

To convert to $\overline{MS}$, we need

$$\mathcal{Z}(\mu,a)_{\overline{MS},RI'} = \frac{Z_\Gamma^{\overline{MS}}(\mu)}{Z_\Gamma^{RI'}(\mu,a)} \tag{13.38}$$

and we compute this ratio in perturbation theory. Three loop formulas are available for the conversion (Franco and Lubicz, 1998; Chetyrkin and

Retey, 2000).

13.2.3 *Schrödinger functional methods*

The partition function in the Feynman path integral, $\mathrm{Tr}\exp(-LH)$ (for inverse temperature L), requires a bosonic periodic or fermionic antiperiodic boundary condition in Euclidean time to generate the trace (Ch. 3). The related "Schrödinger functional" $Z(\phi_b, \phi_a) = \langle\phi_b|\exp(-LH)|\phi_a\rangle$ instead has fixed boundary conditions at time 0 and L, specified by the states ϕ_a and ϕ_b. Expectation values of Schrödinger functional observables are determined in a numerical simulation in the same manner as with the partition function.

The combination of the "Schrödinger functional method" and renormalization group can produce matching factors or determine a running coupling constant over a large range of scales. This is accomplished by measuring the dependence of Schrödinger functional observables on the box size L.

We illustrate the method by sketching the determination of the running coupling $\alpha(q)$ of QCD (Lüscher *et al.*, 1992, 1993, 1994), although it was first introduced for the $d = 2$ sigma model (Lüscher *et al.*, 1991). The basic idea is to define a coupling through the response of a system in a box of size L to its environment

$$\alpha(q) = \frac{g^2(L)}{4\pi}; \qquad q = \frac{1}{L}. \tag{13.39}$$

In the gauge theory this is done by fixing the value of the spatial link variables on the faces of the box at Euclidean time $t = 0$ and $t = L$. This fixing involves a free parameter η. Call the resulting Schrödinger functional $Z(\eta)$. The coupling is then defined through the variation of the effective action Γ, the negative logarithm of the partition function $Z(\eta)$: $\Gamma = -\ln Z(\eta)$. In lowest order perturbation theory, Γ is equal to the classical action, which can be computed since the link variables simply interpolate between their boundary values. At this order the action is proportional to the inverse squared bare coupling $1/g_0^2$. One then defines the renormalized coupling $g^2(L)$ through

$$\frac{\partial\Gamma}{\partial\eta} = \frac{k}{g^2(L)}, \tag{13.40}$$

where the constant k is adjusted so that $g^2(L) = g_0^2$ in lowest order. Observe that $\frac{\partial\Gamma}{\partial\eta}$ is an expectation value of plaquettes on the boundaries.

Now we take take the beta function

$$\beta(g) = -L\frac{dg^2}{dL}, \qquad (13.41)$$

stretch the length of the compact dimension from L_0 to sL_0, and compute the change in the coupling

$$\int_{L_0}^{sL_0} \frac{dL}{L} = \int_{g^2(L_0)}^{g^2(sL_0)} \frac{dg^2}{\beta(g^2)} \equiv \int_u^{\sigma(s,u)} \frac{dv}{\beta(v)}, \qquad (13.42)$$

where the "step scaling function", $\sigma[s, u = g^2(L)] = g^2(sL)$, is the new coupling constant.

The running coupling is found by doing simulations with the same bare coupling on systems of size L_0 and sL_0, and, by measuring u and $\sigma(s,u)$, to see how the new coupling depends on the original one. An example (Lüscher *et al.*, 1993) of how this works for $SU(2)$ pure gauge theory is shown in Table 13.1. Each horizontal pair are the coupling and its step function (for $s = 2$). The output coupling on the first line is used as the input coupling on the second line (or rather, since these couplings are derived, not bare, couplings, one must interpolate calculations done with different bare couplings to begin with the old output coupling as the new input coupling). When the matching of couplings to set the scales is not perfect, perturbation theory is used for the small amount of running that is required. After four steps the physical length scale has increased by a factor of $2^4 = 16$.

Now we can unfold the couplings. We measure distances in terms of the largest L in the simulation and show the coupling constant in Table 13.2.

Table 13.1 Data for the running coupling constant in pure SU(2) gauge theory.

u	$\sigma(2,u)$
2.037	2.45(4)
2.380	2.84(6)
2.840	3.54(8)
3.55	4.76

We can introduce a scale in GeV by measuring something physical (a mass, or the string tension) with the bare parameters corresponding to one of the L's. An example of a running coupling constant determined via this prescription is shown in Fig. 13.1. This is a coupling constant

Table 13.2 The Schrödinger functional running coupling constant in pure SU(2) gauge theory.

L/L_{max}	$g^2(L)$
1.0	4.765
0.500(23)	3.55
0.249(19)	2.840
0.124(13)	2.380
0.070(8)	2.037

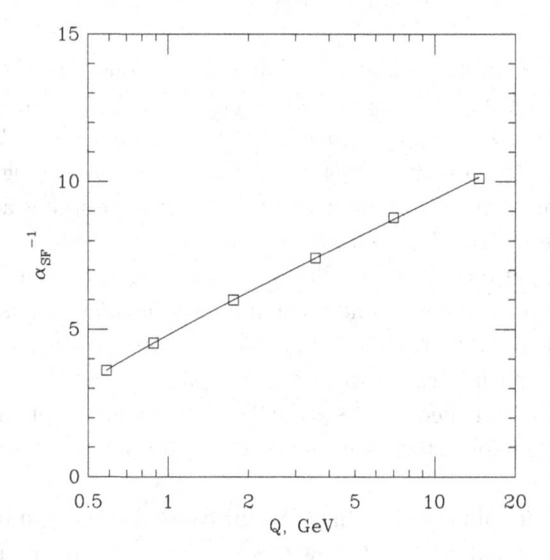

Fig. 13.1 The pure gauge $SU(3)$ coupling constant from the Schrödinger functional method of Lüscher *et al.* (1994), with superimposed three-loop prediction.

in a particular prescription; when it is small enough it could be matched to any other prescription using perturbation theory. All the simulations are done in small lattice volumes. This calculation checks perturbation theory; it does not use it overtly (apart from the small amount of running we described).

Schrödinger functional methods can be applied to find operator matching factors. Usually the calculation combines a simultaneous construction of an improved operator with a determination of the Z factor. For a recent example involving the strange quark mass, see Della Morte *et al.* (2005a,c).

Chapter 14

Chiral symmetry and lattice simulations

14.1 Minimal introduction to chiral perturbation theory

We return to a discussion of chiral effective theories for the low energy behavior of QCD. This is a vast subject (probably bigger than all of lattice gauge theory) and so we cannot begin to do it justice. For an introduction, see Scherer (2002). Our goal is more modest: to explore connections between lattice calculations and the predictions of chiral Lagrangians. These connections play a key role in chiral extrapolations, *i.e.* the extrapolation of results from simulations at unphysical values of the up and down quarks to yield predictions at their physical values. If one could do simulations at physical masses, they could also provide functional forms to fit correlation functions involving light quarks. Chiral models are available for several variants of lattice QCD, including quenched and partially quenched QCD and the staggered fermion formulation with its extra taste degree of freedom.

As far as the chiral Lagrangians are concerned, their parameters are just bare parameters. As far as QCD is concerned, the parameters of chiral Lagrangians are quantities that can be determined from a suitable simulation. Then the comparison of lattice simulations of QCD with experiment can go through an intermediate stage: First, determine the parameters of the chiral Lagrangian, and then use it to make predictions.

The low energy effective theory is constructed with all the symmetries and degrees of freedom of QCD. They are organized as an expansion in powers of the meson momenta and masses. The parameters of the theory include the quark masses, which are matched to QCD quark masses, and a set of other low energy constants, starting with the chiral condensate and pseudoscalar decay constants. At any particular order in the expansion parameter, it is not altogether clear in advance, what is the range of quark masses or momenta over which the predictions of chiral Lagrangian are reliable. It is first necessary to know what the predictions are, then to try to confront lattice data with them. When the lattice results fit the functional form predicted by the chiral Lagrangian, its low energy constants can be determined. For this purpose it is not necessary to simulate at the physical values of the quark masses.

A classic example of a chiral prediction is the functional dependence of the meson mass on the quark mass. Here we sketch the derivation up to next-to-leading order and in the process show how chiral perturbative power counting is organized.

As in Ch. 2, we introduce the Goldstone boson fields π^k through the $N_f \times N_f$ matrix field

$$U = \exp\left(\frac{i}{f}\sum_{k=1}^{N_f^2-1} \pi^k(x)\tau_k\right), \tag{14.1}$$

where the τ_k are the generators of the flavor group, and f is the tree-level coupling to the axial vector current, $A^i_\mu = -f\partial_\mu \pi^i + \ldots$. The leading-order effective Lagrangian is

$$\mathcal{L}_{\text{eff}} = \frac{f^2}{4}\text{Tr}(\partial_\mu U \partial^\mu U^\dagger). \tag{14.2}$$

Expanding U in powers of the π field, we find

$$\mathcal{L}_{\text{eff}} = \frac{1}{2}\partial_\mu \vec{\pi}\cdot\partial^\mu\vec{\pi} + \frac{1}{48f^2}\text{Tr}([\partial_\mu\pi,\pi][\partial^\mu\pi,\pi]) + \ldots. \tag{14.3}$$

where $\pi = \vec{\pi}\cdot\vec{\tau}$. The (global) flavor symmetry transformation is $U \to V_L U V_R^\dagger$. Quark masses explicitly break this symmetry; if parity is conserved and the quark mass matrix $\mathcal{M}$ is assumed to be real and diagonal, symmetry breaking is encoded in the term

$$\mathcal{L}_{\text{sb}} = \frac{f^2 B}{2}\text{Tr}[\mathcal{M}(U + U^\dagger)], \tag{14.4}$$

which can be expressed in terms of the Goldstone boson fields as

$$\mathcal{L}_{\text{sb}} = f^2 B\,\text{Tr}\mathcal{M} - \frac{B}{2}\text{Tr}[\mathcal{M}\pi^2] + \ldots. \tag{14.5}$$

The first term shifts the vacuum energy, and the second term generates a squared mass for the Goldstone boson that is proportional to the quark mass. For $N_f = 3$, and in the limit $m_u = m_d = m_\ell$,

$$M_\pi^2 = 2m_\ell B; \qquad M_K^2 = 2B(m_\ell + m_s); \qquad M_\eta^2 = 2B\left(\frac{1}{3}m_\ell + \frac{2}{3}m_s\right). \tag{14.6}$$

In QCD the derivative of the Hamiltonian with respect to a quark mass gives the operator $\bar{q}q$ for that quark. Evaluating this derivative with the first term in Eq. (14.5) gives

$$\langle\bar{q}q\rangle = -f^2 B, \tag{14.7}$$

and when this relation is combined with Eq. (14.6), we arrive (again) at the Gell-Mann, Oakes, Renner relation, which for the pion and kaon is

$$f^2 M^2 = -(m_i + m_j)\langle\bar{q}q\rangle + O(\mathcal{M}^2). \tag{14.8}$$

The sum $\mathcal{L}_{\text{eff}} + \mathcal{L}_{\text{sb}}$ is only the leading term in an expansion of the effective Lagrangian in a double series in powers of derivatives and powers of the quark mass from the symmetry breaking term. We can unite the two expansions by realizing that the matrix elements we will compute are to be found at momenta $p^2 \sim (m_i + m_j)B$. This means that we can count quark masses as two powers of the momenta in a derivative expansion. The leading terms are then both order p^2 contributions. If we denote the power of momentum by a superscript, the full effective Lagrangian is the series (Gasser and Leutwyler, 1985)

$$\mathcal{L}_{\text{eff}} = \mathcal{L}^{(2)} + \mathcal{L}^{(4)} + \mathcal{L}^{(6)} + \ldots \tag{14.9}$$

The term $\mathcal{L}^{(4)}$ contains all terms with four powers of derivatives, or two derivatives and a mass term, or terms quadratic in the quark mass. It consists of the following invariants:

$$\begin{aligned}
P_1 &= \langle\partial_\mu U \partial^\mu U^\dagger\rangle^2 \\
P_2 &= \langle\partial_\mu U \partial_\nu U^\dagger\rangle\langle\partial^\mu U \partial^\nu U^\dagger\rangle \\
P_3 &= \langle\partial_\mu U \partial^\mu U^\dagger \partial_\nu U \partial^\nu U^\dagger\rangle \\
P_4 &= \langle\partial_\mu U \partial^\mu U^\dagger\rangle\langle\chi U^\dagger + U\chi^\dagger\rangle \\
P_5 &= \langle\partial_\mu U \partial^\mu U^\dagger(\chi U^\dagger + U\chi^\dagger)\rangle \\
P_6 &= \langle\chi U^\dagger + U\chi^\dagger\rangle^2 \\
P_7 &= \langle\chi U^\dagger - U\chi^\dagger\rangle^2 \\
P_8 &= \langle\chi U^\dagger \chi U^\dagger + U\chi^\dagger U\chi^\dagger\rangle \\
P_9 &= \langle\chi^\dagger\chi\rangle
\end{aligned} \tag{14.10}$$

Here the notation $\langle W \rangle$ stands for TrW and $\chi = 2B\mathcal{M}$.

If there are fewer than four flavors, not all of these invariants are independent. Experts will note that there is also a Wess-Zumino term, proportional to $\epsilon^{\mu\nu\rho\sigma}\partial_\mu U \partial_\nu U \partial_\rho U \partial_\sigma U$, which we will not consider further. The fourth-order Lagrangian is a linear combination of all these terms,

$$\mathcal{L}^{(4)} = \sum_i L_i P_i. \tag{14.11}$$

The leading order Lagrangian is characterized by two low energy constants (f and B), but in fourth order we have an additional nine coefficients $L_1, \ldots, L_9$, all dimensionless. The higher order $\mathcal{L}^{(n)}$'s have increasingly more coupling constants.

So far, we have been discussing only tree graphs. It is not correct to so limit our discussion, because keeping only the tree graphs in a scattering amplitude violates unitarity. (In fact, we need to keep all the Feynman graphs of a local field theory to satisfy unitarity.) However, we are interested in constructing scattering amplitudes order by order in p^2, and for each order we do not need an infinite set of graphs to satisfy unitarity. For example, graphs containing ℓ loops with vertices coming only from $\mathcal{L}^{(2)}$ scale as $p^{2+2\ell}$ (in four dimensions), and loop graphs containing terms from $\mathcal{L}^{(4)}$ and higher scale as higher powers of p. Thus to work to $O(p^4)$, we need consider only tree level graphs from $\mathcal{L}^{(4)}$ and one loop graphs from $\mathcal{L}^{(2)}$. The one loop graphs are divergent, but their divergent part can be absorbed into a renormalization of the coefficients L_i in $\mathcal{L}^{(4)}$.

Let us illustrate this procedure by sketching a calculation of the pion mass to order m^2 (or p^4). Let us take all the quark masses to be degenerate. We will call the lowest order result $M_1^2 = 2mB$. In next order, for tree graphs, we have to collect all the terms that are quadratic in the pion field. This excludes $P_1 - P_3$. The contributions P_4 and P_5 include terms proportional to $M_1^2 \partial_\mu \vec{\pi} \partial^\mu \vec{\pi}$. The contributions P_6 and P_8 include terms proportional to $M_1^4 \vec{\pi}^2$. Thus, the relevant part of the effective Lagrangian is

$$\begin{aligned}\mathcal{L}^{(2)} + \mathcal{L}^{(4)} &= \frac{1}{2}\partial_\mu \vec{\pi} \partial^\mu \vec{\pi} \left[1 + \frac{8M_1^2}{f^2}(N_f L_4 + L_5)\right] \\ &\quad -\frac{1}{2} M_1^2 \vec{\pi}^2 \left[1 + \frac{16 M_1^2}{f^2}(N_f L_6 + L_8)\right]. \end{aligned} \tag{14.12}$$

Next, we have the one-loop term. The vertex is proportional to $\mathrm{Tr}\{[\partial_\mu \pi, \pi][\partial^\mu \pi, \pi]\}$ [as we recall from Eq. (14.3)]. The one-loop graph is the usual single-bubble tadpole, proportional to derivatives of the propagator $\Delta(x = 0)$. To evaluate it, we use dimensional regularization (since it preserves the symmetry). Using

$$\partial_\mu \partial^\mu \Delta(x) = \delta(x) - M_1^2 \Delta(x) \tag{14.13}$$

plus the fact that $\delta(0) = 0$ in dimensional regularization, we combine the one-loop graph with the tree-level terms in Eq. (14.12) to find

$$M^2 = M_1^2 \left[1 - \frac{8M_1^2}{f^2}(N_f L_4 + L_5 - 2N_f L_6 - 2L_8) + \frac{1}{N_f f^2}\Delta(0) \right] \tag{14.14}$$

where the (dimensionally-regularized) propagator is

$$\begin{aligned} \Delta(0) &= \int \frac{d^D k}{(2\pi)^D} \frac{\mu^{D-4}}{k^2 + M_1^2} \\ &= -\frac{M_1^2}{16\pi^2}\left(\frac{1}{\epsilon} - \gamma_E + 1 + \ln 4\pi + \ln \frac{M_1^2}{\mu^2} \right) \end{aligned} \tag{14.15}$$

in $D = 4 - 2\epsilon$ dimensions. The divergence and constants can be absorbed into a renormalization of the L_i's. The mass is

$$M^2 = M_1^2 \left[1 - \frac{8M_1^2}{f^2}(N_f L_4^r + L_5^r - 2N_f L_6^r - 2L_8^r) + \frac{M_1^2}{16\pi^2 N_f f^2} \ln \frac{M_1^2}{\mu^2} \right]. \tag{14.16}$$

The formula shows that the expansion in powers of the quark mass is not analytic; instead, it contains a "chiral logarithm," proportional to $m_q^2 \log m_q$.

Chiral logarithms are ubiquitous in quantities computed in chiral perturbation theory. Notice that apart from the scale μ the coefficient of the chiral logarithm is completely determined by the order p^2 parameters of the theory (in contrast to the coefficient of M_1^4, which is a combination of the L_i's). This fact can be used by the lattice practitioner to test whether or not one's simulation is in the chiral regime by whether or not it exhibits chiral-logarithm behavior, with a known coefficient.

14.2 Quenching, partial quenching, and unquenching

In the quenched approximation the fermion determinant is simply set equal to unity in the functional integral, and only valence quark lines are retained in correlators. In partial quenching some part of the fermion determinant is kept, but retained sea quarks do not match the valence quarks in all respects (number and mass). A convenient form of partial quenching is in fact to compute hadronic properties for one set of sea quark masses (because each set is expensive) and many values of valence quark masses (because each set is cheap).

It has proven very useful to extend the low energy effective theory accordingly. The necessary modifications are interesting. For example, recall that in full QCD, because of the anomaly, the flavor singlet meson, the eta prime, is not a Goldstone boson. In the low energy limit it can be decoupled from the interactions of the ordinary Goldstone bosons, because it is massive and they are not. In quenched QCD the eta prime is not really a particle. The would-be eta prime gives rise to "hairpin insertions" that pollute essentially all predictions.

14.2.1 *The eta prime correlator*

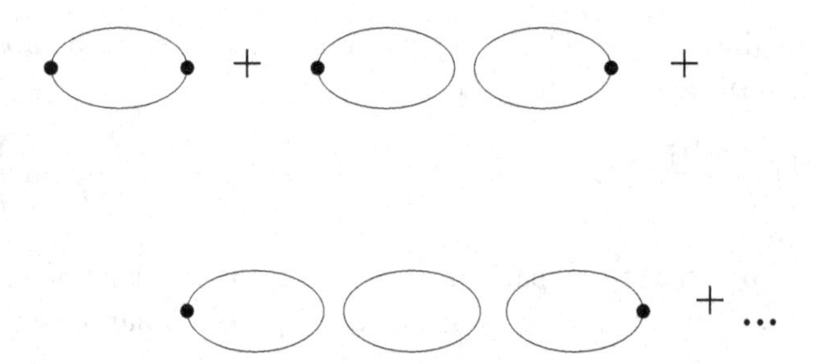

Fig. 14.1 The eta-prime propagator in terms of a set of annihilation graphs summing into a geometric series to shift the eta-prime mass away from the masses of the flavor non-singlet pseudoscalar mesons. In the quenched approximation, only the first two terms in the series survive as the "direct" and "hairpin" graphs.

Let us consider the eta prime channel in full QCD, partially quenched, and quenched QCD in various extreme limits. In ordinary QCD the eta prime propagator includes a series of terms in which the flavor singlet $q\bar{q}$ pair alternately annihilates into some quarkless state and reappears. This process is shown in Fig. 14.1. Coupling to the quarkless state in the pseudoscalar channel is associated with the flavor singlet axial anomaly, which has strength proportional to $1/N_c$ in the large N_c limit, as we recall from Sec. 2.7. Let us assume we have N_v degenerate valence quarks, N_s sea quarks of the same mass, and N_c colors. (The case $N_v \neq N_s$ with $N_s \neq 0$ is called "partial quenching", as opposed to the $N_s = 0$ quenched approximation.) Let us sum the geometric series for the eta prime propagator:

$$\eta'(q) = C(q) - H_0(q) + H_1(q) + \ldots, \tag{14.17}$$

where $C(q) = 1/d$ for $d = q^2 + M_\pi^2$ is the "connected" meson propagator, the same as for any other Goldstone boson. (We normalize the meson color and flavor singlet wave function to 1.) The term H_n is the nth order hairpin (with n internal fermion loops). Let us explore the flavor and color dependence of the hairpin contributions in the large N_c limit (with $g^2 N_c$ fixed). With the same normalization we find the lowest order hairpin is

$$H_0(q) = \frac{1}{d}\frac{N_v}{N_c}\lambda^2 \frac{1}{d} \tag{14.18}$$

where λ^2/N_c comes from the anomaly. Each extra loop gives another factor of $-N_s\lambda^2/(N_c d)$. Summing the geometric series, we find

$$\eta'(q) = \frac{1}{d}\left(1 - \frac{N_v}{N_s}\right) + \frac{N_v}{N_s}\frac{1}{d + \lambda^2 N_s/N_c}. \tag{14.19}$$

This formula has a number of interesting cases. First, if $N_v = N_s = N_f \neq 0$, we see that we have generated a massive η' propagator with $m_{\eta'}^2 = m_\pi^2 + \lambda^2 N_f/N_c$. Next, if $N_v < N_s$, we have both the Goldstone mode and the eta prime propagator. This result is also as expected. Suppose $N_s = 3$. In the $SU(3)$ flavor symmetry limit, there is an octet of Goldstone bosons. The source $\bar{u}u + \bar{d}d$ couples to a mixture of the Goldstone octet η and the singlet η'. We could take N_c to infinity at fixed N_v and N_s. With N_c-scaling of the vertex, the eta-prime mass falls to zero, and it becomes a ninth Goldstone boson as the anomaly disappears and $U(1)_A$ is restored. At any finite N_c, the η' is not a Goldstone boson.

The quenched limit $N_s = 0$ is different, however. In this case only the first two terms contribute to Eq. (14.17):

$$\eta'(q) = \frac{1}{d} - \frac{1}{d}\frac{N_v}{N_c}\lambda^2 \frac{1}{d}, \tag{14.20}$$

In the eta prime channel there is an ordinary (but flavor singlet) Goldstone boson and a new contribution–a double-pole ghost (negative norm) state. In the $N_c = \infty$ limit, the double pole decouples, but finite-N_c quenched QCD remains different from finite-N_c full QCD.

The double pole would also appear if the sea quarks and valence quarks had different masses. Its residue would be proportional to the difference of the sea and valence pseudoscalar squared masses.

The quenched double pole has dramatic consequences for chiral perturbation theory. We have seen that radiative corrections involve processes with internal Goldstone boson loops, contributing terms like

$$\int d^4k\, G(k,m) \sim \left(\frac{m}{4\pi}\right)^2 \log\left(\frac{m^2}{\Lambda^2}\right) \tag{14.21}$$

(plus cutoff effects). The eta-prime hairpin can appear in these loops, replacing $G(k,m) \to -G(k,m)\lambda^2 G(k,m)$ and altering the chiral logarithm. Thus, in a typical observable, with a small mass expansion

$$Q(m_{PS}) = A\left(1 + C\frac{m_{PS}^2}{f_{PS}^2}\log m_{PS}^2\right) + \ldots \tag{14.22}$$

quenched and $N_f = 3$ QCD can have different coefficients [different C's in Eq. (14.22)]. For example, it happens that quenched f_π has no chiral logarithm, whereas it does in full QCD. By contrast the coefficients of O_+, the operator measured for B_K, are identical in quenched and full QCD. The differences can go farther than mere coefficients. Even worse, one can find a different functional form. For example, the relation between pseudoscalar mass and quark mass in full QCD is given by Eq. (14.16). In quenched QCD, the analogous relation is

$$M^2 = M_1^2\{1 - \delta[\ln(M_1^2/\mu^2) + 1]\} + \ldots \tag{14.23}$$

where $\delta = \lambda^2/(8\pi^2 N_c f_\pi^2)$ is expected to be about 0.2, if we use the physical η' mass to fix λ^2. Note that the ratio of the squared pseudoscalar mass to the quark mass now diverges in the chiral limit.

14.2.2 *Chiral Lagrangian with partial or complete quenching*

Formulating a chiral Lagrangian for quenched or partially-quenched QCD presents an interesting problem. There are several solutions in the literature. It would take us too far from the lattice to describe any of them in detail, but one commonly seen one is the "supersymmetric" formulation of Morel (1987) and Bernard and Golterman (1992, 1994). If we have n fermions, of which k are quenched, the QCD action after integrating out the fermions would be

$$Z = \int [dU]\, e^{-S_g} \det(D+m)^{n-k} = \int [dU]\, e^{-S_g} \det(D+m)^n \det[(D+m)^{-1}]^k. \tag{14.24}$$

The same combination of determinants is produced by n quarks and k pseudoquarks, *i.e.*, bosonic spinor fields entering the Lagrangian through the same Dirac bilinear form. In that case, we could have bound states of two quarks, of two pseudoquarks, or of a quark and a pseudoquark. The symmetry transformation among these states is "almost" a $U(n+k) \times U(n+k)$ invariance, except that these rotations mix fermions and bosons, so what we have is called a "graded symmetry." The group is called $U(n|k)$ and we define our chiral fields

$$\Phi = \begin{pmatrix} \phi & \chi^\dagger \\ \chi & \tilde{\phi} \end{pmatrix}, \tag{14.25}$$

where ϕ is an $n \times n$ matrix (of ordinary numbers) representing ordinary mesons made of quarks and antiquarks, $\tilde{\phi}$ is a $k \times k$ matrix (again of ordinary numbers) representing bound states of the pseudoquarks, and χ is a $k \times n$ matrix of anticommuting numbers representing bound states of quarks and pseudoquarks. The chiral Lagrangian is constructed in terms of

$$\Sigma = \exp\left(i\frac{\Phi}{f}\right). \tag{14.26}$$

Because Σ contains both commuting and anticommuting numbers, the trace in the usual chiral Lagrangian is replaced by the "supertrace: If

$$U = \begin{pmatrix} A & C \\ D & B \end{pmatrix}, \tag{14.27}$$

then $\mathrm{Str}U = \mathrm{Tr}A - \mathrm{Tr}B$.

The η' is (again) a complication. The anomaly should break $U(n|k) \times U(n|k)$ to $SU(n|k)_L \times SU(n|k)_R \times U(1)$. The anomalous field is $\Phi_0 = (\eta' - \tilde{\eta}')/\sqrt{2}$ (the minus sign from the relative minus sign between fermion and boson loops). This field is proportional to $\mathrm{Str}\ln\Sigma$. We can add arbitrary functions of Φ_0 to our chiral Lagrangian, but the simplest choice is

$$\mathcal{L} = \frac{f^2}{4}\mathrm{Str}(\partial_\mu\Sigma\partial^\mu\Sigma) - \frac{f^2 B}{2}\mathrm{Str}\mathcal{M}(\Sigma + \Sigma^\dagger) + \frac{m_0^2}{6}\Phi_0^2. \tag{14.28}$$

The $\frac{m_0^2}{6}\Phi_0^2$ term generates a two-point coupling of the Goldstone bosons, the H_0 term of Eq. (14.17). Sometimes a $\frac{\alpha}{6}(\partial_\mu\Phi_0)^2$ term is added. The neutral meson propagators resemble the expressions we have derived from our bubble-sum example, with $m_0^2 + \alpha p^2$ playing the role of the parameter λ^2.

14.3 Chiral perturbation theory for staggered fermions

As we have seen in Ch. 6, staggered fermions suffer from residual fermion doublers. Each intentional quark flavor comes in a "taste" multiplet of four. So each $\bar{q}q$ meson comes in a multiplet of 16. The fourth-root approximation is used to suppress their contributions to the fermion determinant. Without taking the fourth root we have a well-defined, local field theory with four times the number of desired fermion species. The taste symmetry is broken at nonzero lattice spacing, but is restored to $SU(4)$ in the continuum limit. The interesting question is whether it is possible to write a low energy chiral Lagrangian for such a theory. The question was answered by Lee and Sharpe (2003) for a single flavor and by Aubin and Bernard (2003) for multiple flavors. This development is especially important, since it makes it possible to model small lattice artifacts in a low-energy effective theory, thus facilitating the extrapolation of low-energy lattice quantities simultaneously to physical quark masses and the continuum limit. The modified low energy theory is called staggered chiral perturbation theory (SXPT).

The most interesting question is whether there is a SXPT formulation that matches the rooted theory at low energies (Aubin and Bernard, 2003; Bernard, 2006). We first present SXPT without the fourth root and then comment about extensions to the rooted theory.

14.3.1 *Staggered chiral perturbation theory*

It is beyond our scope to give details of the Lee and Sharpe (2003) and Aubin and Bernard (2003) construction of the the SXPT Lagrangian. Essentially, it proceeds in two steps. First they analyzed the lattice symmetries of the staggered fermion action. Then, following Symanzik, they proceeded to construct an intermediate continuum action involving only quarks with additional four-fermi interaction terms representing corrections to $\mathcal{O}(a^2)$ in the lattice spacing. These most general additional terms were designed to break the conventional continuum symmetry to that of the lattice action. Finally, they made a translation from quark language to the language of chiral fields. The resulting chiral Lagrangian models the taste-symmetry breaking to $\mathcal{O}(a^2)$.

The chiral field Φ appears as usual in the Lagrangian as $\Sigma = \exp(i\Phi/f)$. But because of the additional taste degree of freedom, for N flavors, it has the expansion

$$\Phi = \sum_{a=1}^{16} t_a \phi^a \tag{14.29}$$

where t_a are generators of the taste group and $\phi^a_{ff'}$ is the chiral field corresponding to the quark flavors $\bar{f}f'$ for meson taste a. For the $SU(2)$ flavor of the previous sections $\phi^a = \pi^{ak}\tau_k$. The taste multiplet is conveniently labeled in the Dirac basis $t_a \in \{\xi_5, \xi_\mu, i\xi_{\mu 5} = i\xi_\mu\xi_5, i\xi_{\mu\nu} = i\xi_\mu\xi_\nu, \xi_I = I\}$ or, sometimes labeled more succinctly as P, V, A, T, S. (To compensate for this normalization, extra factors of 1/4 appear with traces.) The Euclidean Lagrangian is then

$$\mathcal{L} = \frac{f^2}{16}\mathrm{Tr}\left(\partial_\mu\Sigma^\dagger\partial_\mu\Sigma\right) - \frac{Bf^2}{8}\mathrm{Tr}[\mathcal{M}(\Sigma^\dagger + \Sigma)] + \frac{m_0^2}{6}\phi_{0I}^2 + a^2\mathcal{V}(\Sigma) \tag{14.30}$$

where the taste-symmetry-breaking term

$$-\mathcal{V}(\Sigma) = \sum C_i \mathcal{O}_i \tag{14.31}$$

is a linear combination of the operators

$$\begin{aligned}
\mathcal{O}_1 &= \mathrm{Tr}\left(t_5\Sigma t_5\Sigma^\dagger\right)\\
\mathcal{O}_{2V} &= \frac{1}{4}\left[\mathrm{Tr}(t_\mu\Sigma)\mathrm{Tr}(t_\mu\Sigma) + \mathrm{h.c.}\right]\\
\mathcal{O}_{2A} &= \frac{1}{4}\left[\mathrm{Tr}(t_{\mu 5}\Sigma)\mathrm{Tr}(t_{5\mu}\Sigma) + \mathrm{h.c.}\right]\\
\mathcal{O}_3 &= \frac{1}{2}\left[\mathrm{Tr}(t_\mu\Sigma t_\mu\Sigma) + \mathrm{h.c.}\right] \qquad (14.32)\\
\mathcal{O}_4 &= \frac{1}{2}\left[\mathrm{Tr}(t_{\mu 5}\Sigma t_{5\mu}\Sigma) + \mathrm{h.c.}\right]\\
\mathcal{O}_{5V} &= \frac{1}{2}\left[\mathrm{Tr}(t_\mu\Sigma)\mathrm{Tr}(t_\mu\Sigma^\dagger)\right]\\
\mathcal{O}_{5A} &= \frac{1}{2}\left[\mathrm{Tr}(t_{\mu 5}\Sigma)\mathrm{Tr}(t_{5\mu}\Sigma^\dagger)\right]\\
\mathcal{O}_6 &= \sum_{\mu<\nu}\mathrm{Tr}\left(t_{\mu\nu}\Sigma t_{\nu\mu}\Sigma^\dagger\right) \ . \qquad (14.33)
\end{aligned}$$

The mass matrix $\mathcal{M}$ is diagonal and contains the quark masses. Note that the anomaly term in m_0^2 is a taste singlet as well as a flavor singlet.

Let us specialize to three flavors (u, d, and s) with $m_u = m_d = m_\ell \neq m_s$ and consider the spectrum of pseudoscalar mesons in this theory.

At zero lattice spacing the flavor nonsinglet meson masses are the usual ones from chiral perturbation theory.

$$
\begin{aligned}
m_\pi^2 &= 2Bm_\ell \\
m_K^2 &= B(m_\ell + m_s)
\end{aligned}
\tag{14.34}
$$

Recall, however, that both of these species appear with taste multiplicity of 16.

The flavor singlet mesons must be treated with special care. Because the anomaly is a taste singlet, the flavor-singlet taste multiplets are split, even at zero lattice spacing. The taste singlet receives the usual contribution from the anomaly and behaves in the usual way. With a large anomaly mass m_0^2 we have

$$
m_\eta^2 = B(m_\ell + 2m_s)/3 \tag{14.35}
$$

$$
m_{\eta'}^2 = m_0^2, \tag{14.36}
$$

but the remaining 15 taste nonsinglet members do not see the anomaly and have quite different masses,

$$
m_\eta^2 = 2Bm_\ell \tag{14.37}
$$

$$
m_{\eta'}^2 = 2Bm_s. \tag{14.38}
$$

(Nonetheless, we choose to call them η and η', since they are the lightest two flavor-singlet pseudoscalar mesons.) These additional states appear as a consequence of making taste a physical quantum number.

At a small nonzero lattice spacing the taste multiplets are split further. The splitting is independent of the flavor assignment, depending only on taste. It is not difficult to calculate the splitting from the taste-breaking term $a^2\mathcal{V}$. There are two steps in the process. First, there are diagonal terms in the mass matrix whose contributions we write down immediately. For taste a it is

$$
\Delta m_a^2 = a^2 \Delta_a, \tag{14.39}
$$

where

$$\Delta_P = 0 \tag{14.40}$$

$$\Delta_V = \frac{16}{f^2}(C_1 + C_3 + 3C_4 + 3C_6) \tag{14.41}$$

$$\Delta_A = \frac{16}{f^2}(C_1 + 3C_3 + C_4 + 3C_6) \tag{14.42}$$

$$\Delta_T = \frac{16}{f^2}(2C_3 + 2C_4 + 4C_6) \tag{14.43}$$

$$\Delta_I = \frac{16}{f^2}(4C_3 + 4C_4) \; . \tag{14.44}$$

Next, there are axial-taste and vector-taste hairpin diagrams like the anomaly-generated singlet-taste hairpin. One mixes the η_A and η'_A, and the other, the η_V and η'_V. This mixing is parameterized by

$$\delta'_A = 16(C_{2A} - C_{5A})/f^2 \tag{14.45}$$

$$\delta'_V = 16(C_{2V} - C_{5V})/f^2. \tag{14.46}$$

From simulations these mixings are found to be fairly small. We see that the pseudoscalar-taste member is protected from additive mass renormalization by the remnant chiral symmetry of the theory. The others are split, but not completely. Some degeneracy remains. We see, also, that most of the low-energy constants of the SXPT Lagrangian can be determined immediately from a measurement of the meson mass spectrum.

14.3.2 *Rooted staggered chiral perturbation theory*

To suppress the doublers in the fermion determinant, we take its fourth root. What is the appropriate SXPT Lagrangian, then? As far as we know, there is no comparably rigorous mapping of rooted lattice QCD to a chiral theory. But there is a rule of thumb that seems to work in all cases we know of. It is called the "replica" trick (Damgaard and Splittorff, 2000; Aubin and Bernard, 2004). Instead of taking a fourth root, we take an arbitrary integer power n_r of the determinant. This procedure is tantamount to increasing the multiplicity of each fermion flavor by a factor of n_r. It is trivial to generalize the SXPT Lagrangian to this case. The rule then proceeds as follows. Calculate any quantity of interest in replicated SXPT, keeping n_r as a parameter. In the final result set $n_r = 1/4$. In Green's functions some restrictions on acceptable external hadrons must also be imposed.

As a nontrivial example of how this trick works, consider the a_0 correlator generated by the flavor nonsinglet combination $\bar{u}d$. This is a scalar meson ($J^P = 0^+$). We consider only the taste singlet. The correlator for this meson receives contributions from a pair of intermediate pseudoscalar mesons. Indeed, the physical $a_0(980)$ meson is found predominantly in the $\eta\pi$ channel but also seen in $K\bar{K}$. In the taste-singlet channel, these intermediate mesons may have any taste, but the tastes are paired. That is, any pion taste is allowed, but it must be paired with an eta of the same taste. In view of our discussion of the masses of the eta taste multiplets, we see a potential disaster. The taste nonsinglet etas are degenerate (or nearly so in the case of the vector and axial tastes) with the pion. So it appears that there is a series of unphysical, low-lying two-particle intermediate states at approximately twice the mass of each member of the pion taste multiplet, a possibility first pointed out by Prelovsek (2006). If n_r is an integer, as we have noted above, we accept them as the expected consequence of an enlarged flavor group. But when we set $n_r = 1/4$ in addition to the unwanted low-lying two-particle intermediate states, we find a ghost threshold with a negative weight, where we had hoped to find only states that appear in nature. The disaster is averted, however, when we take the continuum limit. A detailed analysis of the two-particle states shows that precisely when $n_r = 1/4$ all the unwanted intermediate thresholds cancel, and we are left with only the physical taste-singlet $\eta\pi$ and $K\bar{K}$ channels. Still, at nonzero lattice spacing, these unphysical channels appear, and their contributions are apparently seen in simulations (Aubin *et al.*, 2004; Gregory *et al.*, 2005).

In this example we see that contributions to SXPT diagrams from internal meson loops seem to be as desired. Clearly, to preserve unitarity, we do not put states on external lines that do not also appear internally in the limit of zero lattice spacing. This exclusion applies to the taste nonsinglet etas, for example.

These considerations have led some practitioners to speculate that taking the fourth root at a nonzero lattice spacing is no more perilous than partial quenching with a mismatch of valence and sea quark masses (Bernard, 2006).

14.4 Computing topological charge

We conclude this chapter with a discussion of the measurement of topological charge or winding number ν on the lattice. Particularly, the related topological susceptibility $\langle \nu^2 \rangle / V$ can be predicted from chiral models at small quark mass. In that regime it can be shown in leading order chiral perturbation theory that the susceptibility obeys

$$\chi_{\rm top} = \frac{\langle \nu^2 \rangle}{V} = \frac{\Sigma}{(1/|m_u| + 1/|m_d| + 1/|m_s|)} \tag{14.47}$$

in terms of the quark masses and chiral condensate (Leutwyler and Smilga, 1992). That is, as any quark mass vanishes, topological fluctuations are suppressed, as we remarked in Sec. 2.6.3. It is important to test this prediction in lattice simulations.

Topological quantities on a lattice are always ambiguous because the lattice is not a manifold. For a given gauge configuration, it is necessary to specify an interpolation between neighboring lattice sites, usually coupled with a smoothing process, in order to establish a definite result.

A distinction is usually made between semiclassical gauge field configurations that might or might not be instanton-like and ultraviolet fluctuations. Unfortunately, in lattice QCD, as in any field theory, the dominant feature of the QCD vacuum is ultraviolet fluctuations, which diverge as a large power of the inverse lattice spacing. There is no clear cut separation of a "fluctuation" from a "small instanton." Nevertheless, people try. We will examine how this is done with the various actions.

We begin by reviewing the implications of the index theorem that connects topological charge and fermionic zero modes. Recall that in continuum QCD the flavor singlet axial current is not conserved,

$$\partial_\mu J^{\mu 5} = -N_f Q_T(x), \tag{14.48}$$

where the right hand side is the topological charge density

$$Q_T(x) = \frac{g^2}{16\pi^2} \epsilon^{\alpha\beta\mu\nu} F^a_{\alpha\beta}(x) F^a_{\mu\nu}(x). \tag{14.49}$$

The integral of Eq. (14.48) gives the change in axial charge over the time extent of integration:

$$\Delta Q_5 = \int dt \, \frac{dQ}{dt} = \int d^4x \, \partial_\mu J^{\mu 5} = -N_f \int d^4x \, Q_T(x). \tag{14.50}$$

The integral takes on integer values, the winding number or topological charge of the gauge configuration ν. It is related to the number of fermionic zero modes, which have definite chirality (eigenstates of γ_5) with eigenvalue ± 1, through the Atiyah-Singer "index theorem," $\nu = n_- - n_+$, where $n_\pm$ counts the number of zero modes of positive or negative chirality. The presence of zero modes means that gauge configurations with nonzero ν are suppressed in the functional integral for QCD with n_f flavors by a factor $m_q^{|\nu| n_f}$. This is the QCD origin of the suppression predicted by chiral models in Eq. (14.47).

Chiral lattice fermions capture much of this physics at nonzero lattice spacing, but not all of it: there are anomalous Ward identities and chiral zero modes, but the gauge configuration whose integral gives ν only becomes the $Q_T(x)$ of Eq. (14.49) for smooth gauge fields. Actions without a full $SU(N_f) \times SU(N_f)$ chiral symmetry do not have any of the connections of chirality, zero modes, and winding number as exact relations, but generally do realize them approximately.

If we generate artificial gauge field configurations that are lattice discretized versions of the classical self-dual instanton solution and compute the spectrum of Dirac eigenmodes, we discover that a generic non-chiral lattice Dirac operator has a small Dirac eigenvalue, the corresponding eigenfunction of which has an expectation value of γ_5 ("chirality") approaching ± 1. Instanton solutions are parameterized by a size, r_0, and as $r_0/a \gg 1$ the solutions become more and more "chiral" (γ_5 expectation value closer to ± 1) and the Dirac eigenvalue approaches zero. As r_0/a falls to be $O(1)$ the would-be zero eigenvalue of a Wilson-type fermion moves away from zero (while remaining real) and the chirality of the mode drops. At some value of r_0/a the real mode typically becomes degenerate with another real mode and both disappear from the spectrum, replaced by a complex-conjugate pair. The eigenvalues of the Dirac operator for staggered fermions are purely imaginary. For large r_0/a, four low eigenvalues cluster near zero. As r_0/a falls, the eigenmodes move away from zero and become less degenerate and less chiral. Chiral lattice fermions (such as the overlap) have exact zero modes that are chiral for all r_0 greater than some minimum value, at which their spectrum changes discontinuously.

Different fermionic actions – even different chiral fermionic actions – can disagree over the topological charge of a gauge configuration because they have different cutoff effects. Thinking about our lattice discretized instanton configuration, the difference will occur for excitations of the gauge field whose size variation is $\delta r_0 \sim a$. It is believed that the disagreement of different actions is a cutoff effect, vanishing as $a \to 0$.

With chiral actions, it is customary to turn the index theorem around and to define the topological charge of a configuration through the difference of positive and negative chirality zero modes. This choice certainly has the correct naive continuum limit. At finite a, it is arbitrary, but it does have the advantage that the fermionic functional integral suppresses topology proportional to $|m_q|^{n_f|\nu|}$ at all a. This will probably not be the case for any action without full $SU(N_f) \times SU(N_f)$ chiral symmetry. It is also unlikely to occur in partially-quenched simulations, when the valence and sea quarks are described with different discretizations. Of course, chiral lattice predictions require taking the $a \to 0$ limit, and it becomes a practical question, whether undesirable behavior at nonzero a contaminates this limit.

One could, alternatively, define topology through a quantity involving only the lattice gauge fields. A popular choice is a discrete version of the topological charge density Eq. (14.49). The appropriate closed path of gauge links runs around two plaquettes connected to the site x. The plaquettes are aligned in the disjoint $\alpha\beta$ and $\mu\nu$ planes. With a suitable linear combination of orientations of the plaquettes, one obtains an operator that reduces to Eq. (14.49) in the continuum limit. Improved operators include paths other than plaquettes (DeGrand *et al.*, 1997).

The winding number can also be determined geometrically starting from a smooth interpolation of the gauge field across each hypercube (Phillips and Stone, 1986). The interpolation is then used to count the number of times and with which orientation, positive or negative, the map crosses a selected point in the group manifold. The net crossing over all hypercubes is the winding number.

It is customary with either of these definitions first to smooth out the ultraviolet fluctuations. There are two methods in common use. One of them involves filtering the gauge fields. One can use fat links as local averaging operators, and measure observables made of them. In this case the filtering process can be viewed as merely an extension of the definition of the topological charge density operator. As an alternative, one can quench the gauge fields by re-equilibrating the configuration at a very large β value. (This procedure is called "cooling.") The idea is that UV fluctuations will equilibrate quickly, but longer range structures would persist.

None of these definitions is unique. One expects that all sensible definitions agree for instantons much larger than the lattice scale [see, for example, Cundy *et al.* (2002)]. But for smaller structures the result depends on the smoothing process. Even the Dirac zero mode method, in effect, smooths the gauge field through the overlap Dirac operator. The instantons are the objects that the fermions can "see" as instantons. Depending on the construction of the operator, they may not see small structures. The other methods count instantons that those operators can see. The principal advantage of dynamical chiral theories is that "instantons" suppressed by the chiral fermionic determinant can be the same instantons that are counted. With the other methods the fermions may not see the same instantons that the operator sees. Thus one would expect a better accounting for Eq. (14.47) with the zero mode method.

Chapter 15

Finite volume effects

In this chapter we consider the effects of finite simulation volume on correlation functions. In nearly all sections below we will assume that the volume is large compared with the Compton wavelength of the particles ($ML \gg 1$). The one exception will be our discussion of the so-called "epsilon regime," where $m_\pi L$ is assumed to be very small, while $ML \gg 1$ for all other excitations. We will begin by examining this problem in the context of chiral perturbation theory, where the discussion is particularly simple, and then proceed to a more general treatment of finite volume effects. As we will see, finite volume effects, rather than being a problem, can be used as a diagnostic to study physics beyond spectroscopy.

15.1 Finite volume effects in chiral perturbation theory

Our discussion of finite volume effects is that of Hasenfratz *et al.* (1991). It uses a formula for the propagator that is particularly convenient for exposing finite volume effects in the large volume limit. To introduce it, we recall the Poisson resummation formula. The function

$$f(x) = \sum_{m=-\infty}^{\infty} \delta(x - mx_0) \tag{15.1}$$

is periodic with period x_0. Thus it has a Fourier series expansion with equal Fourier coefficients:

$$f(x) = \frac{1}{x_0} \sum_{n=-\infty}^{\infty} \exp\left(i\frac{2\pi}{x_0}nx\right). \tag{15.2}$$

Now in a one-dimensional periodic box of length L, a propagator $\Delta_L(x)$ has a discrete Fourier transform ($k_m = 2\pi m/L$), which can be rewritten in terms of the infinite volume propagator $G(k)$ as follows:

$$\begin{aligned}\Delta_L(x) &= \frac{1}{L}\sum_{m=-\infty}^{\infty} e^{ixk_m} G(k_m) \\ &= \frac{1}{L}\int_{-\infty}^{\infty} dk\, G(k) e^{ikx} \sum_{m=-\infty}^{\infty} \delta\left(k - \frac{2\pi m}{L}\right).\end{aligned} \tag{15.3}$$

Substituting Eq. (15.2) for the delta-function gives

$$\Delta_L(x) = \int \frac{dk}{2\pi} G(k) \sum_n e^{ik(x+nL)}, \tag{15.4}$$

which in D-dimensions generalizes obviously to

$$\Delta_L(x) = \int \frac{d^D k}{(2\pi)^D} G(k) \sum_{n_\mu} e^{ik_\mu(x_\mu + n_\mu L_\mu)} = \sum_{n_\mu} \Delta(x + n_\mu L_\mu). \tag{15.5}$$

Thus propagation in a periodic box is equivalent to propagation via $\Delta(x)$ in an infinite system with a direct contribution from the source ($n_\mu = 0$) plus a sum of contributions from the infinite set of images of the source.

For the propagator in an infinite system the literature makes extensive use of the "heat kernel" form,

$$\Delta(x) = \int \frac{d^4 p}{(2\pi)^4} \frac{e^{ipx}}{p^2 + m^2} = \frac{m^2}{16\pi^2} \int_0^\infty d\alpha \exp\left(-\frac{1}{\alpha} - \frac{\alpha m^2 x^2}{4}\right). \tag{15.6}$$

To derive this improbable formula, write

$$\frac{1}{p^2 + m^2} = \int_0^\infty dy \exp[-y(p^2 + m^2)] \tag{15.7}$$

and, specializing to four dimensions, substitute into the middle expression in Eq. (15.6) to get

$$\Delta(x) = \int_0^\infty dy\, e^{-ym^2} \prod_i \frac{dp_i}{2\pi} e^{-yp_i^2 + ip_i x_i}. \tag{15.8}$$

Completing the square, integrating over the p_i's, and changing variables to $\alpha = 1/(m^2 y)$ gives Eq. (15.6). This formula has a simple saddle point approximation at large mx,

$$\Delta(x) \simeq \frac{m^2}{(4\pi)^2}\left(\frac{8\pi}{(mx)^3}\right)^{1/2}\exp(-mx)+\ldots, \tag{15.9}$$

which can be used with Eq. (15.5) to investigate large-finite-volume corrections.

Now the way is clear to describe volume dependence in chiral perturbation theory. At one loop level, we have seen [compare Eq. (14.16)] how infinite volume observables show a typical behavior

$$O(L=\infty) = O_0\left[1 + C_0\frac{1}{f^2}I_1(m)\right] \tag{15.10}$$

where C_0 is a constant, $G(k,m) = 1/(k^2+m^2)$ is the pion propagator, and

$$I_1(m) = \int \frac{d^4k}{(2\pi)^4}G(k,m) = \Delta(0). \tag{15.11}$$

The integral must be regularized. For present purposes we introduce a simple hard momentum cutoff $k \leq \Lambda$ to give

$$I_1(m) = \frac{\Lambda^2}{16\pi^2} + \frac{m^2}{16\pi^2}\log\frac{m^2}{\Lambda^2} + O\left(\frac{m^2}{\Lambda^2}\right). \tag{15.12}$$

As before, the divergence is absorbed into a redefinition of bare parameters, and the logarithm is the famous chiral logarithm.

Let $I_1(m,L)$ denote the corresponding finite volume tadpole term. The difference

$$\overline{I}_1(m,L) = I_1(m,L) - I_1(m), \tag{15.13}$$

can be expanded in contributions from the images points. So to lowest order in chiral perturbation theory, we correct Eq. (15.10) with

$$O(L) = O(L=\infty)\left[1 + C_0\frac{1}{f^2}\overline{I}_1(m,L)\right] \tag{15.14}$$

where

$$\overline{I}_1(m,L) = \sum_{n_\mu\neq 0}\Delta(n_\mu L). \tag{15.15}$$

From Eq. (15.9), $\overline{I}_1(m,L) \sim L^{-3/2}\exp(-mL)$. Thus, all observables experience an extra correction due to the finite volume, which is exponentially small in the size of the box.

As an example, consider the volume-dependent pion mass, which we can compute by merely recycling our earlier $O(m^2)$ calculation. In the language of this section, it is

$$m_\pi(L) = m_\pi(L=\infty)\left[1 + \frac{1}{N_f f^2}\bar{I}_1(m,L)\right]. \tag{15.16}$$

This formula has two uses to the lattice practitioner. First of all, it shows us how large a simulation volume we need to assure that finite volume effects are small (in this case in the determination of m_π). Second, if the lattice volume is small, but the L dependence is given by Eq. (15.14), then given a sufficiently accurate measurement of $O(L)$, one could determine f from a measurement of $O(L)$. For a very readable discussion of these points, with applications, see Sharpe (1992).

15.2 The ϵ-regime

With a suitable choice of quark masses the pion can be made so much lighter than the other hadrons that there is a range of box sizes for which $ML \gg 1$ for all hadrons, but $m_\pi L \ll 1$. This is called the epsilon regime (Gasser and Leutwyler, 1987a,b). Here the partition function is dominated by pions; everything else is massive, and their contribution to the partition function is exponentially suppressed. (The finite volume case we discussed in the last section is called the "p-regime," $m_\pi L \gg 1$.) The epsilon regime provides another way to match lattice measurements with chiral predictions and permits insightful contact with random matrix theory.

15.2.1 *Banks-Casher formula*

To see how results are altered in the epsilon regime we first examine the Banks and Casher (1980) formula, a beautiful relation between the level density of the spectrum of the Dirac operator and the chiral condensate Σ.

Call the mean number of eigenvalues in an interval $d\lambda$, per unit volume, $\rho(\lambda)d\lambda$. Next, write the fermion propagator for a particular background gauge configuration as an eigenmode expansion,

$$\langle q(x)\bar{q}(y)\rangle = \sum_n \frac{u_n(x)u_n^\dagger(y)}{m - i\lambda_n}. \tag{15.17}$$

Apart from zero modes, the eigenvalues come in complex-conjugate pairs. The condensate is computed by setting $x = y$, integrating over volume, and averaging over all gauge configurations,

$$\frac{1}{V}\int dx\,\langle\bar{q}(x)q(x)\rangle = -\frac{2m}{V}\sum_{\lambda_n>0}\frac{1}{\lambda_n^2+m^2}. \tag{15.18}$$

Taking the infinite volume limit and introducing the spectral density $\rho(\lambda)$ through $(1/V)\sum_n \to \int d\lambda\,\rho(\lambda)$, we arrive at

$$\langle 0|\bar{q}q|0\rangle = -2m\int_0^\infty d\lambda\,\frac{\rho(\lambda)}{\lambda^2+m^2}, \tag{15.19}$$

and taking the limit of zero mass, where $m/(\lambda^2+m^2) \to (\pi/2)\delta(\lambda)$, we get

$$\langle 0|\bar{q}q|0\rangle = -\pi\rho(0). \tag{15.20}$$

Thus the condensate is determined by the Dirac eigenmode density at $\lambda = 0$.

15.2.2 *Chiral Lagrangian in the epsilon regime*

In the Banks-Casher formula, the infinite volume limit plays an essential role. Indeed, at finite volume the behavior of the quark condensate must be different, because symmetries do not break spontaneously if V is finite. The spectrum is discrete, and the singularity does not appear. If the limit of vanishing m is taken at fixed volume, the condensate disappears, and the symmetry is restored. We got a nonzero condensate because we first took the limit $V \to \infty$, then took the mass to zero. Clearly, the two limits do not commute.

Nonetheless, the effect of the infinite volume condensate persists at finite volume, as do the effects of other parameters of the low energy effective Lagrangian, and they can be observed with appropriate simulations. Let us see how the observed finite volume- and mass-dependent chiral condensate provides this information.

Because the partition function is dominated by the pions, it is described by an effective Lagrangian of the type we have discussed previously. In principle, these Lagrangians have an infinite number of terms. However, in the limit we have just described, the pion contribution is dominated by zero modes, field configurations that are flat across the volume. Then the effective Lagrangian is dominated by a single term,

$$Z = \int_{SU(N_f)} dU\,\exp[V\Sigma\mathrm{Re}\mathrm{Tr}(MU^\dagger)], \tag{15.21}$$

where we introduced the constant $\Sigma = -f^2 B$.

The mass or volume-dependent condensate $\Sigma_i(m, V)$ for quark flavor i can be computed from $\Sigma_i(m, V) = 1/VZ^{-1}\partial Z/\partial m_i$. Notice that Z depends on the quark masses m_i, the parameter Σ, and the volume through the single combination $\mu_i = m_i \Sigma V$. Thus $\Sigma_i(m, V) = \Sigma f(\mu_i)$. At small μ it scales as $\Sigma_i(m, V) \sim m\Sigma^2 V$. Even though the condensate measured in a simulation vanishes in the epsilon regime in the zero quark mass limit, its behavior at small quark mass still gives the low energy constant Σ.

Leutwyler and Smilga (1992) combined the limit of the epsilon regime with fixed topology of the gauge field. Consider QCD with a vacuum angle θ. Equation (15.21) generalizes to

$$Z(\theta) = \int_{SU(N_f)} dU \, \exp[V\Sigma \mathrm{ReTr}(\exp(i\theta/N_f)MU^\dagger)]. \tag{15.22}$$

The partition function $Z(\theta)$ has an expansion in terms of partition functions in sectors of fixed winding number ν,

$$Z(\theta) = \sum_\nu e^{i\nu\theta} Z_\nu. \tag{15.23}$$

Because of the anomaly the partition function $Z(\theta)$ does not depend separately on θ and on the quark mass matrix M, but on the combination $\exp(i\theta/N_f)M$. Since U enters through the combination $U\exp(-i\theta/N_f)$, the projection onto sectors of fixed topological charge converts the $SU(N_f)$ integral over the zero mode U into a $U(N_f)$ integral, and

$$Z_\nu = \int_{U(N_f)} dU \, (\det U)^\nu \exp[V\Sigma \mathrm{ReTr}(MU^\dagger)]. \tag{15.24}$$

The mass or volume-dependent condensate in a sector of fixed topology can be computed by differentiation as before.

15.2.3 *Random matrix theory*

Equation (15.24) allows for a direct and rather surprising connection with random matrix theory. This was first described by Shuryak and Verbaarschot (1993), and has been the subject of a large literature since. We cannot do it justice here. The review of Damgaard (2002) provides a reasonably gentle introduction. We briefly summarize the high points.

The original idea was to consider the integral over a set of $N \times (N+\nu)$ complex matrices, W,

$$\tilde{Z}_\nu = \int dW \exp\left[-\frac{N}{2}\mathrm{Tr}V(M^2)\right] \prod_{i=1}^{N_f} \det(iM + \tilde{m}_i), \tag{15.25}$$

where

$$M = \begin{pmatrix} 0 & W^\dagger \\ W & 0 \end{pmatrix}. \tag{15.26}$$

The $2N+\nu$ dimensional square matrix M has ν zero modes and its other eigenmodes come in complex conjugate pairs. We denote its eigenvalue density with $\tilde{\rho}(\tilde{\lambda})$. The $\tilde{m}$'s are just numbers, and $V(M^2)$ is unspecified.

Intuitively, this expression looks like its epsilon-regime chiral analog. The W variables play the role of the "flat across the box" sigma model fields U, and the determinant will give a factor of m_i for every zero mode of each flavor. The careful derivation of the result (which we admit, seems very improbable at a first reading) is given by Shuryak and Verbaarschot (1993).

The connection between random matrix theory and QCD in the epsilon regime is not directly at the level of the variables in the functional integral. Instead, it involves taking the so-called "microscopic limit," where N is taken to infinity while $\tilde{\mu}_i = 2\pi N\tilde{\rho}(0)m_i$ is held fixed. In that limit,

$$Z_\nu(\{\mu_i\}) = \tilde{Z}_\nu(\{\tilde{\mu}_i\}) \tag{15.27}$$

when $\tilde{\mu}_i = \mu_i$. The limit is universal in that it can be shown to be true for any choice of $V(M)$ in large classes of random matrices, characterized by "Dyson indices". These classes are associated with the pattern of symmetry breaking in the original chiral model (Verbaarschot, 1994). If we regard $Z_\nu(\{\mu_i\})$ as a generating function for the chiral condensate, this result implies that in the infinite volume limit, at fixed $m_i\Sigma V$, the mass dependence of the chiral condensate can be predicted using random matrix theory.

While the partition functions of the two theories coincide in the microscopic limit, in general, the eigenvalues of the Dirac operator do not coincide with the eigenvalues of M. (After all, at large λ, asymptotic freedom says that $\rho(\lambda) \sim \lambda^3$). However, near $\lambda = 0$ both densities approach constants (as the Banks-Casher relation requires for QCD). In that case, one can define rescaled variables, $\zeta = \lambda\Sigma V$ for QCD, and $\tilde{\zeta} = \tilde{\lambda}2\pi N\tilde{\rho}(0)$. The "microscopic spectral density" of QCD, $\rho_s(\zeta) \equiv \rho(\lambda)/V$ can then be shown to be equivalent to the spectral density in random matrix theory in an equivalent limit. Random matrix theory predicts the functional form of the probability distribution of individual eigenvalues of the Dirac operator. These distributions depend on N_f [notice the N_f in Eq. (15.25)] and ν, and are functions of the combinations $\lambda\Sigma V$ and $m_q\Sigma V$. Fitting the probability distribution of individual Dirac eigenvalues to the random matrix theory formula then gives a measurement of Σ.

15.2.4 *Further applications*

The practical consequence of these observations is that it may be possible to compute chiral parameters of QCD via simulations in the epsilon regime in fixed sectors of topological charge. One would measure a set of "chirally interesting" correlators, such as the pseudoscalar or axial vector correlator, and compare their asymptotic shapes with formulas derived from epsilon-regime chiral perturbation theory, fitting parameters such as the condensate or the pseudoscalar decay constant. Most of the relevant expressions from chiral Lagrangians have been computed in full QCD and in quenched approximation. At the time this section was written. little had been done with simulations of full QCD in the epsilon regime, but pioneering studies in the quenched approximation are promising enough [for one example, see Giusti *et al.* (2004)] that we feel that the extended introduction to the subject we have given is warranted.

For further reading, see Hansen (1990); Hansen and Leutwyler (1991); Damgaard *et al.* (2002, 2003).

15.3 Finite volume, more generally

15.3.1 *Single particle states*

In Sec. 15.1 we examined finite-volume mass shifts in lowest order chiral perturbation theory and developed an image source expansion. Here we consider a more general treatment that makes contact with the Lüescher formula. We consider a spatial volume of finite extent L and infinite Euclidean time extent. The energy of a particle is determined from the zero of its inverse propagator as a function the Euclidean energy p_0 at fixed spatial momentum $\vec{p}$, *i.e.*, $G(p)^{-1} = G(p_0 = i\omega_L(\vec{p}), \vec{p})^{-1} = 0$. As usual we separate contributions to the L-dependent mass from the bare mass and L-dependent self energy and write $G(p)^{-1} = p_0^2 + (\vec{p})^2 + m^2(L) = p^2 + m_0^2 + \Sigma(L, p)$.

For example, ϕ^4 field theory resembles the chiral model; its $\Sigma(L,p)$ is given by the single bubble and, just as in Eq. (15.11),

$$m^2(L) - m^2(\infty) = \Sigma(L,p) - \Sigma(\infty,p) = \frac{g}{2}\sum_{n_\mu\neq 0}\Delta(n_\mu L_\mu). \qquad (15.28)$$

Rather than proceeding with the heat kernel expression Eq. (15.6), the literature customarily continues with a slightly different integral expression for the propagator,

$$\Delta(x) = \frac{1}{8\pi^2|x|}\int_{-\infty}^{\infty} dy\, \exp(-\sqrt{y^2+m^2}|x|), \qquad (15.29)$$

where $|x| = \sqrt{x^2}$, and we neglect lattice discretization effects. Again, a quick derivation: Start from

$$\Delta(x) = \int \frac{d^4k}{(2\pi)^4}\frac{e^{ikx}}{k^2+m^2}. \qquad (15.30)$$

Pick x along the $\hat{t}$ axis, let $k = (k_0, \vec{y})$ and integrate over the directions of $\vec{y}$ to give

$$\Delta(x) = \frac{1}{4\pi^2}\int\frac{dk_0}{2\pi}\int_{-\infty}^{\infty}\frac{y^2 dy}{y^2+k_0^2+m^2}e^{ik_0 x}. \qquad (15.31)$$

Perform the contour integral over k_0 and integrate the result by parts in y. Counting only the nearest image points in the six spatial directions, and approximating $m^2(L) - m^2(\infty) = 2m(\infty)[m(L) - m(\infty)]$, we have

$$m(L) - m(\infty) = \frac{g}{16\pi^2}\frac{3}{mL}\int dy\, \exp(-\sqrt{y^2+m^2}L), \qquad (15.32)$$

where $m = m(\infty)$ on the right hand side.

Rather than thinking about the mass shift as a self-energy effect, we can think of it as a scattering effect. The incoming particle emits a particle that travels to an image point on the right, say, where it is absorbed. Because they are images, all of them do the same thing to their neighbors, so the incoming particle also absorbs a particle from its image on the left. Like the direct term, this scattering amplitude is just $-g$, and the mass shift is proportional to it. This idea was generalized by Lüscher (1986a) to think of the scattering not only in lowest order, but in all orders, and to replace g by the scattering amplitude. His formula is

$$m(L) - m(\infty) = -\frac{1}{16\pi^2}\frac{3}{mL}\int dy\, \exp(-\sqrt{y^2+m^2}L)F(iy) \qquad (15.33)$$

where $F(iy)$ is the analytic continuation $\nu \to iy$ of the scattering amplitude $F(\nu) = T(s,t,u)$, evaluated at forward scattering, $t = 0$, and ν is the "crossing variable," $4m\nu = s - u$. If we assign the momentum p to the initial particle and k to the particle it sends to its neighboring image, it must absorb an identical particle of momentum k from its opposite neighboring image, leaving it with its original momentum and no net momentum transfer, $t = (p-p)^2 = 0$. Then $s = (p+k)^2$ and $u = (p-k)^2$ and $s - u = 4pk$. In the rest frame of the particle, $k_0 = \nu$ is the zeroth component of k. The variable y in the integral is the continuation of k_0 to Euclidean space.

The above treatment generalizes to other field theories and requires only a four-point scattering amplitude with two legs that can be connected to form a self-energy diagram. In a $\lambda\phi^3$ theory, for example, the lowest order four-point forward scattering amplitude has poles at $s = m^2$ and $u = m^2$. Taking only the singular terms, we have

$$\lim_{\nu \to \frac{1}{2}m} \left(\nu^2 - \frac{1}{4}m^2\right) F(\nu) = \frac{1}{2}\lambda^2, \tag{15.34}$$

and we find

$$\delta m = -\frac{3}{16\pi^2}\lambda^2 \exp\left(-\frac{\sqrt{3}}{2}mL\right). \tag{15.35}$$

This result is – rather surprisingly – a slower fall off that the term we get with no singularity, Eq. (15.33), which typically shows a $1/L^{3/2}\exp(-mL)$ fall off [compare Eq. (15.9)].

It is interesting to make the connection between Lüscher's formula and the current algebra formula for the finite volume mass shift, Eq. (15.16). This is easily done if we know $F(\nu) = T(s,0,u)$. In this case the scattering amplitude is

$$T(s,t,u) = A(s,t,u) + 3A(t,s,u) + A(u,s,t) \tag{15.36}$$

with $A(s,t,u) = (m_\pi^2/f_\pi^2)(s/m_\pi^2 - 1)$ in lowest order. Thus $s + u = 4m_\pi^2$ and $F(\nu) = -m_\pi^2/f_\pi^2$, just a constant, again. The Lüscher formula reduces to Eq. (15.16) when we restrict the sum over images in the latter expression to the 6 nearest neighbors (images at $\pm L$ in each direction).

15.3.2 *Two particle states*

Finite size effects in two-particle states can be exploited to obtain some scattering information.

15.3.2.1 *Scattering lengths from the two-particle ground state*

Suppose we consider two-particle states in a finite box of size L. The energy of two free particles is just the sum of their individual energies, but if they interact, their energy is different. Lüscher (1986b) showed that the energy shift of the lowest two-particle state is proportional to $-a_0/(mL^3)$ where a_0 is the scattering length.

It is easiest to sketch the derivation by considering not quantum field theory, but two identical non-relativistic bosons of mass m, in the box. The free particle wave function is

$$\psi(p,q) = \langle xy|pq\rangle = \frac{1}{\sqrt{2L^3}}[e^{i(px+qy)} + e^{i(py+qx)}]. \tag{15.37}$$

Assume that our state has total momentum $P = p+q = 0$, or $q = -p$, and relabel $|pq\rangle$ as just $|p\rangle$. These states are eigenstates of the free Hamiltonian with $H_0|p\rangle = E(p)|p\rangle$ with $E(p) = p^2/m$ and $\langle p'|p\rangle = \delta_{p,p'} + \delta_{p,-p'}$. In the finite box, components of p are quantized in multiples of $2\pi/L$.

Next, let's assume that we have a short range potential $V(x-y)$. (It should be periodic, $\sum_n V(x-y+nL)$, but in leading order, that doesn't matter.) Now consider the matrix element

$$\begin{aligned}
\langle p'|V|p\rangle &= \frac{1}{2L^6}\int dxdy\, V(x-y)[e^{ip'(x-y)} + e^{-ip'(x-y)}][e^{ip(x-y)} + e^{-ip(x-y)}] \\
&= \frac{1}{2L^3}\int dz\, V(z)[e^{ip'z} + e^{-ip'z}][e^{ipz} + e^{-ipz}] \\
&\equiv \frac{1}{L^3}\hat{V}(p',p) \equiv \frac{1}{L^3}\langle p'|\hat{V}|p\rangle.
\end{aligned} \tag{15.38}$$

We are interested in the energy shift of the lowest $p = 0$ state, but we'll keep the variable p a bit longer. The perturbative expansion is

$$\begin{aligned}
E &= E(p) + \langle p|V|p\rangle + \sum_{\hat{k}}\langle p|V|k\rangle\frac{1}{E(p)-E(k)}\langle k|V|p\rangle + \dots \\
&= E(p) + \frac{1}{L^3}\hat{V}(p,p) + \frac{1}{L^3}\frac{1}{L^3}\sum_{\hat{k}}\hat{V}(p,k)\frac{1}{E(p)-H_0}\hat{V}(k,p) + \dots
\end{aligned} \tag{15.39}$$

Now if the box is large, we can replace $\frac{1}{L^3}\sum_{\hat{k}}$ by $\int d^3k/(2\pi)^3$, and, adding an $i\epsilon$ to regulate the denominator, we get

$$E = E(p) + \frac{1}{L^3}[\hat{V}(p,p) + \int \frac{d^3k}{(2\pi)^3}\hat{V}(p,k)\frac{1}{E(p) - H_0 + i\epsilon}\hat{V}(k,p) + \ldots]$$
$$= E(p) + \frac{1}{L^3}\langle p|[\hat{V} + \hat{V}\frac{1}{E(p) - H_0 + i\epsilon}\hat{V} + \ldots]|p\rangle. \tag{15.40}$$

The object in brackets is the familiar T matrix of quantum mechanics.

Now go to very low p. The low-energy limit of the T-matrix is

$$T = -\frac{4\pi a_0}{m}, \tag{15.41}$$

where a_0 is the S-wave scattering length. Thus the energy shift is

$$\Delta E = -\frac{4\pi a_0}{mL^3}. \tag{15.42}$$

The $1/L^3$ is easy to understand. The wave functions of the particles are flat across the box, and only the parts of the wave functions of the two particles that are close together can interact.

For two relativistic particles, the energy becomes $W = 2\sqrt{m^2 + m\Delta E}$. This would be the relevant formula to use for (say) the energy shift of a $\pi\pi$ state. The measurement of W would still give the scattering length.

This simple formula is valid in large volumes. Lüscher has calculated correction terms to it,

$$\Delta E = -\frac{4\pi a_0}{mL^3}\left[1 + c_1\frac{a_0}{L} + c_2\left(\frac{a_0}{L}\right)^2 + \ldots\right], \tag{15.43}$$

where $c_1 = -2.837\ldots$ and $c_2 = 6.375\ldots$. These terms can be important in practice. The coefficients are universal (at least for cubic lattices); the scattering lengths are (of course) not.

15.3.2.2 *Decay widths from level repulsion*

As we move up in the spectrum of two-particle states we can encounter more interesting behavior. Again, a quantum mechanical model tells the story most easily. Imagine that we have two spinless bosons, this time in one spatial dimension. If their total momentum is zero, their wave function $\psi(x,y) = f(x-y)$ is an even function of its argument. The wave function is, of course, a solution of the one-dimensional Schrödinger equation,

$$\left[-\frac{1}{m}\frac{d^2}{dz^2} + V(z)\right] f(z) = Ef(z). \tag{15.44}$$

If we look for scattering solutions, the energy is positive, and far away from the center of the potential $f(z) \to \cos[k|z| + \delta(k)]$, where the energy is $E = k^2/m$ and $\delta(k)$ is the scattering phase shift. So much for L infinite; if the range of z is finite $(-L/2 < z < L/2)$ and periodic, the wave function and its derivative have to match at the joining point, which means that

$$kL + 2\delta(k) = 0 \quad \mod 2\pi. \tag{15.45}$$

Thus, a measurement of the energy of the two particle state can reveal scattering information (the phase shift). Physically, what is happening is that the potential either pulls in or pushes out the wave function, modifying the quantization condition that the finite length of the box imposes.

The three-dimensional formula has been worked out, again by Lüscher (1986b, 1991). Its physics is identical to the one dimensional case, but the formula for the shift in the wave number is much messier than Eq. (15.45). In an infinite volume, the scattering amplitude has its easiest expression in the familiar partial wave series, and yet the unperturbed wave functions in a box separate in rectangular coordinates. The high part of their spectrum becomes dense (though still discrete), and the quantization condition for several modes can become entangled.

The most interesting situation occurs when the two particles can form a resonance. Near a continuum resonance, $\tan\delta = \Gamma/(2\Delta E)$ where $\Delta E = E_R - E$. In a periodic box the resonance appears as a distinct eigenstate that mixes with the two-particle states. As the energy of the two-particle state approaches that eigenstate, kL suffers a big shift, essentially an avoided level crossing with the resonance. The shape of the variation of the energy with L of either of the would-be crossing states will give information about the decay width Γ.

A little model illustrates this more cleanly. Suppose that we have "ordinary" energy levels ϵ_i where $i = 1, 2, \ldots, N$. Also let us have one "special" energy level that we denote by ϵ_0. The "ordinary" energy levels are also assumed to be decoupled from one another, but not from the "special" energy level. The Hamiltonian matrix is

$$\hat{H} = \begin{bmatrix} \epsilon_0 & V_{01} & V_{02} & V_{03} & \cdots \\ V_{10} & \epsilon_1 & 0 & 0 & \cdots \\ V_{20} & 0 & \epsilon_2 & 0 & \cdots \\ V_{30} & 0 & 0 & \epsilon_3 & \cdots \\ \vdots & \vdots & \vdots & \vdots & \ddots \end{bmatrix}. \tag{15.46}$$

One can find the energy eigenvalues of this system by solving

$$\det[\hat{H} - \delta_{ij} E] = 0 \tag{15.47}$$

or

$$0 = (\epsilon_0 - E)\prod_i(\epsilon_i - E) - \sum_i |V_{0i}|^2 \prod_{j\neq i}(\epsilon_j - E). \tag{15.48}$$

There are $N+1$ roots to this characteristic equation.

Let's focus on the case where $E \approx \epsilon_0$ and rewrite the secular equation as

$$E = \epsilon_0 + \sum_i \frac{|V_{0i}|^2}{(E - \epsilon_i)}. \tag{15.49}$$

When the resonance and the ordinary levels are far apart, there is a slight repulsion of the two levels. Replacing E with ϵ_0 on the RHS gives the familiar second order energy shift. If ϵ_0 and one ϵ_i are close, then we have two-state mixing. The two nearby levels mix and split: the energy eigenvalues come no closer together than $2|V_{0i}|$.

In the case of a continuum the sum is promoted to an integral, and

$$\frac{1}{E-\epsilon} \to \mathcal{P}\frac{1}{E-\epsilon} + i\pi\delta(E-\epsilon). \tag{15.50}$$

The eigenvalue equation can be broken into real and imaginary parts, with

$$\mathrm{Re}(E) = \epsilon_0 + \mathcal{P}\int d\epsilon \frac{|V_{0\epsilon}|^2}{(E-\epsilon)} \tag{15.51}$$

giving the location of the resonance and

$$\mathrm{Im}(E) = \int d\epsilon |V_{0\epsilon}|^2 \pi\delta(E-\epsilon) \tag{15.52}$$

giving the decay width. Both expressions involve weighted sums over $|V_{0i}|^2$ just as in the behavior of the avoided level crossing.

The physical situation we have in mind is that the "ordinary" states are the free two-particle states in the box, for example, two-pion states, and the special state is a resonance like the rho meson. Since lattice calculations are done in Euclidean space, we do not actually observe the decay of a resonance into its constituents; rather, we hope to observe connected Euclidean correlation functions that include exchanges of the resonance and other states, and the coupling of the resonance to the other states. We can do this by varying the locations of either kind of state, either by tuning the quark mass, or varying the box size, or both, and it may happen that we can bring an ordinary state and the special state close together. Then the level repulsion gives us information about the coupling of the resonance to its decay products. For applications to spin models, see (Gattringer and Lang, 1993; Rummukainen and Gottlieb, 1995).

15.4 Miscellaneous comments

Of course, we have not given a complete description of finite volume effects. If the size of the box is really very tiny in physical units, then the coupling constant is never large and the spectrum of states can be attacked analytically. Much beautiful work, *e.g.*, by Koller and van Baal (1986) was done with this limit in the earlier days of lattice gauge theory.

An all-too-common occurrence is that the volume of the box is small compared with the size of the bound state. Let us think (again) in Schrödinger wave function language, and imagine that the quarks are trapped in a bag of size R. As R is varied, their energies vary inversely with R, and if the size of the simulation region is on the order of the size of the bound state, the state is squeezed and its energy rises. The scattering length formula in the last section models a version of this effect (the particle interacts with its images), but it is not the whole story. Your authors have encountered this situation in their own research, more often than they would care to admit. If the offending spectroscopy's variation with box size cannot be fit to one of the functional forms we have already described, then the data set is probably hopelessly compromised.

A simple argument suggests (just as data supports) the idea that volume effects are more important in full QCD than in quenched QCD. In full QCD, particles can exchange pions, while in quenched QCD, unless we explicitly include quark exchange diagrams in the correlators, they can exchange only glueballs or quenched eta-primes.

Chapter 16

Testing the standard model with lattice calculations

16.1 Overview

One of the major goals of lattice calculations is to provide hadronic matrix elements that test QCD or other components of the Standard Model. A wide variety of hadronic matrix elements can be computed on the lattice and used to extract CKM mixing angles from experimental data. In Fig. 16.1 we show the CKM matrix in Wolfenstein parameterization and, paired with each entry, a physical process that is sensitive to it. Each process involves both strong and electroweak interactions. The strong part must be determined nonperturbatively, whereas perturbation theory suffices for the electroweak part. So to determine the CKM parameter from experimental data requires knowledge of the hadronic component of the interaction. Because of the vast difference in scale between hadronic and electroweak interactions, a pure QCD lattice calculation of the hadronic matrix element suffices.

This field is enormous and rapidly evolving, representing a large fraction of the contemporary effort devoted to lattice calculations. To do justice to the field would require a separate book that would be rapidly outdated. Instead, we will be content to provide some theoretical background and then give a brief introduction to the lattice methodology.

16.2 Strong renormalization of weak operators

16.2.1 *Effective Hamiltonian*

The scale of the weak interactions is set by the masses of the W and Z vector bosons, *i.e.*, at distances almost two orders of magnitude smaller than the sizes of hadrons containing u, d, s, c, or b quarks. Thus, once again, we turn naturally to effective field theories, first, in writing an effective weak Hamiltonian and second, in evolving it through strong renormalization to the few-GeV scale μ of lattice computations. For this purpose we start with the continuum theory and use the well-developed technology of the continuum operator product expansion (OPE) and renormalization group. The lattice enters only at the last stage. The effective Hamiltonian then reduces to a sum of four-fermion interactions.

For example, single-W exchange processes are described through the OPE by

$$H_W^{\text{eff}} = \frac{G_F}{\sqrt{2}} \sum_i c_i(\mu) O_i(\mu), \tag{16.1}$$

where the O_i's are the four fermion operators and the c_i's are (known) Wilson coefficients. The physics content of Eq. (16.1) is that the $c(\mu)$'s encode all physics on momentum scales greater than μ. They include the full effects of W's, Z's and top quarks, plus any beyond-Standard-Model physics, plus short-distance effects of QCD (which give them their μ dependence). The scale μ is arbitrary; we have introduced it only to separate short and long distance physics. Quantum chromodynamics introduces a set of corrections to the tree-level graphs that scale as $\alpha_s^n[\ln(M_W/\mu)]^n$. They are small when $\mu \sim M_W$, but as we push the boundary between short and long distance physics down to a scale of μ on the order of a few GeV, these contributions grow and must be summed. Until we reach scales where the strong interactions become nonperturbative, the OPE and renormalization group do the work for us. In this construction, the initial and final states are not relevant and can be chosen for convenience. Thus the $c(\mu)$'s are

$$\begin{pmatrix} V_{ud} = 1 - \frac{\lambda^2}{2} & V_{us} = \lambda & V_{ub} = A\lambda^3(\rho - i\eta) \\ \pi \to l\nu & K \to l\nu & B \to \pi l\nu \\ V_{cd} = -\lambda & V_{cs} = 1 - \frac{\lambda^2}{2} & V_{cb} = A\lambda^2 \\ D \to l\nu & D_s \to l\nu & B \to Dl\nu \\ D \to \pi l\nu & D \to Kl\nu & \\ V_{td} = A\lambda^3(1 - \rho - i\eta) & V_{ts} = -A\lambda^2 & V_{tb} = 1 \\ \langle B_d|\overline{B}_d\rangle & \langle B_s|\overline{B}_s\rangle & \end{pmatrix}$$

Fig. 16.1 Matrix elements that can be computed reasonably reliably with lattice methods, and their impact on the CKM matrix.

process-independent.

Armed with $H_W^{\rm eff}$, we then wish to compute a matrix element between hadronic states $|I\rangle$ and $|F\rangle$

$$A(I \to F) = \frac{G_F}{\sqrt{2}} \sum_i c_i(\mu)\langle F|O_i(\mu)|I\rangle. \tag{16.2}$$

In this expression, the matrix element $\langle F|O_i(\mu)|I\rangle$ must be computed outside perturbation theory. There are a variety of approaches: models, large-N expansions, and, of course, the lattice. The lattice calculation $\langle F|O_i(\mu)|I\rangle$ involves replacing the continuum operator O_i by an appropriate lattice discretization, and then computing some Euclidean correlation function involving the operator.

The construction of $H_W^{\rm eff}$ is a technical problem as difficult as any we have discussed so far. Fortunately, there are excellent reviews [our favorite is Buras (1998)]. Let us be content with an oversimplified introduction.

16.2.2 *An example:* $c \to s\bar{d}u$

Consider the quark transition $c \to s\bar{d}u$. With only electroweak interactions the tree-level W-exchange amplitude is

$$\begin{aligned} M &= -\frac{G_F}{\sqrt{2}} V_{cs}^* V_{ud} \frac{M_W^2}{k^2 - M_W^2} (\bar{s}_\alpha c_\alpha)_L (\bar{u}_\beta d_\beta)_L \\ &= \frac{G_F}{\sqrt{2}} V_{cs}^* V_{ud} (\bar{s}c)_L (\bar{u}d)_L + O(\frac{k^2}{M_W^2}). \end{aligned} \tag{16.3}$$

We have explicitly included the quark colors as subscripts in the first line and have written $(\bar{s}c)_L$ for $\bar{s}\gamma_\mu(1-\gamma_5)c$. An "R" subscript would indicate a $\gamma_\mu(1+\gamma_5)$ coupling. Because the momentum carried by the W is much smaller than its mass, it makes sense to neglect it. The amplitude can also be obtained from the tree-level effective Hamiltonian

$$H_W^{\rm eff} = \frac{G_F}{\sqrt{2}} V_{cs}^* V_{ud} (\bar{s}c)_L (\bar{u}d)_L + \ldots \tag{16.4}$$

where the ... are higher-dimensional operators that produce the momentum dependence in the original expression. This result is an example of an operator product expansion. The original expression is replaced by a sum of operators of progressively higher dimension, whose contributions are weighted by effective coupling constants. In this simple example, there is one $c(\mu)$, which we take to be unity.

Now we consider QCD radiative corrections. As we have already seen many times, the connection between the full and effective field theory involves computing an equivalent process in both the original and effective theories and tuning parameters in the effective theory to match the results. The relevant graphs generating the lowest order QCD corrections are shown in Fig. 16.2. We first compute the corrections in the full theory. We discover that the weak decay, which involved a matrix element of the operator

$$O_2 = (\bar{s}_\alpha c_\alpha)_L(\bar{u}_\beta d_\beta)_L, \tag{16.5}$$

picks up a new contribution from the "color-mixed operator"

$$O_1 = (\bar{s}_\alpha c_\beta)_L(\bar{u}_\beta d_\alpha)_L. \tag{16.6}$$

The low energy effective Hamiltonian must be a linear combination of these two operators. We write it as

$$H_W^{\text{eff}} = \frac{G_F}{\sqrt{2}} V_{cs}^* V_{ud}[c_1(\mu)O_1 + c_2(\mu)O_2]. \tag{16.7}$$

A few pages of algebra give [for color $SU(3)$ and at one-loop order]

$$c_1(\mu) = -3\frac{\alpha_s}{4\pi}\ln\frac{M_W^2}{\mu^2}; \qquad c_2(\mu) = 1 + \frac{\alpha_s}{4\pi}\ln\frac{M_W^2}{\mu^2} \tag{16.8}$$

(notice that at $\mu = M_W$ we recover the original operator). The operators O_1 and O_2 have mixed under renormalization. (This is a general feature of these calculations.) It will prove convenient to eliminate the mixing with a change of basis,

$$O_\pm = \frac{O_2 \pm O_1}{2}; \qquad c_\pm = c_2 \pm c_1 \tag{16.9}$$

and then

$$c_\pm(\mu) = 1 + (1 \mp 3)\frac{\alpha_s}{4\pi}\ln\frac{M_W^2}{\mu^2}. \tag{16.10}$$

Unfortunately, at $\mu \sim 1$ GeV, the radiative correction is as large as the zeroth order term. The naive perturbative expansion has broken down because of the large logarithms, which come from the large ratio of scales, M_W/μ. We must sum the large logs; the renormalization group does that

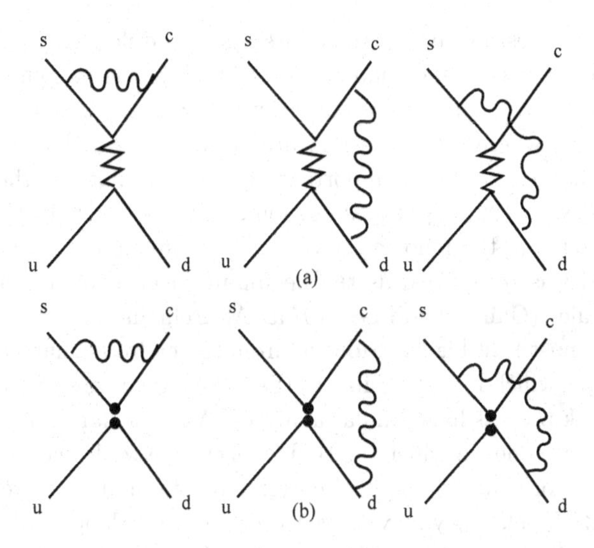

Fig. 16.2 Radiative corrections (a) in the full theory and (b) in the effective field theory.

for us. That is easy to do, once we realize that we can write $c_\pm(\mu) = Z_\pm(\mu)c_\pm^{(0)}$ and that the bare $c_\pm^{(0)}$ does not depend on μ:

$$\frac{dc_\pm(\mu)}{d\ln\mu} = \frac{1}{Z_\pm}\frac{dZ_\pm}{d\ln\mu}c_\pm(\mu) = \gamma_\pm(g)c_\pm(\mu) \tag{16.11}$$

where γ is the anomalous dimension of the operator $O_\pm$, $\gamma_\pm = \gamma_\pm^0\alpha_s/(4\pi)$. The solution to this equation is

$$c_\pm(\mu) = c_\pm(M_W)\exp\left[\int_{g(M_W)}^{g(\mu)} dg' \, \frac{\gamma_\pm(g')}{\beta(g')}\right], \tag{16.12}$$

where $\beta(g)$ is the usual beta function, $\beta(g) = -\beta_0\alpha_s/(4\pi) + \ldots$. Working to lowest nontrivial order, we can do the integral and find

$$c_\pm(\mu) = c_\pm(M_W)\left[\frac{\alpha_s(M_W)}{\alpha_s(\mu)}\right]^{\gamma^0/(2\beta_0)} . \tag{16.13}$$

At the scale M_W there are no large logarithms and the coefficient can be computed perturbatively; in fact $c_\pm(M_W) = 1$. For five flavors, the explicit result is

$$c_+(\mu) = \left[\frac{\alpha_s(M_W)}{\alpha_s(\mu)}\right]^{\frac{6}{23}}; \qquad c_-(\mu) = \left[\frac{\alpha_s(M_W)}{\alpha_s(\mu)}\right]^{-\frac{12}{23}} . \tag{16.14}$$

Already, this expression reveals some interesting physics. At low scales c_- has grown to be larger than than c_+. If we analyze the isospin content of $O_\pm$, we find that O_- contributes only to weak decays in which the isospin changes by 1/2, whereas O_+ is a mixture of $\Delta I = 3/2$ and 1/2 processes. The result that $c_- > c_+$ is important, because in nature, the ratio of $\Delta I = 3/2$ to $\Delta I = 1/2$ amplitudes in kaon decays (specifically $K \to \pi\pi$) is about 21. Early on, the renormalization group running argument we have just sketched was recognized as the beginning of an explanation for this "$\Delta I = 1/2$ rule" (Gaillard and Lee, 1974a; Altarelli and Maiani, 1974), but clearly there must be a big contribution from the hadronic matrix elements of O_+ and O_-, which is inaccessible to the theory we have so far discussed.

Other weak process have similar analyses. A particularly important set of transitions are ones in which quark flavors change while their charges do not. These flavor-changing neutral currents are forbidden at tree level in the Standard Model. They may, however, occur as one-loop or higher order electroweak radiative processes. The fact that they they appear only as loop effects makes them particularly useful for testing the Standard Model or for looking for physics beyond the Standard Model. Again, it is the low energy effective field theory that is important, not the particular structure of the amplitude in terms of W's, Z's and t quarks. At zeroth order in the strong interactions, there is an H_W^{eff} that superficially resembles Eq. (16.1). It is a sum over a set of four-fermion operators with a set of coefficient functions that now are second order (proportional to G_F^2 or αG_F) and that also depend parametrically on the masses of the heavy particles through ratios $x_i = m_i^2/M_W^2$, where $i = Z$, t, and even c if charm quarks are integrated out, too. Light quark contributions can be removed by exploiting the GIM mechanism (Donoghue *et al.*, 1992). Of course, as we run down from the weak scale, operators are mixed and renormalized.

16.2.3 *Lattice vs continuum renormalization of the effective Hamiltonian*

Lattice calculations of matrix elements use essentially every ingredient we have discussed in previous chapters. The choice of lattice discretization can be crucial to the success or failure of the project. Even if one is considering processes with a heavy quark, it is still almost always necessary to perform a chiral extrapolation. Moreover, we are working with an effective field theory that we have run down from some high momentum scale. In order to do the running, we had to pick a renormalization scheme and a scale.

The renormalization scheme was almost certainly not the lattice, so we must convert the resultant lattice-regularized matrix element into a continuum scheme, such as $\overline{MS}$. Schematically, the steps are as follows:

$$\begin{array}{lcl} \text{continuum SM} \stackrel{\text{OPE}}{\rightarrow} C(M_W)O^{\text{cont}}(M_W) \stackrel{\text{RG}}{\rightarrow} & C(\mu)O^{\text{cont}}(\mu) \\ & \text{matching} \updownarrow \\ & C(\mu)Z(\mu a)O^{\text{latt}}(a) \end{array}$$

The combination $C(\mu)Z(\mu a)$ must be independent of μ, at least to the order in g one is working.

One might ask, why not eliminate the matching and simulate the whole standard model on the lattice? This is not done for two reasons. The first is practical. We want to capture the hadron in our simulation, so the box size we are simulating should be bigger than a fermi or so. For computational reasons (we do not want to have too many lattice points), the inverse lattice spacing should not be bigger than a few GeV. This is a much bigger distance than M_W^{-1}, so the W will "fall through" the lattice. The second reason is that the standard model is a theory of chiral fermions, which we must discretize. This is a difficult theoretical problem. One could, in principle, do the OPE and RG running on the lattice. This would avoid the theoretical problem of a lattice chiral fermion, but would still leave the great disparity in scales. In addition, we would have to compute the running from a lattice calculation. That could be done in principle (use of the Schrödinger functional comes to mind), but if we make the easier choice of working in perturbation theory, doing the running in the continuum allows us to take two loop (or sometimes three loop) OPE and RG calculations from the literature. Lattice perturbation theory is very cumbersome, so it is easier to use it in only one place, at the matching point.

16.2.4 *Mixing with operators of higher and lower dimension*

Weak matrix elements present a particularly cruel arena for lattice fermions without exact chiral symmetry, making the operator matching delicate. For example, suppose that we want to do a lattice calculation of the operator responsible for the $\bar{K}_0 - K_0$ mass difference, $O_+ = O_1 + O_2$ where $O_1 = (\bar{s}_\alpha d_\alpha)_L(\bar{s}_\beta d_\beta)_L$ and $O_2 = (\bar{s}_\alpha d_\beta)_L(\bar{s}_\beta d_\alpha)_L$. When the lattice action is chiral, the lattice operators O_+ match to the continuum operators O_+ without further mixing, $O_\pm^{\text{cont}} = Z_\pm O^{\text{latt}}\pm$. When the lattice action is not chiral, the mixing involves additional terms. We can write the mixing expression generically as

$$O_{\pm}^{\rm cont} = Z_{\pm} O_{\pm}^{\rm latt} + Z'_{\pm} O'_{\pm}. \tag{16.15}$$

As we have previously remarked, the diagonal mixing coefficients contain logarithms that exponentiate into the anomalous dimensions of the operators, plus constant terms:

$$Z_{\pm} = 1 + \frac{g^2}{16\pi^2}[\gamma_{\pm}^{(1)} \ln(a\mu) + b_{\pm}] + \ldots \tag{16.16}$$

The $Z'_{\pm}$ terms can have no logarithms because the $O'_{\pm}$'s do not appear in the continuum, and anomalous dimensions are scheme independent at one loop.

Now several things can happen, and some of them are bad. First, consider a mismatch in scaling dimensions between $O'_{\pm}$ and $O_{\pm}$.

If $O'_{\pm}$ has a dimension higher than $O_{\pm}$, then $Z'_{\pm}$ must include compensating scale factors proportional to a positive power of the lattice spacing a. Then we might do a lattice simulation in which we measure a matrix element $\langle O_{\pm}^{\rm latt} \rangle$ and set it equal to $\langle O_{\pm}^{\rm cont} \rangle / Z_{\pm}$. This procedure, of course, ignores the contribution of $\langle O'_{\pm} \rangle$ and introduces a scaling violation that vanishes as the lattice spacing goes to zero.

If $O_{\pm}$ and $O'_{\pm}$ have the same dimension, the situation is more serious but not desperate (at least in principle). Perturbatively, $Z'_{\pm}$ is proportional to g^2, and because of asymptotic freedom, g^2 vanishes slowly as a becomes small.

Finally, it could happen that $O'_{\pm}$ has a lower dimension than $O_{\pm}$. Now we are in real trouble. In principle $Z'_{\pm}$ must grow as $1/a^p$ for $p > 0$. And yet, something delicate must be happening, because in the continuum limit, the operators do not mix. In practice, we try to avoid this problem by choosing to measure processes in which matrix elements of the lower dimensional operators vanish due to some symmetry.

Second, suppose the operators $O'_{\pm}$ and $O_{\pm}^{\rm cont}$ have a different chiral behavior. Then in the chiral limit the need for a cancellation of lattice artifacts among terms on the rhs of Eq. (16.15) may degrade the signal-to-noise ratio in the result.

Consider how mixing occurs for Wilson-type fermions. Here $O_{\pm}$ mixes off shell with operators whose chiral structure is not LL. The massless

Wilson quark propagator behaves like

$$S(p) \sim \frac{1}{\not{p} - ap^2} = \frac{\not{p} + ap^2}{p^2 - a^2p^4} \tag{16.17}$$

for small momentum. The ap^2 term violates chiral symmetry in loops. These extra mixed operators can dominate the contribution of the LL operators at small quark mass. In the example of $O_\pm$, chiral symmetry forces the pseudoscalar meson matrix element of $O_+^{\rm cont}$ to scale as the square of the meson mass. The matrix element of the non-LL O'_+ remains a constant as the meson mass vanishes. Then, as Eq. (16.15) shows, there must be a cancellation between the matrix elements of O'_+ and $O_+^{\rm latt}$. Thus, $O_+^{\rm latt} \sim$ const $+O(m^2)$. The constant is a lattice artifact mixing term that overwhelms the $O(m^2)$ answer.

Broadly speaking, there are two solutions to this problem. The conceptually simplest one is to choose a lattice discretization for which the unwanted operator mixing does not occur, even at nonzero lattice spacing. This can typically be done by choosing fermions with some amount of chiral symmetry: staggered fermions, domain wall fermions, or overlap fermions. The other choice is to measure a particular combination of lattice operators that reduces to the continuum operator and has the correct chiral limit at nonzero lattice spacing [explicitly, the complete right hand side of Eq. (16.15)]. This usually requires a separate calculation of the ratio $Z'_\pm/Z_\pm$, which might be done perturbatively or nonperturbatively.

For now, let us neglect all these issues and consider some new ones, more closely related to simulations. We will first look at some lattice symmetries that can be usefully applied to enhance signals, and then look at some simple examples.

16.3 Lattice discrete symmetries

The lattice has a set of discrete symmetries that are very useful for matrix element calculations. We will list them for Wilson-type (four component) fermions; there are corresponding ones for staggered fermions.

The parity transformation connects variables at site x with its parity-inverted partner $x^p = (x_0, -\vec{x})$. It reverses the direction of the spatial links and has its usual effect on the fermions:

$$
\begin{aligned}
P : \psi(x) &\to \gamma_0 \psi(x^P) \\
\bar{\psi}(x) &\to \bar{\psi}(x^P)\gamma_0 \\
U_0(x) &\to U_0(x^P) \\
U_j(x) &\to -U_{-j}(x^P)
\end{aligned}
\tag{16.18}
$$

We use $U_{-\mu}(x)$ to denote the link leaving point x in the $-\mu$ direction, equal to $U_\mu^\dagger(x-\hat{\mu})$. Time reversal is

$$
\begin{aligned}
T : \psi(x) &\to \gamma_0\gamma_5 \psi(x^t) \\
\bar{\psi}(x) &\to \bar{\psi}(x^t)\gamma_5\gamma_0 \\
U_j(x) &\to U_j(x^t) \\
U_0(x) &\to -U_{-0}(x^t)
\end{aligned}
\tag{16.19}
$$

where $x^t = (-x_0, \vec{x})$ is the time-reversed point. In Euclidean space T is a linear transformation, and by Euclidean invariance it must be just like inversion about any of the other directions. Charge conjugation is

$$
\begin{aligned}
C : \psi(x) &\to C\bar{\psi}(x)^T \\
\bar{\psi}(x) &\to -\psi(x)^T C^{-1} \\
U_\mu(x) &\to U_\mu^*(x).
\end{aligned}
\tag{16.20}
$$

The superscript T means "transpose," and C obeys $C\gamma_\mu C = -\gamma_\mu^* = \gamma_\mu^T$. An explicit realization in the Euclidean Bjorken-Drell convention for gamma matrices is $C = \gamma_0\gamma_2$.

Wilson-type Dirac operators obey the property that $\gamma_5 D_{xy}^\dagger \gamma_5 = D_{yx}$, which means that the propagator $S(x, y, [U]) = D_{xy}^{-1}$ obeys

$$\gamma_5 S^\dagger(x, y, [U])\gamma_5 = S(y, x, [U]). \tag{16.21}$$

For staggered fermions the analogous formula is obtained by replacing γ_5 by $(-)^{x_0+x_1+x_2+x_3}$ (remember, staggered fermions have no spin indices). We will call this "H-invariance."

The discrete symmetries can be implemented on quark propagators:

$$\begin{aligned}
P &: S(x,y,[U]) = \gamma_0 S(x^p, y^p, [U]^p)\gamma_0 \\
T &: S(x,y,[U]) = \gamma_0\gamma_5 S(x^t, y^t, [U]^t)\gamma_5\gamma_0 \\
C &: S(x,y,[U]) = CS(y,x,[U]^c)^T C^{-1} \\
H &: S(x,y,[U]) = \gamma_5 S^\dagger(y,x,[U])\gamma_5 \\
CH &: S(x,y,[U]) = C\gamma_5 S^*(x,y,[U]^c)\gamma_5 C^{-1}.
\end{aligned} \tag{16.22}$$

By $[U]^p$, $[U]^c$, and $[U]^t$ we specify the P, C, or T-transformed links.

Given $S(x,y,[U])$, the propagator on a particular configuration, Eq. (16.22) tells us the propagator on configurations $[U]^p$, $[U]^c$, and $[U]^t$. We can then imagine that these configurations have also been generated in our data set and average (virtually) over all of them. Often this averaging can be done analytically. For example, including configurations $[U]$ and $[U]^p$ picks out the correct parity part of an operator (the $VV + AA$ part of a LL operator) in a two-meson matrix element. The symmetry CH implies that the real part of all lattice amplitudes is the physical part. [To prove this, use Eq. (16.22) plus the identity $C\gamma_5\gamma_\mu\gamma_5 C^{-1} = \gamma_\mu^*$.] This might seem like good news, in that the fluctuating imaginary parts of amplitudes can be discarded. However, it is also bad news, in that phases are not directly available from a Euclidean lattice calculation. In particular, it is not possible to compute phase shifts for scattering amplitudes directly on the lattice. (It can be done indirectly, as we saw in Ch. 15.)

Another discrete symmetry, CPS (Bernard *et al.*, 1985), is also valid in the continuum. It affects almost all weak operators. The S stands for "switch:" it interchanges s and d quarks. For example, consider $(\bar{s}u\bar{u}d)_{LL}$. Under CP this is

$$\bar{s}\gamma_\mu(1-\gamma_5)u\bar{u}\gamma_\mu(1-\gamma_5)d \rightarrow \bar{u}\gamma_\mu(1-\gamma_5)s\bar{d}\gamma_\mu(1-\gamma_5)u, \tag{16.23}$$

which, in turn, under S goes to

$$\rightarrow \bar{u}\gamma_\mu(1-\gamma_5)d\bar{s}\gamma_\mu(1-\gamma_5)u = \bar{s}\gamma_\mu(1-\gamma_5)u\bar{u}\gamma_\mu(1-\gamma_5)d, \tag{16.24}$$

which is our starting operator. The action is not invariant under CPS unless $m_s = m_d$. When they are equal, CPS plus other symmetries (Bose statistics) can be used to eliminate Feynman graphs and hence simplify calculations. When the masses are unequal, CPS is only softly broken, meaning that its violations appear multiplied by factors of $(m_s - m_d)$. It may still be a good approximation to neglect CPS-violating graphs.

16.4 Some simple examples

Let us consider (in no particular order) some of the simpler matrix elements that are accessible to determination by lattice simulations, and describe the strategies for computing them.

16.4.1 *Leptonic decay constants of mesons*

Consider first the weak decay of a pseudoscalar meson, composed of a $\bar{q}_1 q_2$ pair, to a lepton ℓ plus an anti-neutrino. The interaction Hamiltonian is

$$H_I = \frac{G_F}{\sqrt{2}} V_{12} J^5_\mu \bar{\nu}(p)\gamma_\mu(1-\gamma_5)\ell(q), \tag{16.25}$$

where V_{12} is a CKM matrix element and the hadronic current is

$$J^5_\mu = \bar{q}_1\gamma_\mu(1-\gamma_5)q_2. \tag{16.26}$$

The matrix element of the hadronic current between a pseudoscalar meson state carrying four momentum k_μ and the vacuum is parameterized as

$$\langle 0|J^5_\mu|P(k)\rangle = -ik_\mu f_P e^{-ikx} \tag{16.27}$$

The quantity f_P is the decay constant for the meson. For the spinless meson, its four momentum k_μ is the only four vector available to dot into the leptonic current.[1]

This matrix element can be computed from the two-point correlator

$$C_{JO}(t) = \sum_x \langle 0|J(x,t)O(0,0)|0\rangle, \tag{16.28}$$

where $O(0,0)$ is an interpolating operator that excites the desired meson from the vacuum. Inserting a complete set of correctly normalized momentum eigenstates

$$1 = \frac{1}{L^3}\sum_{A,\vec{p}} \frac{|A,\vec{p}\rangle\langle A,\vec{p}|}{2E_A(p)}, \tag{16.29}$$

using translational invariance, and going to large t gives

$$C_{JO}(t) = e^{-m_A t}\frac{\langle 0|J|A\rangle\langle A|O|0\rangle}{2m_A}. \tag{16.30}$$

[1] With this definition, $f_\pi \sim 137$ MeV. An alternate definition often seen in the current algebra literature, and which we used ourselves, introduces an additional factor of $\sqrt{2}$ on the left hand side and then the charged pion decay constant is $f_\pi \sim 93$ MeV.

A second calculation of

$$C_{OO}(t) = \sum_x \langle 0|O(x,t)O(0,0)|0\rangle \to e^{-m_A t}\frac{|\langle 0|O|A\rangle|^2}{2m_A} \tag{16.31}$$

is needed to extract $\langle 0|J|A\rangle$, which is accomplished by fitting the two correlators with three parameters.

Measurements of decay constants have a long lattice history. They are one of the simplest matrix elements to compute. They also play an important role in determining Standard Model parameters. For example, the full leptonic decay amplitude is proportional to a product of the CKM matrix V_{12} from Eq. (16.25) and f_P from Eq. (16.27), so an experimental measurement of a leptonic decay can be combined with a lattice measurement of f_P to determine the CKM element.

16.4.2 *Leptonic decay constants in the heavy quark limit*

Decay constants of heavy mesons of mass M_H scale as $f_P \sim 1/\sqrt{M_H}$. This can be seen (somewhat formally) by evaluating a two-point function of J^5_μ's in the context of heavy-quark effective theory. In HQET we define the current and matrix element so as to remove all explicit quark-mass dependence as the quark mass is sent to infinity. (On the lattice, this makes sense – we replace the heavy quark propagator by a product of links, with no explicit factors of the heavy quark mass ever present.) We can do this by replacing the momentum by the velocity, $k_\mu/M_Q = v_\mu$, which is independent of the heavy quark mass M_Q, and by replacing the current J^5_μ by an effective current $\hat{J}^5_\mu$, where in Euclidean space

$$J^5_\mu(x) = \hat{J}^5_\mu(x)\exp(-M_Q v_\nu x_\nu). \tag{16.32}$$

The factor of $2E(k)$ in the denominator of Eq. (16.29) becomes $2M_H v_0$, where the mass of the hadron M_H differs from the mass of the heavy quark by a constant in leading order in the heavy quark limit. Then

$$C_{JJ}(t) = \frac{(f_P M_H)^2}{2M_H v_0} e^{-(M_H - M_Q)t}, \tag{16.33}$$

which now must be finite in the infinite mass limit. The difference $M_H - M_Q$ is finite, but to remove all the heavy quark mass dependence, the prefactor $f_P^2 M_H$ must also be finite in the limit, giving what we wanted to show, $f_P \sim 1/\sqrt{M_H}$. In a nonrelativistic quark model, $f_p \sim \psi(0)/\sqrt{M_H}$, where $\psi(0)$ is just the wave function at the origin. In lattice calculations of heavy meson decay constants, one often sees the data plotted as $f_P\sqrt{M_H}$ vs $1/M_H$, with the idea that this quantity has the smoothest dependence on quark mass. In addition to simulations with heavy or nonrelativistic quarks and light quarks the lattice calculation could include a determination in the static (infinite mass) limit to help "anchor" an interpolation through the physical values of the charmed or bottom meson masses.

16.4.3 *Electromagnetic widths of vector mesons*

Matrix elements of vector currents can be used to predict the electromagnetic widths of vector mesons, for example, the decay width for $\phi \to e^+e^-$. The lattice calculation is a variation on the decay constant calculation we just described. There is no standard notation for the matrix element, but one commonly used choice parameterizes the matrix element between the vacuum and a vector meson of polarization ν as

$$\langle 0|J_\mu|V_\nu\rangle = \frac{1}{f_V} m_V^2 \delta_{\mu\nu}; \tag{16.34}$$

where J_μ is some choice for a vector current. Here f_V is the decay constant.

16.4.4 *Form factors*

Many processes require the calculation of a three-point function. One example is that of the rate for an exclusive semileptonic weak decay such as $B \to \pi\ell\nu$. As in the actual experimental measurement of such a quantity, the amplitude is parameterized in terms of a set of form factors, each multiplied by its own kinematic factor. For example, in $B \to \pi\ell\nu$,

$$\langle\pi(p)|V_\mu|B(p')\rangle = f^+(q^2)\left[p' + p - \frac{m_B^2 - m_\pi^2}{q^2}q\right]_\mu + f^0(q^2)\frac{m_B^2 - m_\pi^2}{q^2}q_\mu. \tag{16.35}$$

Although the kinematic factors are dictated by symmetry, the particular functional form of the form factor is usually only loosely constrained by current algebra in special kinematic limits and by unitarity. Often the form factor is fit to a model-dependent function with a set of fit parameters, which are then compared with experiment.

Sometimes the model dependence is robust, and can be used to check the lattice simulation. One example is in the decay of a meson containing a heavy quark into another heavy quark. At the particular kinematic point where the velocities of the initial and final heavy quarks are equal, the spectator light quarks and gluons are unaware that the heavy quark's flavor has changed, and so their wave function is unaffected by the decay. The form factor at that kinematic point would be unity (Isgur and Wise, 1989, 1990). Of course, away from that point, the form factor would have to be parameterized, and away from the heavy quark limit (say for the physical b or c quarks) the form factor would have a parameterization in terms of functions scaled as inverse powers of the quark mass.

The actual lattice calculation involves evaluating lattice correlators of three-point functions (two mesons and the current). They can be obtained from a correlator of the form

$$\begin{aligned} C_{AB}(t,t',\vec{p},\vec{q},\vec{p}') &= \sum_z \langle 0|O_A(p,t)J(z,t')e^{i\vec{q}\cdot\vec{z}}O_B(p',0)|0\rangle \\ &= \sum_{xyz} \langle 0|O_A(x,t)e^{i\vec{p}\cdot\vec{x}}J(z,t')e^{i\vec{q}\cdot\vec{z}}O_B(y,0)e^{i\vec{p}'\cdot\vec{y}}|0\rangle \end{aligned} \tag{16.36}$$

by stretching the source and sink operators O_A and O_B far apart on the lattice, letting the lattice project out the lightest states, and then measuring and dividing out $\langle 0|O_A|h\rangle$ and $\langle 0|O_B|h\rangle$. The relevant Feynman graph is given in Fig. 16.3.

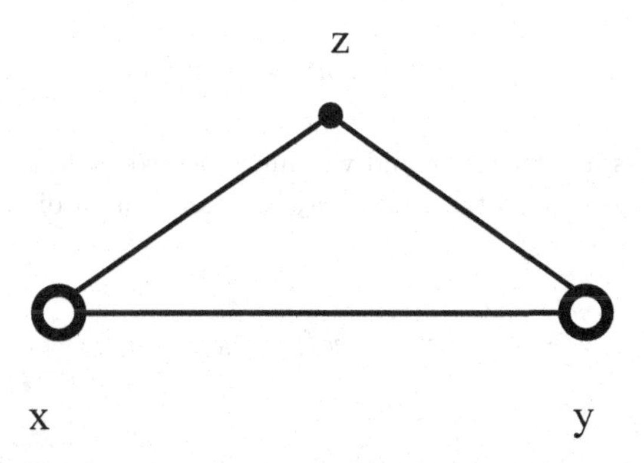

Fig. 16.3 Feynman graph for computing a form factor. The open circles are hadronic wave functions, while the closed circle is the current operator.

In the actual calculation one would need many different values of momenta to map the form factor. (Remember, momentum is quantized to be a multiple of $2\pi/L$ in a box of size L.) The best signals come when the momenta of the initial and final hadron are small. Efficiently computing the actual three point function is nontrivial. Let us neglect all Dirac algebra and focus only on the locations of the vertices. We assume that the meson on the left hand side of the graph is created at space-time location x, the meson the right is annihilated at y, and the current is inserted at z. Then the amplitude is $M \sim \mathrm{Tr} S(x,y)S(y,z)S(z,x)$. Now we see a problem. $S(z,x)$ and $S(x,y) = \gamma_5 S^\dagger(y,x)\gamma_5$ can be computed with a single source (a delta-function at x), at a cost that is proportional to the volume of the lattice. But $S(y,z)$ involves many source points and many sink points. Its cost is apparently volume-squared. The trick for handling this problem is called the "sequential-source" or "exponentiation" method. One way to implement it is to write $M = \mathrm{Tr} S(x,y)T(t,x)$ where $T(y,x) = S(y,z)S(z,x)$. Then $D(w,y)T(y,x) = S(w,x)$. Use the first propagator as the source for the second one. The cost of the matrix element now scales as twice the volume, not its square.

16.4.5 *Meson B parameters*

As a last simple case consider matrix elements that determine the $\bar{K} - K$ or $\bar{B} - B$ mass differences and consequently particle-antiparticle oscillations between them. The hadronic matrix element for the first of these processes is characterized by

$$M = \langle \bar{K} | \bar{s}_a \gamma_\mu (1-\gamma_5) d_a \bar{s}_b \gamma_\mu (1-\gamma_5) d_b | K \rangle. \tag{16.37}$$

The subscript is a color index and we sum repeated indices. It has been customary to write M in terms of decay constants and a dimensionless B parameter:

$$M = \frac{8}{3}(m_K f_K)^2 B_K. \tag{16.38}$$

The motivation for this breakup goes back to pre-lattice days and an approximation called "vacuum saturation" (Gaillard and Lee, 1974b). The idea was to approximate the amplitude by inserting the lowest contributing state, the vacuum, in all possible ways between the bilinear currents that make up the four fermion operator. A Fierz identity

$$\bar{\psi}_1 \gamma_\mu (1-\gamma_5) \psi_2 \bar{\psi}_3 \gamma_\mu (1-\gamma_5) \psi_4 = \bar{\psi}_1 \gamma_\mu (1-\gamma_5) \psi_4 \bar{\psi}_2 \gamma_\mu (1-\gamma_5) \psi_3 \tag{16.39}$$

lets us write the vacuum-saturated expression as

$$M = 2(\langle \bar{K}|\bar{s}_a\gamma_\mu(1-\gamma_5)d_a|0\rangle\langle 0|\bar{s}_b\gamma_\mu(1-\gamma_5)d_b|K\rangle + \langle \bar{K}|\bar{s}_a\gamma_\mu(1-\gamma_5)d_b|0\rangle\langle \bar{K}|\bar{s}_b\gamma_\mu(1-\gamma_5)d_a|0\rangle). \tag{16.40}$$

To get this result, note that there are four possible ways to assign the quarks in the four-fermion operator to the quarks in the mesons: two for the s quark (left and right), two for the d quark. When the assignments mix quarks from both Dirac inner products in the four-fermion operator, we used the Fierz rearrangement to re-write them before saturating with the vacuum.

Now we use

$$\langle 0|\bar{s}_b\gamma_\mu(1-\gamma_5)d_b|K\rangle = -if_K q^\mu e^{-iqx} \tag{16.41}$$

and

$$\langle 0|\bar{s}_b\gamma_\mu(1-\gamma_5)d_a|K\rangle = -if_K q^\mu e^{-iqx}\delta_{ab}/3 \tag{16.42}$$

(the kaons are color singlets, carrying momentum q), to get a weighting factor for the matrix element

$$M = 2\left[1 \times 1 + (1/3) \times (1/3)\sum_{a,b}\delta_{ab}^2\right](f_K m_K)^2 = 8/3(f_K m_K)^2. \tag{16.43}$$

Thus in the vacuum saturation approximation, $B_K = 1$. This analysis is oversimplified, of course. The matrix element M carries an anomalous dimension, whereas the vacuum saturation expression has no μ dependence, since it is written in terms of (partially) conserved currents. The true B_K is not unity, but it is a dimensionless number of order unity.

In a real lattice calculation, one might want to measure

$$B_K = \frac{\langle \bar{K}|(\bar{s}d)_{LL}(\bar{s}d)_{LL}|K\rangle}{8/3(f_K m_K)^2} \tag{16.44}$$

as a ratio of correlators, because fluctuations between the numerator and denominator may be highly correlated, and the ratio would tend to suppress them. The denominator would be a product of two-point functions, each of which has an axial vector current at the sink.

The Feynman graph for M is given in Fig. 16.4(a). Imagine that the two mesons are produced using interpolating fields at lattice time slices $t = t_1$ and $t = t_2$, and the four-point operator is evaluated at $t = t_4$, where

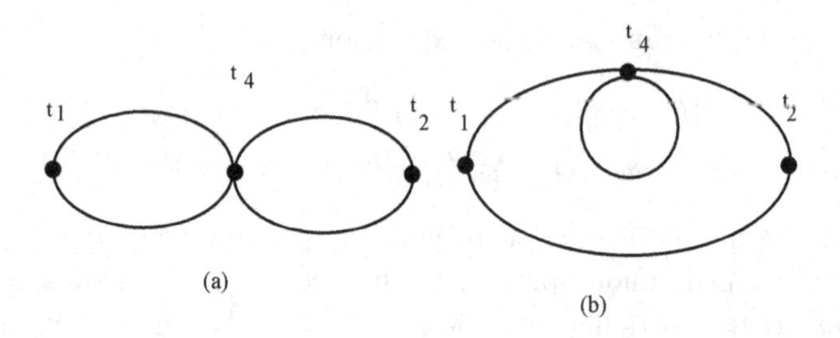

Fig. 16.4 "Eight" (a) and "eye" (b) graphs for four-fermion operators.

$t_2 \gg t_4 \gg t_1$ is required to make sure that the matrix element is evaluated between the lowest mass mesons.

There are obviously many ways to organize the calculation. One way to proceed is to use the space-time point of the four-fermion operator as the source for all the propagators. Then one would join the propagators to make mesons and average the meson propagators over spatial sites to project on $\vec{p} = 0$ (or any other momentum eigenstate). This choice is nearly always used when one is computing B_B. It is the cheapest choice (it requires only one set of propagators per lattice), but it is probably not the cleanest. The operator is measured at one point per configuration.

Another choice is to use two sources per lattice, one for each meson, and to join the propagators at the operator. This procedure averages the operator over many lattice locations per configuration, giving better statistics. However, the initial and final states are not momentum eigenstates, and the zero-momentum signal (behaving like $\exp[-m(t_2 - t_1)]$) has a contamination from nonzero momentum states, going like $\exp[-E(p)(t_2 - t_1)]$. On the lattice p is quantized in multiples of $2\pi/L$, so this might not be a problem for small hadron mass, since $E(p) \gg m$ for all p. Difficulties arise when the mass of the meson becomes large, as $E(p) \sim m + p^2/(2m) + \ldots$, $E(p) - m$ becomes small, and the contamination term does not fall off very fast with t. So this would not be a good choice for B_B. It works well for B_K, though.

The B parameters are one class of lattice graphs for measuring matrix elements of four-fermion operators. In the lattice literature they are called "figure-eight graphs." Other weak processes can involve graphs in which two of the fermion lines at the vertex are tied together. These are called "eye graphs." An example is shown in Fig.16.4(b). They can be computed using the sequential source trick.

In Sec. 16.2.4 we alluded to problems associated with operator mixing.

Let us return to that discussion for the specific case of B_K. For lattice discretizations that preserve chiral symmetry, operator mixing is as in the continuum, plus additional mixing with higher dimensional operators. This mixing gives rise to scale violations.

With Wilson fermions the operator $O_{LL} = (\bar{s}d)_{LL}(\bar{s}d)_{LL}$ mixes with a sum of operators of the form $(\bar{s}\Gamma d)(\bar{s}\Gamma d)$ where Γ is a Dirac matrix. These operators have a different chiral structure from O_{LL}. For Wilson or clover fermions, the perturbative mixing coefficient goes as $Z' \sim r^2 g^2$ where r is the coefficient of the Wilson term. This result is understandable, since mixing is due to explicit chiral symmetry breaking – the mixing does not occur for naive fermions ($r = 0$). (A good description of the perturbative calculation can be found in Bernard *et al.* (1987).) Our vacuum saturation argument tells us that the matrix element M scales quadratically with the pseudoscalar mass $M \sim m_{PS}^2$ in the chiral limit. This does not happen for the opposite chirality operators. For example, a $\gamma_5 - \gamma_5$ term would have a matrix element that would go to a nonzero constant in the chiral limit (since $\langle 0|\bar{\psi}\gamma_5\psi|PS\rangle \sim (m_{PS}^2/m_q)f_{PS}$. Thus the desired part of the lattice operator for O_{LL} becomes a vanishingly small part of the lattice measurement in the chiral limit. It is possible to compute the coefficient of operator mixing nonperturbatively [for one example, see Conti *et al.* (1998)], and subtract it, but our impression (not having tried to do it ourselves) is that this is not an easy procedure.

As for staggered fermions, because of the residual $U(1)$ chiral symmetry, they do not suffer from a chiral symmetry-violating mixing. To construct operators with the correct flavor and spin involves combining contributions from one-component fields distributed over the sites of the hypercube. They must be made gauge-invariant by including products of link fields, or the operators must be measured after gauge-fixing. The construction of the operators is a bit technical, so we refer the reader to the original literature. A good beginning are the papers (Kilcup and Sharpe, 1987; Sharpe *et al.*, 1987).

At the time we are writing, B_K measurements with chiral lattice fermions (staggered, overlap, domain wall) have smaller error bars than ones done with Wilson-type fermions. We believe that this is a direct consequence of their better chiral properties.

16.4.6 *Other purely hadronic weak interactions*

Other than the B parameters, the only purely hadronic weak processes with an extensive literature are matrix elements for the decay $K \to \pi\pi$ that probe either the $\Delta I = 1/2$ rule or measure ϵ'/ϵ, the ratio of direct to indirect CP violation in K decays. Rather than working directly with the two-pion state, practitioners typically first use chiral perturbation theory to convert the decay amplitude $K \to \pi\pi$ to off-shell matrix elements of the axial vector current, $K \to \pi$ and $K \to$ vacuum. To make matters worse, the expression for ϵ'/ϵ happens to be dominated by the difference of two of the two operators with the largest matrix elements, the QCD penguin

$$Q_6 = (\bar{s}d)_L \left[(\bar{u}u)_R + (\bar{d}d)_R + (\bar{s}s)_R\right], \tag{16.45}$$

and the electroweak penguin

$$Q_8 = \frac{1}{2}(\bar{s}d)_L \left[(2\bar{u}u)_R - (\bar{d}d)_R - (\bar{s}s)_R\right]. \tag{16.46}$$

There is a lot of cancellation in the actual numerical simulation data. Your authors have no experience with these calculations and suggest that you go directly to the journal literature if you want to learn about them.

16.5 Evading a no-go theorem

So far we have considered initial and final states with zero or one particle in them. What about processes like $K \to \pi\pi$? It turns out that a direct attack on this problems from the lattice is impossible. This impossibility is called the "Maiani-Testa no-go theorem" (Maiani and Testa, 1990).

There are two parts to the theorem. First, in any calculation performed in Euclidean space-time, there are no asymptotic noninteracting states, so no distinction between "in" and "out" states and no energy conservation. All one can measure are real numbers, which are averages of the two cases,

$$\langle\pi\pi|H_W|K\rangle_E = \frac{1}{2}[\langle\pi\pi|H_W|K\rangle + \langle K|H_W|\pi\pi\rangle]. \tag{16.47}$$

Any phases in the amplitudes (due to final-state interactions, for example) are lost. Second, matrix elements are extracted on the lattice from the long-time behavior of correlation functions. Maiani and Testa showed that in this limit the correlations functions are dominated by off shell amplitudes in which all the final state particles are at rest, like $\langle\pi(\vec{p}=0)\pi(\vec{p}=0)|H_W|K\rangle$.

At the time we are writing there are solutions to this problem that are theoretically sound; however, no large scale tests have been done to show whether or not they are feasible. The reader who has just read the chapter on finite volume effects can guess the answer. In a finite lattice there is no continuum of free particle states. Instead, there is a set of discrete modes, "particles in the box." When we introduce a special state and tune its energy toward one of these levels, there is an avoided level crossing, and the repulsion of the levels gives information about a finite-volume matrix element. Lellouch and Lüscher (2001) showed how to relate this matrix element to the infinite-volume matrix element characterizing the physical decay. The constant of proportionality involves the function that parameterizes the avoided level crossing.

Chapter 17

QCD at high temperature and density

At high temperatures and densities strongly interacting matter becomes a quark gluon plasma. Lattice simulations are well suited for the study of high temperature equilibrium properties of the plasma. High density has proven far more difficult. The study of nonequilibrium processes and transport phenomena is an even more challenging problem.

17.1 Simulating high temperature

As we have noted in Ch. 3, thermal equilibrium properties are determined from the thermal expectation values of operators

$$\langle \mathcal{O} \rangle = \frac{\mathrm{Tr}\ (\mathcal{O} e^{-H/T})}{\mathrm{Tr}\ e^{-H/T}}. \tag{17.1}$$

As we saw, the corresponding fields in the Feynman path integral are defined over a Euclidean space-time with time extent $N_t a = 1/T$. Simulations are typically carried out at a fixed number of lattice sites in the time direction N_t. The temperature is adjusted by varying the gauge coupling g^2. Decreasing g^2 decreases the lattice spacing and increases the temperature. Of course, at any given temperature, approaching the continuum limit requires increasing N_t.

The link variables obey periodic boundary conditions in time. The fermions are antiperiodic, as described in Ch. 3.

17.2 Introducing a chemical potential

To simulate at nonzero baryon number density one introduces a chemical potential coupled to the conserved baryonic charge B. In the continuum the expression is

$$\langle \mathcal{O} \rangle = \frac{\mathrm{Tr}\,(\mathcal{O} e^{-(H-\mu B)/T})}{\mathrm{Tr}\, e^{-(H-\mu B)/T}}. \tag{17.2}$$

For naive fermions the coupling of the chemical potential to the conserved charge density on the lattice takes the form

$$\frac{1}{2a}[\bar{\psi}_x e^{\mu a/3} \gamma_0 U_{0x} \psi_{x+\hat{t}} - \bar{\psi}_{x+\hat{t}} e^{-\mu a/3} U^\dagger_{0,x} \gamma_0 \psi_x] \tag{17.3}$$

which reduces in the continuum limit to

$$\bar{\psi}_x \gamma_0 (\partial_t - i A_{x,0} + \mu/3) \psi_x. \tag{17.4}$$

For other actions one simply replaces each forward gauge link by $e^{\mu a/3} U_{0,x}$ and each backward link by $e^{-\mu a/3} U^\dagger_{0,x}$. Thus the baryon chemical potential appears as an imaginary vector potential. Unfortunately, as a consequence, the corresponding fermion determinant is no longer real and the determinant can no longer be used directly as a probability weight. (An imaginary chemical potential, however, produces a positive weight.) This problem is equivalent to the the notorious sign problem in condensed matter physics that arises in the Monte Carlo simulation of strongly correlated electron systems doped away from a half-filled electron band. The solution of one problem is the solution of the other.

17.3 High quark mass limit and chiral limit

The phase diagram as a function of quark mass, temperature, and density is particularly interesting. As a function of the quark masses the phase transition from ordinary matter to the quark-gluon plasma has two important limits. When all quark masses are infinite we have, in effect, a pure Yang-Mills theory, which for the $SU(3)$ gauge group has a first order high temperature phase transition from a confined to a deconfined phase. The order parameter is the asymptotic value of the Polyakov loop correlator introduced in Sec. 5.4.3. That is, for

$$\exp[-\beta F(R)] = \langle P(x) P^*(x+R) \rangle\,, \tag{17.5}$$

with $\beta = 1/T$, the order parameter $2F_0 = \lim_{R\to\infty} F(R)$, can be interpreted as (twice) the free energy of a point charge. In the confined phase that free energy is infinite. In the deconfined phase it is finite. It is common in simulations to replace this definition of the order parameter by the simpler expedient $\langle|P(x)|\rangle$.

As we noted in Sec. 5.4.3 the Euclidean action of the pure $SU(3)$ Yang-Mills theory is unchanged if we multiply all time-like links on a single time slice by $z \in Z(3)$, the center of the gauge group. Under this transformation the Polyakov loop becomes $zP(x)$. It is useful to study the analogy with a three-dimensional $Z(3)$ Potts model [a generalization of the Ising model with $Z(3)$ spins]. The Polyakov loop plays the role of the spin variable and the physical interpretation of temperature is inverted. That is, high temperature QCD corresponds to low temperature Potts. The spin system is ferromagnetically coupled, so the QCD deconfined phase corresponds to a ferromagnetic phase. The mapping of QCD onto a such three-dimensional, three-state Potts model is a reasonable approximation in the high temperature, strong coupling, large quark mass limit (Yaffe and Svetitsky, 1982; DeGrand and DeTar, 1983). The Potts Hamiltonian is

$$J \sum_{r,j} \mathrm{Re}(z_r z_{r+j}^*) + h \sum_r \mathrm{Re} z_r + ih' \sum_r \mathrm{Im} z_r \qquad (17.6)$$

for sites r and directions j on a three-dimensional lattice. The spin variables z_r are in $Z(3)$ and correspond to the Polyakov loop, reduced to the center of $SU(3)$. The real coupling h drops to zero with increasing quark mass. The imaginary coupling h' vanishes with vanishing chemical potential μ and increasing quark mass. The order parameter is now $\langle z \rangle \propto \exp(-F_0/T)$. As long as the quark mass is not infinite, the phase of the order parameter is strictly real.

We see that at nonzero chemical potential in this spin basis $h' > 0$, the Hamiltonian is imaginary so $\exp(-H/T)$ cannot be used as a probability weight for a Monte Carlo simulation. The Boltzmann factor has a phase with fluctuations that grow with the lattice volume. This problem carries over to full QCD, making it extremely difficult to carry out simulations at nonzero baryon number density.

As the quark mass is decreased or, equivalently, the magnetic field h is increased, the first order phase transition is weakened, and we eventually reach a critical point beyond which there is only a crossover.

In the opposite, chiral limit of vanishing quark masses the high temperature phase transition restores the spontaneously broken chiral symmetry.

In this case the order parameter is the chiral condensate $\langle\bar{\psi}\psi\rangle$. The order of the phase transition depends on the number of quark flavors and whether the $U_A(1)$ symmetry is involved. If it is not, for two flavors the transition is second order. For three or more it is first order (Pisarski and Wilczek, 1984).

17.4 Locating and characterizing the phase transition

In the intermediate regime of physical quark masses that are neither zero nor infinite it is necessary to carry out a numerical simulation to determine whether there is a phase transition at all, to locate its temperature and determine its order. In addition to looking for an inflection point in Polyakov loop parameter and chiral condensate as a function of temperature, it is often easier to look for peaks in corresponding susceptibilities

$$\chi_P = \langle P^2 \rangle - \langle P \rangle^2 \tag{17.7}$$

or

$$\chi_m = \langle (\bar{\psi}\psi)^2 \rangle - \langle \bar{\psi}\psi \rangle^2 . \tag{17.8}$$

A finite size scaling analysis studies the peak height as a function of increasing lattice volume V. If the peak height is asymptotically constant, there is no phase transition. If it rises linearly with the volume, the phase transition is first order. If it rises less rapidly, it is a critical point. The Binder cumulant B_4 is often used to help determine the universality class of a critical point (Binder, 1981):

$$B_4(0) = \frac{\langle (\delta\bar{\psi}\psi)^4 \rangle}{\langle \delta\bar{\psi}\psi \rangle^2}, \tag{17.9}$$

where $\delta\bar{\psi}\psi = \bar{\psi}\psi - \langle\bar{\psi}\psi\rangle$.

17.5 Simulating in a nearby ensemble

Particularly in the study of high temperature and density QCD, it has proven useful to find ways to simulate in one ensemble and infer results for a different ensemble. This trick is accomplished through "reweighting". For a simple illustration, suppose we wanted to get the expectation value of an observable $\mathcal{O}$ at an inverse gauge coupling $\beta + \delta\beta$ when we have generated an ensemble for inverse gauge coupling β. Suppose, further that the effective action depends on β only as $S_{\rm eff}(U) = \beta H(U)$, as it does, for example, in an unimproved Yang-Mills theory. We would take a weighted average of the observable as follows:

$$\begin{aligned}
\langle \mathcal{O} \rangle (\beta + \delta\beta) &= \frac{\int [dU] \mathcal{O} e^{-(\beta+\delta\beta)H(U)}}{\int [dU] e^{-(\beta+\delta\beta)H(U)}} \\
&= \frac{\left\langle \mathcal{O} e^{-\delta\beta H(U)} \right\rangle |_\beta}{\left\langle e^{-\delta\beta H(U)} \right\rangle |_\beta}
\end{aligned} \tag{17.10}$$

where $\langle\rangle |_\beta$ means averaging with respect to the ensemble at β. More generally we simply weight the operator by a factor that corrects for the change in the exponential of the effective action when the parameter is changed. As we see, it is easiest to do this when the action has a simple dependence on the parameter.

This method works if the "overlap" between the simulated ensemble β and the unsimulated shifted ensemble $\beta + \delta\beta$ is significant. If we actually simulated both ensembles, this difference could be measured by looking at the probability distribution of the values of $H(U)$ on each ensemble. For either ensemble the distribution is typically peaked with a rapid fall off on either side. As we shift β the peak typically shifts. If the two distributions have a strong overlap, the method succeeds. When they do not overlap significantly, the variance of the weighted averages is large and the signal to noise ratio is poor. Since the width of the distribution narrows with increasing volume, the accessible parameter range shrinks.

A generalization of the reweighting method to several ensembles was developed by Ferrenberg and Swendsen (1988). A typical application selects a series of parameter values β_i and generates one ensemble for each. One then interpolates the expectation values of the operators of interest in β by reweighting. When two ensembles are involved, the reweighting method must establish the free energy difference between them. This is possible if the two simulated ensembles overlap in the sense described above. As the volume is increased the density of β_i values must increase. Eventually, it is probably more cost effective to resort to a parametric interpolation of the unreweighted results.

17.6 Dimensional reduction and nonperturbative behavior

In the high temperature limit the time dimension of the lattice shrinks and we have an effective three-dimensional theory. The gluon field satisfies a periodic boundary condition. The squared four momentum for a free gluon becomes

$$p^2 = (2\pi n T)^2 + |\vec{p}|^2 \tag{17.11}$$

where n is the Matsubara quantum number for the periodic time dimension. For $n > 0$ and high T, the gluon modes are very massive, so we ignore all but the $n = 0$ mode. By contrast the quark fields must satisfy an antiperiodic boundary condition, so their squared momentum is

$$p^2 = [(2n+1)\pi T]^2 + |\vec{p}|^2. \tag{17.12}$$

They are massive for any n. The time component of the color vector potential becomes an effective colored scalar field – a "Higgs" field. The remaining spatial components of the vector potential are interpreted as a three-dimensional Euclidean Yang-Mills field. Thus QCD reduces to a three-dimensional Yang-Mills theory coupled to a Higgs field and a massive quark field. The gauge coupling carries a dimension. In terms of the usual four-dimensional coupling, it is $g^2 T$. This is high temperature dimensional reduction (Ginsparg, 1980; Appelquist and Pisarski, 1981).

If we select one of the large spatial dimensions and call it our new Euclidean time dimension, we may discuss the spectrum of the dimensionally reduced theory. It is confining with an area law for Wilson loops that are spatially oriented in the original four-dimensional world. The quark excitations are charmonium-like with meson masses of order $2\pi T$ and baryon masses, $3\pi T$. These masses translate to "screening masses" in the original four-dimensional world. That is, the asymptotic behavior of correlators of operators with space-like separation z,

$$C(z) = \langle O(z) O(0) \rangle, \tag{17.13}$$

is controlled by the masses of these states. Asymptotic freedom would suggest that as the temperature is increased, the momentum scale of interactions increases, the coupling decreases, and the quark-plasma becomes a free gas. This is true for quanta in the ensemble with momenta of the order T. But from dimensional reduction we see that no matter how high the temperature there is always a residual nonperturbative component coming from quanta with momenta of order g^2T. For example, in a perturbative expansion of a thermodynamic quantity such as the pressure, these nonperturbative effects start at $\mathcal{O}(g^6)$ [for a recent reference, see (Kajantie *et al.*, 2003)].

This same analysis has been applied to electroweak theory by Kajantie *et al.* (1996).

17.7 Miscellaneous observables

Here we mention a few important observables.

17.7.1 *Quark number susceptibilities*

Quark number susceptibilities are of phenomenological interest, since they quantify fluctuations in flavor and baryon number in the plasma, quantities that may be indirectly related to fluctuations in flavor and baryon production in a heavy-ion collision. The quark number susceptibility is defined in terms of conserved quark numbers Q_α as follows

$$\chi_{\alpha,\beta}(T) = \langle (Q_\alpha - \langle Q_\alpha \rangle)(Q_\beta - \langle Q_\beta \rangle) \rangle \,. \tag{17.14}$$

One can either choose the flavor basis $\alpha = u, d, s$ or the somewhat more physical isospin, hypercharge, baryon number basis $\alpha = I, Y, B$. Since the charges are bilinears in the quark fields, the expectation values involve meson-like correlators with quark-line connected and often quark-line disconnected pieces. Typically, one uses a stochastic or random source method to evaluate the latter [see Ch. 11, Eq. (11.15)].

17.7.2 *Quarkonium potential*

The free energy of a pair of static quarks as a function of separation R Eq. (17.5) is used in modeling the production of the J/ψ particle in heavy ion collisions. Thermal effects increase screening of the color charges and weaken the binding of the state (Matsui and Satz, 1986).

17.7.3 *Equation of state*

The equation of state, *i.e.*, energy density or pressure *vs* temperature, is especially important for hydrodynamical models of plasma evolution. The standard determination starts from the thermodynamic identities

$$\begin{aligned}
\varepsilon V &= - \left.\frac{\partial \ln Z}{\partial(1/T)}\right|_V \\
\frac{p}{T} &= \left.\frac{\partial \ln Z}{\partial V}\right|_T \approx \frac{\ln Z}{V} \\
I = \varepsilon - 3p &= -\frac{T}{V}\frac{d \ln Z}{d \ln a}
\end{aligned} \tag{17.15}$$

for energy density ε, pressure p, "interaction measure" I, lattice spatial volume V, temperature T, and lattice spacing a. The quantities of phenomenological importance are the differences $\Delta\varepsilon$, Δp, and ΔI between the energy density, pressure, and interaction measure of the plasma and the corresponding values for the vacuum. So all quantities in Eq. (17.15) must be computed on two ensembles that differ only in the time extent aN_t, one of them at large aN_t to simulate the vacuum (zero temperature) ensemble and the other at the desired high temperature value of aN_t. The paired ensembles are typically simulated over a range of temperatures T.

The interaction measure is easiest to measure, since it is derived from from an isotropic variation of the lattice spacing. The derivative $d \ln Z / d \ln a$ reduces to a linear combination of expectation values of the various operators that enter into the definition of the action. From the large volume approximation $p \approx \ln Z/V$ we can convert the expression for the interaction measure into an integral relation for the pressure

$$\Delta p a^4 = -\int_{\ln a_0}^{\ln a} d \ln a' \, \Delta I a'^4. \tag{17.16}$$

from which the energy density follows from $\Delta\varepsilon = \Delta I + 3\Delta p$. The integration over lattice spacings a is equivalent to an integration over temperature. The lower limit of integration $\ln a_0$ corresponds to a low temperature. It must be sufficiently low that the pressure difference there is essentially zero.

The equation of state is an especially difficult quantity to compute because it involves the difference between high and low temperature observables. The energy density is strongly dominated by ultraviolet fluctuations in both ensembles, so the difference has a large variance. We require high statistics to compensate. The difficulty grows rapidly as the lattice spacing is decreased.

17.8 Nonzero density

The transition from confined matter to a quark-gluon plasma is expected to occur at high baryon number density as well. The phase diagram in temperature and chemical potential is expected to have a line of phase transitions extending from the well-studied high temperature, zero μ point to a possible tricritical point at nonzero μ and T (Halasz *et al.*, 1998). At low temperature and high baryon density even more exotic phases have been suggested, such as a color superconducting phase (Bailin and Love, 1984) or a color-flavor locked phase (Alford *et al.*, 1999). It has been a long standing challenge for lattice QCD to explore this region.

As we mentioned above, when we introduce a nonzero baryon chemical potential, the fermion determinant becomes complex so it cannot be used as a probability weight. With the much simpler Potts model one can evade this problem by a change of basis in the functional integral. Condella and DeTar (2000) showed that if one works in the analog of a Fock basis for quark occupation and color flux quanta one can recast the functional integral into one with a positive definite probability weight. For strong-coupling lattice QCD the monomer-dimer basis of Karsch and Mutter (1989) is similar in spirit, but it is impractical at weaker coupling.

Another series of approaches use reweighting. Early attempts extracted the phase of the determinant and generated an ensemble based on its absolute value (Toussaint, 1990), but the method worked only on uninterestingly small volumes. The Glasgow method is based on a Maclaurin series expansion of the fermionic determinant in the fugacity variable $z = \exp(\mu/3T)$, the coefficients of which define contributions of fixed baryon number (canonical ensemble). Simulations with this method attempted reweighting based on an ensemble at zero chemical potential. This approach did not prove particularly successful (Barbour *et al.*, 1998).

The most successful of the reweighting methods was developed by Fodor and Katz (2002) for the purpose of tracking the transition line from zero μ into the (μ, T) plane. Like the Glasgow method, it works with an ensemble generated at the $\mu = 0$ transition point but it shifts simultaneously in both the inverse gauge coupling β and μ, all the while striving to stay on the transition line. It is claimed that the method owes its success to the fact that the ensembles along the transition line have strong overlap. As of this writing it remains to be seen whether the method can be carried out successfully at larger volumes and smaller lattice spacing.

A more modest but solid approach explores only small chemical potential, *i.e.*, small baryon number densities. This regime is still of relevance in the study of heavy ion collisions. It assumes analyticity in the observable of interest at small μ. There are two versions of this method. One uses the fact that the fermion determinant is real when the chemical potential is imaginary. So one simulates at imaginary chemical potential, fits the result to a power series in μ and performs an analytic continuation to real values of μ (de Forcrand and Philipsen, 2002). Alternatively, by carrying out the necessary derivatives by hand, one can develop the Taylor series expansion of an observable of interest with the explicit coefficients evaluated at zero chemical potential (Allton *et al.*, 2003). In either case the ensemble is generated with a positive definite Boltzmann weight.

17.9 Spectral functions and maximum entropy

Drell-Yan pair production in the quark gluon plasma is governed by the thermal correlator of the electromagnetic current

$$D(\tau,\vec{x}) = \left\langle J_\mu(\tau,\vec{x}) J^\dagger_\mu(0,0)\right\rangle . \tag{17.17}$$

Note that the time variable $\tau \in [0, 1/T]$ is Euclidean and periodic. But only real Minkowski time t is experimentally meaningful. It would be, of course, very interesting to be able to use lattice simulations to predict the rates of lepton pair production. Extending a lattice calculation of the correlator to the real Minkowski region involves an analytic continuation from a finite set of points $t = i\tau$ to the infinite real t axis. Even without the statistical uncertainties in the measurement of the correlator, this is an ill-posed problem requiring additional assumptions to make progress. Still some interesting methods have been proposed.

The analytic continuation to real Minkowski time can be parameterized through the integral transform

$$D(\tau,\vec{k}) = \int_0^\infty d\omega \, \frac{e^{-\tau\omega} + e^{-(1/T-\tau)\omega}}{1 - e^{-\omega/T}} \sigma(\omega,\vec{k}), \tag{17.18}$$

for $0 < \tau < 1/T$, where we have done a spatial Fourier transform to $\vec{k}$. The spectral function $\sigma(\omega,\vec{k})$ has discrete (delta function) contributions from bound states and for a continuum of multiparticle states, it is proportional to the density of states. The pair production rate at lepton pair invariant mass squared $k^2 = \omega^2 - |\vec{k}|^2$ is directly proportional to $\sigma(\omega,\vec{k})$ (Karsch *et al.*, 2002).

So the problem of analytic continuation becomes a problem of inverting Eq. (17.18) from a finite set of LHS data, which is impossible without further constraints. In the absence of other information, it is popular to use a maximum entropy method (Jarrell and Gubernatis, 1996). Here we follow Asakawa *et al.* (2001). The method seeks a spectral function that fits the lattice data through Eq. (17.18) by minimizing the appropriate χ^2, and maximizing the Shannon-Jaynes entropy

$$S = \int_0^\infty d\omega \left[\sigma(\omega) - m(\omega) - \sigma(\omega)\log\left(\frac{\sigma(\omega)}{m(\omega)}\right)\right] \tag{17.19}$$

where $m(\omega)$ is a default model, typically two-body phase space. The two constraints are combined to maximize the combination

$$Q = \alpha S - \chi^2 \tag{17.20}$$

for fixed α. The result is, then, parameterized by α. One then finishes by averaging over α using the likelihood $\exp(Q)$ as a probability weight.

The maximum entropy method is used here as a sort of Occam's razor. It tries to make the spectral function look like the default model in the absence of any constraint from the lattice data. One might presume that any structure that emerges must surely be dictated by the data. However, the method has been known to produce artifact oscillations that require a careful error analysis to exclude.

The maximum entropy method is also being employed in an effort to determine transport coefficients of the plasma (Nakamura and Sakai, 2005).

Clearly the more lattice data, the better, meaning that one should want to simulate at large N_t and small lattice spacing (at least in Euclidean time).

Bibliography

Albanese, M. *et al.* (1987). Glueball masses and string tension in lattice QCD, *Phys. Lett.* **B192**, p. 163.

Alford, M. G., Dimm, W., Lepage, G. P., Hockney, G. and Mackenzie, P. B. (1995). Lattice QCD on small computers, *Phys. Lett.* **B361**, pp. 87–94.

Alford, M. G., Klassen, T. R. and Lepage, G. P. (1997). Improving lattice quark actions, *Nucl. Phys.* **B496**, pp. 377–407.

Alford, M. G., Rajagopal, K. and Wilczek, F. (1999). Color-flavor locking and chiral symmetry breaking in high density QCD, *Nucl. Phys.* **B537**, pp. 443–458.

Allton, C. R. *et al.* (2003). The equation of state for two flavor QCD at non-zero chemical potential, *Phys. Rev.* **D68**, p. 014507.

Altarelli, G. and Maiani, L. (1974). Octet enhancement of nonleptonic weak interactions in asymptotically free gauge theories, *Phys. Lett.* **B52**, pp. 351–354.

Appelquist, T. and Pisarski, R. D. (1981). High-temperature Yang-Mills theories and three-dimensional quantum chromodynamics, *Phys. Rev.* **D23**, p. 2305.

Asakawa, M., Hatsuda, T. and Nakahara, Y. (2001). Maximum entropy analysis of the spectral functions in lattice QCD, *Prog. Part. Nucl. Phys.* **46**, pp. 459–508.

Aubin, C. and Bernard, C. (2003). Pion and kaon masses in staggered chiral perturbation theory, *Phys. Rev.* **D68**, p. 034014.

Aubin, C. and Bernard, C. (2004). Staggered chiral perturbation theory, *Nucl. Phys. Proc. Suppl.* **129**, pp. 182–184.

Aubin, C. *et al.* (2004). Light hadrons with improved staggered quarks: Approaching the continuum limit, *Phys. Rev.* **D70**, p. 094505.

Bailey, J. A. and Bernard, C. (2005). Staggered lattice artifacts in 3-flavor heavy baryon chiral perturbation theory, *PoS* **LAT2005**, p. 047.

Bailin, D. and Love, A. (1984). Superfluidity and superconductivity in relativistic fermion systems, *Phys. Rept.* **107**, p. 325.

Banks, T. and Casher, A. (1980). Chiral symmetry breaking in confining theories, *Nucl. Phys.* **B169**, p. 103.

Barbour, I. M., Morrison, S. E., Klepfish, E. G., Kogut, J. B. and Lombardo, M.-P. (1998). Results on finite density QCD, *Nucl. Phys. Proc. Suppl.* **60A**,

pp. 220–234.
Basak, S. *et al.* (2006). Baryon operators and baryon spectroscopy, *Nucl. Phys. Proc. Suppl.* **153**, pp. 242–249.
Bernard, C. (2006). Staggered chiral perturbation theory and the fourth-root trick, arXiV hep-lat/0603011.
Bernard, C. W. and DeGrand, T. (2000). Perturbation theory for fat-link fermion actions, *Nucl. Phys. Proc. Suppl.* **83**, pp. 845–847.
Bernard, C. W., Draper, T., Soni, A., Politzer, H. D. and Wise, M. B. (1985). Application of chiral perturbation theory to K → 2 pi decays, *Phys. Rev.* **D32**, pp. 2343–2347.
Bernard, C. W. and Golterman, M. F. L. (1992). Chiral perturbation theory for the quenched approximation of QCD, *Phys. Rev.* **D46**, pp. 853–857.
Bernard, C. W. and Golterman, M. F. L. (1994). Partially quenched gauge theories and an application to staggered fermions, *Phys. Rev.* **D49**, pp. 486–494.
Bernard, C. W., Soni, A. and Draper, T. (1987). Perturbative corrections to four fermion operators on the lattice, *Phys. Rev.* **D36**, p. 3224.
Binder, K. (1981). Finite size scaling analysis of Ising model block distribution functions, *Z. Phys.* **B43**, pp. 119–140.
Blatter, M., Burkhalter, R., Hasenfratz, P. and Niedermayer, F. (1996). Instantons and the fixed point topological charge in the two-dimensional O(3) sigma model, *Phys. Rev.* **D53**, pp. 923–932.
Bochicchio, M., Maiani, L., Martinelli, G., Rossi, G. C. and Testa, M. (1985). Chiral symmetry on the lattice with Wilson fermions, *Nucl. Phys.* **B262**, p. 331.
Brodsky, S. J., Lepage, G. P. and Mackenzie, P. B. (1983). On the elimination of scale ambiguities in perturbative quantum chromodynamics, *Phys. Rev.* **D28**, p. 228.
Brown, F. R. and Woch, T. J. (1987). Overrelaxed heat bath and Metropolis algorithms for accelerating pure gauge Monte Carlo calculations, *Phys. Rev. Lett.* **58**, p. 2394.
Buras, A. J. (1998). Weak hamiltonian, CP violation and rare decays, arXiV hep-ph/9806471.
Cabibbo, N. and Marinari, E. (1982). A new method for updating SU(N) matrices in computer simulations of gauge theories, *Phys. Lett.* **B119**, pp. 387–390.
Callaway, D. J. E. and Rahman, A. (1982). The microcanonical ensemble: A new formulation of lattice gauge theory, *Phys. Rev. Lett.* **49**, p. 613.
Callaway, D. J. E. and Rahman, A. (1983). Lattice gauge theory in microcanonical ensemble, *Phys. Rev.* **D28**, p. 1506.
Chetyrkin, K. G. and Retey, A. (2000). Renormalization and running of quark mass and field in the regularization invariant and MS-bar schemes at three and four loops, *Nucl. Phys.* **B583**, pp. 3–34.
Chodos, A., Jaffe, R. L., Johnson, K., Thorn, C. B. and Weisskopf, V. F. (1974). A new extended model of hadrons, *Phys. Rev.* **D9**, pp. 3471–3495.
Clark, M. A. and Kennedy, A. D. (2004a). Accelerating fermionic molecular dynamics, arXiV hep-lat/0409134.
Clark, M. A. and Kennedy, A. D. (2004b). The RHMC algorithm for 2 flavors of

dynamical staggered fermions, *Nucl. Phys. Proc. Suppl.* **129**, pp. 850–852.
Condella, J. and DeTar, C. (2000). Potts flux tube model at nonzero chemical potential, *Phys. Rev.* **D61**, p. 074023.
Conti, L. *et al.* (1998). Lattice B-parameters for Delta(S) = 2 and Delta(I) = 3/2 operators, *Phys. Lett.* **B421**, pp. 273–282.
Creutz, M. (1979). Confinement and the critical dimensionality of space-time, *Phys. Rev. Lett.* **43**, pp. 553–556.
Creutz, M. (1980). Monte Carlo study of quantized SU(2) gauge theory, *Phys. Rev.* **D21**, pp. 2308–2315.
Creutz, M. (1983). *Quarks, gluons and lattices* (Cambridge Univ. Press).
Creutz, M. (1987). Overrelaxation and Monte Carlo simulation, *Phys. Rev.* **D36**, p. 515.
Creutz, M., Jacobs, L. and Rebbi, C. (1979). Experiments with a gauge invariant Ising system, *Phys. Rev. Lett.* **42**, p. 1390.
Cundy, N., Teper, M. and Wenger, U. (2002). Topology and chiral symmetry breaking in SU(N(c)) gauge theories, *Phys. Rev.* **D66**, p. 094505.
Damgaard, P. H. (2002). The microscopic Dirac operator spectrum, *Nucl. Phys. Proc. Suppl.* **106**, pp. 29–37.
Damgaard, P. H., Diamantini, M. C., Hernandez, P. and Jansen, K. (2002). Finite-size scaling of meson propagators, *Nucl. Phys.* **B629**, pp. 445–478.
Damgaard, P. H., Hernandez, P., Jansen, K., Laine, M. and Lellouch, L. (2003). Finite-size scaling of vector and axial current correlators, *Nucl. Phys.* **B656**, pp. 226–238.
Damgaard, P. H. and Splittorff, K. (2000). Partially quenched chiral perturbation theory and the replica method, *Phys. Rev.* **D62**, p. 054509.
Dashen, R. F., Jenkins, E. and Manohar, A. V. (1994). The 1/N(c) expansion for baryons, *Phys. Rev.* **D49**, pp. 4713–4738.
Dashen, R. F. and Manohar, A. V. (1993a). 1/N(c) corrections to the baryon axial currents in QCD, *Phys. Lett.* **B315**, pp. 438–440.
Dashen, R. F. and Manohar, A. V. (1993b). Baryon - pion couplings from large N(c) QCD, *Phys. Lett.* **B315**, pp. 425–430.
de Forcrand, P. (1996). Progress on lattice QCD algorithms, *Nucl. Phys. Proc. Suppl.* **47**, pp. 228–235.
de Forcrand, P. and Philipsen, O. (2002). The QCD phase diagram for small densities from imaginary chemical potential, *Nucl. Phys.* **B642**, pp. 290–306.
De Rujula, A., Georgi, H. and Glashow, S. L. (1975). Hadron masses in a gauge theory, *Phys. Rev.* **D12**, pp. 147–162.
DeGrand, T. (2003). One loop matching coefficients for a variant overlap action and some of its simpler relatives, *Phys. Rev.* **D67**, p. 014507.
DeGrand, T., Jaffe, R. L., Johnson, K. and Kiskis, J. E. (1975). Masses and other parameters of the light hadrons, *Phys. Rev.* **D12**, p. 2060.
DeGrand, T. A. and DeTar, C. E. (1983). Phase structure of QCD at high temperature with massive quarks and finite quark density: A Z(3) paradigm, *Nucl. Phys.* **B225**, p. 590.
DeGrand, T. A., Hasenfratz, A. and Kovacs, T. G. (1997). Topological structure

in the SU(2) vacuum, *Nucl. Phys.* **B505**, pp. 417–441.

Della Morte, M., Hoffmann, R., Knechtli, F., Sommer, R. and Wolff, U. (2005a). Non-perturbative renormalization of the axial current with dynamical Wilson fermions, *JHEP* **07**, p. 007.

Della Morte, M., Shindler, A. and Sommer, R. (2005b). On lattice actions for static quarks, *JHEP* **08**, p. 051.

Della Morte, M. *et al.* (2005c). Non-perturbative quark mass renormalization in two-flavor QCD, *Nucl. Phys.* **B729**, pp. 117–134.

Donoghue, J. F., Golowich, E. and Holstein, B. R. (1992). *Dynamics of the Standard Model* (Cambridge University).

Drouffe, J. M. and Itzykson, C. (1978). Lattice gauge fields, *Phys. Rept.* **38**, pp. 133–175.

Duane, S. and Kogut, J. B. (1985). Hybrid stochastic differential equations applied to quantum chromodynamics, *Phys. Rev. Lett.* **55**, p. 2774.

Duane, S. and Kogut, J. B. (1986). The theory of hybrid stochastic algorithms, *Nucl. Phys.* **B275**, p. 398.

Duncan, A., Roskies, R. and Vaidya, H. (1982). Monte Carlo study of long range chiral structure in QCD, *Phys. Lett.* **B114**, p. 439.

Edwards, R. G. and Heller, U. M. (2001). Domain wall fermions with exact chiral symmetry, *Phys. Rev.* **D63**, p. 094505.

Edwards, R. G., Joo, B., Kennedy, A. D., Orginos, K. and Wenger, U. (2005). Comparison of chiral fermion methods, *PoS* **LAT2005**, p. 146.

Eichten, E. *et al.* (1975). The spectrum of charmonium, *Phys. Rev. Lett.* **34**, pp. 369–372.

Eidelman, S. *et al.* (2004). Review of particle physics, *Phys. Lett.* **B592**, p. 1.

El-Khadra, A. X., Kronfeld, A. S. and Mackenzie, P. B. (1997). Massive fermions in lattice gauge theory, *Phys. Rev.* **D55**, pp. 3933–3957.

Falcioni, M., Paciello, M. L., Parisi, G. and Taglienti, B. (1985). Again on SU(3) glueball mass, *Nucl. Phys.* **B251**, pp. 624–632.

Ferrenberg, A. M. and Swendsen, R. H. (1988). New Monte Carlo technique for studying phase transitions, *Phys. Rev. Lett.* **61**, pp. 2635–2638.

Fodor, Z. and Katz, S. D. (2002). A new method to study lattice QCD at finite temperature and chemical potential, *Phys. Lett.* **B534**, pp. 87–92.

Fodor, Z., Katz, S. D. and Szabo, K. K. (2004). Dynamical overlap fermions, results with hybrid Monte-Carlo algorithm, *JHEP* **08**, p. 003.

Franco, E. and Lubicz, V. (1998). Quark mass renormalization in the MS-bar and RI schemes up to the NNLO order, *Nucl. Phys.* **B531**, pp. 641–651.

Frezzotti, R., Grassi, P. A., Sint, S. and Weisz, P. (2000). A local formulation of lattice QCD without unphysical fermion zero modes, *Nucl. Phys. Proc. Suppl.* **83**, pp. 941–946.

Frezzotti, R., Grassi, P. A., Sint, S. and Weisz, P. (2001). Lattice QCD with a chirally twisted mass term, *JHEP* **08**, p. 058.

Frezzotti, R. and Rossi, G. C. (2004). Chirally improving Wilson fermions. ii: Four-quark operators, *JHEP* **10**, p. 070.

Frommer, A., Hannemann, V., Nockel, B., Lippert, T. and Schilling, K. (1994). Accelerating Wilson fermion matrix inversions by means of the stabilized biconjugate gradient algorithm, *Int. J. Mod. Phys.* **C5**, pp. 1073–1088.

Frommer, A., Nockel, B., Gusken, S., Lippert, T. and Schilling, K. (1995). Many masses on one stroke: Economic computation of quark propagators, *Int. J. Mod. Phys.* **C6**, pp. 627–638.

Furman, V. and Shamir, Y. (1995). Axial symmetries in lattice QCD with Kaplan fermions, *Nucl. Phys.* **B439**, pp. 54–78.

Gaillard, M. K. and Lee, B. W. (1974a). Delta I = 1/2 rule for nonleptonic decays in asymptotically free field theories, *Phys. Rev. Lett.* **33**, p. 108.

Gaillard, M. K. and Lee, B. W. (1974b). Rare decay modes of the K - mesons in gauge theories, *Phys. Rev.* **D10**, p. 897.

Gasser, J. and Leutwyler, H. (1985). Chiral perturbation theory: Expansions in the mass of the strange quark, *Nucl. Phys.* **B250**, p. 465.

Gasser, J. and Leutwyler, H. (1987a). Light quarks at low temperatures, *Phys. Lett.* **B184**, p. 83.

Gasser, J. and Leutwyler, H. (1987b). Thermodynamics of chiral symmetry, *Phys. Lett.* **B188**, p. 477.

Gattringer, C. R. and Lang, C. B. (1993). Resonance scattering phase shifts in a 2-d lattice model, *Nucl. Phys.* **B391**, pp. 463–482.

Gell-Mann, M. and Levy, M. (1960). The axial vector current in beta decay, *Nuovo Cim.* **16**, p. 705.

Gell-Mann, M., Oakes, R. J. and Renner, B. (1968). Behavior of current divergences under SU(3) x SU(3), *Phys. Rev.* **175**, pp. 2195–2199.

Ginsparg, P. H. (1980). First order and second order phase transitions in gauge theories at finite temperature, *Nucl. Phys.* **B170**, p. 388.

Ginsparg, P. H. and Wilson, K. G. (1982). A remnant of chiral symmetry on the lattice, *Phys. Rev.* **D25**, p. 2649.

Giusti, L., Hernandez, P., Laine, M., Weisz, P. and Wittig, H. (2004). Low-energy couplings of QCD from current correlators near the chiral limit, *JHEP* **04**, p. 013.

Gliozzi, F. (1982). Spinor algebra of the one component lattice fermions, *Nucl. Phys.* **B204**, pp. 419–428.

Gockeler, M. *et al.* (1999). Nonperturbative renormalisation of composite operators in lattice QCD, *Nucl. Phys.* **B544**, pp. 699–733.

Golterman, M. and Shamir, Y. (2003). Localization in lattice QCD, *Phys. Rev.* **D68**, p. 074501.

Golterman, M. F. L. and Smit, J. (1984). Selfenergy and flavor interpretation of staggered fermions, *Nucl. Phys.* **B245**, p. 61.

Golterman, M. F. L. and Smit, J. (1985). Lattice baryons with staggered fermions, *Nucl. Phys.* **B255**, p. 328.

Golub, G. H. and Van Loan, C. F. (1996). *Matrix Computations*, 3rd ed. (Johns Hopkins Univ. Press).

Gottlieb, S. A., Liu, W., Toussaint, D., Renken, R. L. and Sugar, R. L. (1987). Hybrid molecular dynamics algorithms for the numerical simulation of quantum chromodynamics, *Phys. Rev.* **D35**, pp. 2531–2542.

Gregory, E. B., Irving, A. C., McNeile, C. C., Miller, S. and Sroczynski, Z. (2005). Scalar glueball and meson spectroscopy in unquenched lattice QCD with improved staggered quarks, *PoS* **LAT2005**, p. 027.

Gribov, V. N. (1978). Quantization of non-abelian gauge theories, *Nucl. Phys.* **B139**, p. 1.

Gupta, R., Guralnik, G., Kilcup, G. W. and Sharpe, S. R. (1991). The quenched spectrum with staggered fermions, *Phys. Rev.* **D43**, pp. 2003–2026.

Halasz, M. A., Jackson, A. D., Shrock, R. E., Stephanov, M. A. and Verbaarschot, J. J. M. (1998). On the phase diagram of QCD, *Phys. Rev.* **D58**, p. 096007.

Hamber, H. and Parisi, G. (1981). Numerical estimates of hadronic masses in a pure SU(3) gauge theory, *Phys. Rev. Lett.* **47**, p. 1792.

Hansen, F. C. (1990). Finite size effects in spontaneously broken SU(N) x su(N) theories, *Nucl. Phys.* **B345**, pp. 685–708.

Hansen, F. C. and Leutwyler, H. (1991). Charge correlations and topological susceptibility in QCD, *Nucl. Phys.* **B350**, pp. 201–227.

Harman, H. H. (1960). *Modern Factor Analysis* (University of Chicago Press), pp. 11–19.

Hart, A., von Hippel, G. M., Horgan, R. R. and Storoni, L. C. (2005). Automatically generating Feynman rules for improved lattice field theories, *J. Comput. Phys.* **209**, pp. 340–353.

Hasenbusch, M. (2001). Speeding up the hybrid-Monte-Carlo algorithm for dynamical fermions, *Phys. Lett.* **B519**, pp. 177–182.

Hasenbusch, M. and Jansen, K. (2003). Speeding up lattice QCD simulations with clover-improved Wilson fermions, *Nucl. Phys.* **B659**, pp. 299–320.

Hasenfratz, A. and Alexandru, A. (2002). Evaluating the fermionic determinant of dynamical configurations, *Phys. Rev.* **D65**, p. 114506.

Hasenfratz, A., Hasenfratz, P. and Niedermayer, F. (2005). Simulating full QCD with the fixed point action, arXiV hep-lat/0506024.

Hasenfratz, A. and Knechtli, F. (2001). Flavor symmetry and the static potential with hypercubic blocking, *Phys. Rev.* **D64**, p. 034504.

Hasenfratz, A. and Knechtli, F. (2002). Simulation of dynamical fermions with smeared links, *Comput. Phys. Commun.* **148**, pp. 81–86.

Hasenfratz, A. *et al.* (1991). Goldstone bosons and finite size effects: A numerical study of the O(4) model, *Nucl. Phys.* **B356**, pp. 332–366.

Hasenfratz, P., Laliena, V. and Niedermayer, F. (1998). The index theorem in QCD with a finite cut-off, *Phys. Lett.* **B427**, pp. 125–131.

Hasenfratz, P. and Niedermayer, F. (1994). Perfect lattice action for asymptotically free theories, *Nucl. Phys.* **B414**, pp. 785–814.

Hasenfratz, P. and Niedermayer, F. (1997). Fixed-point actions in 1-loop perturbation theory, *Nucl. Phys.* **B507**, pp. 399–415.

Heatlie, G., Martinelli, G., Pittori, C., Rossi, G. C. and Sachrajda, C. T. (1991). The improvement of hadronic matrix elements in lattice QCD, *Nucl. Phys.* **B352**, pp. 266–288.

Hernandez, P., Jansen, K. and Lüscher, M. (1999). Locality properties of Neuberger's lattice Dirac operator, *Nucl. Phys.* **B552**, pp. 363–378.

Hornbostel, K., Lepage, G. P. and Morningstar, C. (2003). Scale setting for alpha(s) beyond leading order, *Phys. Rev.* **D67**, p. 034023.

Isgur, N. and Karl, G. (1978). P wave baryons in the quark model, *Phys. Rev.*

D18, p. 4187.

Isgur, N. and Karl, G. (1979). Positive parity excited baryons in a quark model with hyperfine interactions, *Phys. Rev.* **D19**, p. 2653.

Isgur, N. and Wise, M. B. (1989). Weak decays of heavy mesons in the static quark approximation, *Phys. Lett.* **B232**, p. 113.

Isgur, N. and Wise, M. B. (1990). Weak transition form-factors between heavy mesons, *Phys. Lett.* **B237**, p. 527.

Iwasaki, Y. (1983). Renormalization group analysis of lattice theories and improved lattice action. 2. four-dimensional nonabelian SU(N) gauge model, UTHEP-118.

Jarrell, M. and Gubernatis, J. (1996). *Phys. Rep.* **269**, p. 135.

Jegerlehner, B. (1996). Krylov space solvers for shifted linear systems, arXiV hep-lat/9612014.

Jegerlehner, B. (1998). Multiple mass solvers, *Nucl. Phys. Proc. Suppl.* **63**, pp. 958–960.

Kajantie, K., Laine, M., Rummukainen, K. and Schroder, Y. (2003). The pressure of hot QCD up to g**6 ln(1/g), *Phys. Rev.* **D67**, p. 105008.

Kajantie, K., Laine, M., Rummukainen, K. and Shaposhnikov, M. E. (1996). The electroweak phase transition: A non-perturbative analysis, *Nucl. Phys.* **B466**, pp. 189–258.

Kalkreuter, T. and Simma, H. (1996). An accelerated conjugate gradient algorithm to compute low lying eigenvalues: A study for the Dirac operator in SU(2) lattice QCD, *Comput. Phys. Commun.* **93**, pp. 33–47.

Kamleh, W., Adams, D. H., Leinweber, D. B. and Williams, A. G. (2002). Accelerated overlap fermions, *Phys. Rev.* **D66**, p. 014501.

Kaplan, D. B. (1992). A method for simulating chiral fermions on the lattice, *Phys. Lett.* **B288**, pp. 342–347.

Kaplan, D. B. and Schmaltz, M. (1996). Domain wall fermions and the eta invariant, *Phys. Lett.* **B368**, pp. 44–52.

Karsch, F., Laermann, E., Petreczky, P., Stickan, S. and Wetzorke, I. (2002). A lattice calculation of thermal dilepton rates, *Phys. Lett.* **B530**, pp. 147–152.

Karsch, F. and Mutter, K. H. (1989). Strong coupling QCD at finite baryon number density, *Nucl. Phys.* **B313**, p. 541.

Karsten, L. H. and Smit, J. (1981). Lattice fermions: Species doubling, chiral invariance, and the triangle anomaly, *Nucl. Phys.* **B183**, p. 103.

Kennedy, A. D. and Pendleton, B. J. (1985). Improved heat bath method for Monte Carlo calculations in lattice gauge theories, *Phys. Lett.* **B156**, pp. 393–399.

Kilcup, G. W. and Sharpe, S. R. (1987). A tool kit for staggered fermions, *Nucl. Phys.* **B283**, p. 493.

Kluberg-Stern, H., Morel, A., Napoly, O. and Petersson, B. (1983). Flavors of Lagrangian Susskind fermions, *Nucl. Phys.* **B220**, p. 447.

Kokkedee, J. (1969). *The Quark Model* (Benjamin).

Koller, J. and van Baal, P. (1986). A rigorous nonperturbative result for the glueball mass and electric flux energy in a finite volume, *Nucl. Phys.* **B273**, p. 387.

Lee, W.-J. and Sharpe, S. R. (1999). Partial flavor symmetry restoration for chiral staggered fermions, *Phys. Rev.* **D60**, p. 114503.

Lee, W.-J. and Sharpe, S. R. (2002). One-loop matching coefficients for improved staggered bilinears, *Phys. Rev.* **D66**, p. 114501.

Lee, W.-J. and Sharpe, S. R. (2003). Perturbative matching of staggered four-fermion operators with hypercubic fat links, *Phys. Rev.* **D68**, p. 054510.

Leinweber, D. B., Melnitchouk, W., Richards, D. G., Williams, A. G. and Zanotti, J. M. (2005). Baryon spectroscopy in lattice QCD, *Lect. Notes Phys.* **663**, pp. 71–112.

Lellouch, L. and Lüscher, M. (2001). Weak transition matrix elements from finite-volume correlation functions, *Commun. Math. Phys.* **219**, pp. 31–44.

Lepage, G. P. (1980). Vegas: An adaptive multidimensional integration program, CLNS-80/447.

Lepage, G. P. (1999). Flavor-symmetry restoration and Symanzik improvement for staggered quarks, *Phys. Rev.* **D59**, p. 074502.

Lepage, G. P. and Mackenzie, P. B. (1993). On the viability of lattice perturbation theory, *Phys. Rev.* **D48**, pp. 2250–2264.

Lepage, G. P., Magnea, L., Nakhleh, C., Magnea, U. and Hornbostel, K. (1992). Improved nonrelativistic QCD for heavy quark physics, *Phys. Rev.* **D46**, pp. 4052–4067.

Lepage, G. P. *et al.* (2002). Constrained curve fitting, *Nucl. Phys. Proc. Suppl.* **106**, pp. 12–20.

Leutwyler, H. and Smilga, A. (1992). Spectrum of Dirac operator and role of winding number in QCD, *Phys. Rev.* **D46**, pp. 5607–5632.

Lüscher, M. (1986a). Volume dependence of the energy spectrum in massive quantum field theories. 1. stable particle states, *Commun. Math. Phys.* **104**, p. 177.

Lüscher, M. (1986b). Volume dependence of the energy spectrum in massive quantum field theories. 2. scattering states, *Commun. Math. Phys.* **105**, pp. 153–188.

Lüscher, M. (1991). Two particle states on a torus and their relation to the scattering matrix, *Nucl. Phys.* **B354**, pp. 531–578.

Lüscher, M. (1994). A new approach to the problem of dynamical quarks in numerical simulations of lattice QCD, *Nucl. Phys.* **B418**, pp. 637–648.

Lüscher, M. (2000). Chiral gauge theories revisited, arXiV hep-th/0102028.

Lüscher, M. (2003). Lattice QCD and the Schwarz alternating procedure, *JHEP* **05**, p. 052.

Lüscher, M. (2004). Schwarz-preconditioned HMC algorithm for two-flavour lattice QCD, arXiV hep-lat/0409106.

Lüscher, M., Narayanan, R., Weisz, P. and Wolff, U. (1992). The Schrodinger functional: A renormalizable probe for nonabelian gauge theories, *Nucl. Phys.* **B384**, pp. 168–228.

Lüscher, M., Sint, S., Sommer, R., Weisz, P. and Wolff, U. (1997a). Non-perturbative O(a) improvement of lattice QCD, *Nucl. Phys.* **B491**, pp. 323–343.

Lüscher, M., Sint, S., Sommer, R. and Wittig, H. (1997b). Non-perturbative determination of the axial current normalization constant in O(a) improved lattice QCD, *Nucl. Phys.* **B491**, pp. 344–364.

Lüscher, M., Sommer, R., Weisz, P. and Wolff, U. (1994). A precise determination of the running coupling in the SU(3) Yang-Mills theory, *Nucl. Phys.* **B413**, pp. 481–502.

Lüscher, M., Sommer, R., Wolff, U. and Weisz, P. (1993). Computation of the running coupling in the SU(2) Yang-Mills theory, *Nucl. Phys.* **B389**, pp. 247–264.

Lüscher, M. and Weisz, P. (1985a). Computation of the action for on-shell improved lattice gauge theories at weak coupling, *Phys. Lett.* **B158**, p. 250.

Lüscher, M. and Weisz, P. (1985b). On-shell improved lattice gauge theories, *Commun. Math. Phys.* **97**, p. 59.

Lüscher, M. and Weisz, P. (1996). O(a) improvement of the axial current in lattice QCD to one-loop order of perturbation theory, *Nucl. Phys.* **B479**, pp. 429–458.

Lüscher, M., Weisz, P. and Wolff, U. (1991). A numerical method to compute the running coupling in asymptotically free theories, *Nucl. Phys.* **B359**, pp. 221–243.

Ma, S.-K. (1976). *Modern Theory of Critical Phenomena* (Benjamin).

Maiani, L. and Testa, M. (1990). Final state interactions from Euclidean correlation functions, *Phys. Lett.* **B245**, pp. 585–590.

Manohar, A. V. and Wise, M. B. (2000). Heavy quark physics, *Camb. Monogr. Part. Phys. Nucl. Phys. Cosmol.* **10**, pp. 1–191.

Marinari, E., Parisi, G. and Rebbi, C. (1981). Computer estimates of meson masses in SU(2) lattice gauge theory, *Phys. Rev. Lett.* **47**, p. 1795.

Martinelli, G., Pittori, C., Sachrajda, C. T., Testa, M. and Vladikas, A. (1995). A general method for nonperturbative renormalization of lattice operators, *Nucl. Phys.* **B445**, pp. 81–108.

Maschhof, K. and Sorensen, D. (1996). A portable implementation of ARPACK for distributed memory parallel architectures, in *Copper Mountain Conference on Iterative Methods*, `http://www.caam.rice.edu/~kristyn/parpack_home.html`.

Matsui, T. and Satz, H. (1986). J / psi suppression by quark - gluon plasma formation, *Phys. Lett.* **B178**, p. 416.

Metropolis, N., Rosenbluth, A. W., Rosenbluth, M. N., Teller, A. H. and Teller, E. (1953). Equation of state calculations by fast computing machines, *J. Chem. Phys.* **21**, pp. 1087–1092.

Michael, C. and Teper, M. (1989). The glueball spectrum in SU(3), *Nucl. Phys.* **B314**, p. 347.

Montvay, I. and Muenster, G. (1994). *Quantum fields on a lattice* (Cambridge Univ. Press).

Morel, A. (1987). Chiral logarithms in quenched QCD, *J. Phys. (France)* **48**, pp. 1111–1119.

Morningstar, C. and Peardon, M. J. (2004). Analytic smearing of SU(3) link variables in lattice QCD, *Phys. Rev.* **D69**, p. 054501.

Morningstar, C. J. and Peardon, M. J. (1999). The glueball spectrum from an

anisotropic lattice study, *Phys. Rev.* **D60**, p. 034509.
Nakamura, A. and Sakai, S. (2005). Transport coefficients of gluon plasma, *Phys. Rev. Lett.* **94**, p. 072305.
Narayanan, R. and Neuberger, H. (2000). An alternative to domain wall fermions, *Phys. Rev.* **D62**, p. 074504.
Neuberger, H. (1998a). Exactly massless quarks on the lattice, *Phys. Lett.* **B417**, pp. 141–144.
Neuberger, H. (1998b). A practical implementation of the overlap-Dirac operator, *Phys. Rev. Lett.* **81**, pp. 4060–4062.
Nielsen, H. B. and Ninomiya, M. (1981a). Absence of neutrinos on a lattice. 1. proof by homotopy theory, *Nucl. Phys.* **B185**, p. 20.
Nielsen, H. B. and Ninomiya, M. (1981b). Absence of neutrinos on a lattice. 2. intuitive topological proof, *Nucl. Phys.* **B193**, p. 173.
Nielsen, H. B. and Ninomiya, M. (1981c). No go theorem for regularizing chiral fermions, *Phys. Lett.* **B105**, p. 219.
Nobes, M. A. and Trottier, H. D. (2004). Progress in automated perturbation theory for heavy quark physics, *Nucl. Phys. Proc. Suppl.* **129**, pp. 355–357.
Orginos, K., Toussaint, D. and Sugar, R. L. (1999). Variants of fattening and flavor symmetry restoration, *Phys. Rev.* **D60**, p. 054503.
Patel, A. and Sharpe, S. R. (1993). Perturbative corrections for staggered fermion bilinears, *Nucl. Phys.* **B395**, pp. 701–732.
Peskin, M. and Schroeder, D. (1995). *An introduction to quantum field theory* (Addison-Wesley).
Phillips, A. and Stone, D. (1986). Lattice gauge fields, principal bundles and the calculation of topological charge, *Commun. Math. Phys.* **103**, pp. 599–636.
Pisarski, R. D. and Wilczek, F. (1984). Remarks on the chiral phase transition in chromodynamics, *Phys. Rev.* **D29**, pp. 338–341.
Polchinski, J. (1992). Effective field theory and the Fermi surface, arXiV hep-th/9210046.
Polyakov, A. M. (1975). Compact gauge fields and the infrared catastrophe, *Phys. Lett.* **B59**, pp. 82–84.
Prelovsek, S. (2006). Effects of staggered fermions and mixed actions on the scalar correlator, *Phys. Rev.* **D73**, p. 014506.
Press, W. H., Teukolsky, S. A. and William T. Vetterling, B. P. F. (2002). *Numerical Recipes in C++ : The Art of Scientific Computing*, 2nd ed. (Cambridge Univ. Press).
Ramond, P. (1981). *Field Theory: a Modern Primer* (Benjamin).
Regge, T. (1959). Introduction to complex orbital momenta, *Nuovo Cim.* **14**, p. 951.
Rummukainen, K. and Gottlieb, S. A. (1995). Resonance scattering phase shifts on a nonrest frame lattice, *Nucl. Phys.* **B450**, pp. 397–436.
Schafer, T. and Shuryak, E. V. (1998). Instantons in QCD, *Rev. Mod. Phys.* **70**, pp. 323–426.
Scherer, S. (2002). Introduction to chiral perturbation theory, arXiV hep-ph/0210398.
Sexton, J. C. and Weingarten, D. H. (1992). Hamiltonian evolution for the hybrid

Monte Carlo algorithm, *Nucl. Phys.* **B380**, pp. 665–678.
Shamir, Y. (1993). Chiral fermions from lattice boundaries, *Nucl. Phys.* **B406**, pp. 90–106.
Sharpe, S. R. (1992). Quenched chiral logarithms, *Phys. Rev.* **D46**, pp. 3146–3168.
Sharpe, S. R., Patel, A., Gupta, R., Guralnik, G. and Kilcup, G. W. (1987). Weak interaction matrix elements with staggered fermions. 1. theory and a trial run, *Nucl. Phys.* **B286**, p. 253.
Sheikholeslami, B. and Wohlert, R. (1985). Improved continuum limit lattice action for QCD with Wilson fermions, *Nucl. Phys.* **B259**, p. 572.
Shifman, M. A., Vainshtein, A. I. and Zakharov, V. I. (1979a). QCD and resonance physics: Applications, *Nucl. Phys.* **B147**, pp. 448–518.
Shifman, M. A., Vainshtein, A. I. and Zakharov, V. I. (1979b). QCD and resonance physics. sum rules, *Nucl. Phys.* **B147**, pp. 385–447.
Shuryak, E. V. and Verbaarschot, J. J. M. (1993). Random matrix theory and spectral sum rules for the Dirac operator in QCD, *Nucl. Phys.* **A560**, pp. 306–320.
Skyrme, T. H. R. (1961). A nonlinear field theory, *Proc. Roy. Soc. Lond.* **A260**, pp. 127–138.
Sommer, R. (1994). A new way to set the energy scale in lattice gauge theories and its applications to the static force and alpha-s in SU(2) Yang-Mills theory, *Nucl. Phys.* **B411**, pp. 839–854.
Susskind, L. (1977). Lattice fermions, *Phys. Rev.* **D16**, pp. 3031–3039.
Symanzik, K. (1980). *Recent Developments in Gauge Theories*, vol. 16 (Plenum), p. 313.
Symanzik, K. (1983a). Continuum limit and improved action in lattice theories. 1. principles and phi**4 theory, *Nucl. Phys.* **B226**, p. 187.
Symanzik, K. (1983b). Continuum limit and improved action in lattice theories. 2. O(N) nonlinear sigma model in perturbation theory, *Nucl. Phys.* **B226**, p. 205.
't Hooft, G. (1974). A two-dimensional model for mesons, *Nucl. Phys.* **B75**, p. 461.
't Hooft, G. (1976). Computation of the quantum effects due to a four- dimensional pseudoparticle, *Phys. Rev.* **D14**, pp. 3432–3450.
Takaishi, T. (1996). Heavy quark potential and effective actions on blocked configurations, *Phys. Rev.* **D54**, pp. 1050–1053.
Toussaint, D. (1990). Simulating QCD at finite density, *Nucl. Phys. Proc. Suppl.* **17**, pp. 248–251.
van der Vorst, H. (1992). BiCGStab: A fast and smoothly converging variant of Bi-CG for the solution of nonsymmetric linear systems, *SIAM J. Sc. Stat. Comp.* **13**, p. 631.
Veneziano, G. (1979). U(1) without instantons, *Nucl. Phys.* **B159**, pp. 213–224.
Verbaarschot, J. J. M. (1994). The spectrum of the QCD Dirac operator and chiral random matrix theory: The threefold way, *Phys. Rev. Lett.* **72**, pp. 2531–2533.
Wegner, A. M. (1971). Duality in generalized Ising models and phase transitions

without local order parameters, *J. Math. Phys.* **12**, pp. 2259–2272.
Weingarten, D. H. and Petcher, D. N. (1981). Monte Carlo integration for lattice gauge theories with fermions, *Phys. Lett.* **B99**, p. 333.
Weisz, P. (1983). Continuum limit improved lattice action for pure Yang-Mills theory. 1, *Nucl. Phys.* **B212**, p. 1.
Wilson, K. G. (1974). Confinement of quarks, *Phys. Rev.* **D10**, pp. 2445–2459.
Wilson, K. G. and Kogut, J. B. (1974). The renormalization group and the epsilon expansion, *Phys. Rept.* **12**, pp. 75–200.
Witten, E. (1979a). Baryons in the 1/N expansion, *Nucl. Phys.* **B160**, p. 57.
Witten, E. (1979b). Current algebra theorems for the U(1) 'Goldstone boson', *Nucl. Phys.* **B156**, p. 269.
Witten, E. (1983). Current algebra, baryons, and quark confinement, *Nucl. Phys.* **B223**, pp. 433–444.
Wohlert, R. (1987). Improved continuum limit lattice action for quarks, DESY 87/069.
Yaffe, L. G. and Svetitsky, B. (1982). First order phase transition in the SU(3) gauge theory at finite temperature, *Phys. Rev.* **D26**, p. 963.

Index